FM 1-0.2.2
Military Symbols

Department of the Army

NIMBLE BOOKS LLC: THE AI LAB FOR BOOK-LOVERS

~ FRED ZIMMERMAN, EDITOR ~

Humans and AI making books richer, more diverse, and more surprising.

Publishing Information

ISBN: 978-1-60888-250-2

AI-generated Keyword Phrases

Control Measure Symbols; Military Symbol Construct Process; Military Unit and Organizational Symbols; Unit and Organization Symbols; Unit and Organization Frame Shapes; Amplifier Fields for Units; Main Function Symbols for Units; Activity and Installation Symbols; Activity Frame Shapes; Installation Frame Shapes; Main Function Symbols for Activities and Installations; Equipment Symbols; Framed and Unframed Equipment Symbols; Equipment Symbol Frame Shapes; Main and Modifier Symbols and Amplifiers for Equipment; Main Function Symbols for Equipment; Tactical Mission Tasks; Task Organization

Publisher's Notes

Military symbols are an enduring source of fascination for students of military history.

This annotated edition illustrates the capabilities of the AI Lab for Book-Lovers to add context and ease-of-use to manuscripts. It includes several types of abstracts, building from simplest to more complex: TLDR (one word), ELI5, TLDR (vanilla), Scientific Style, and Action Items; essays to increase viewpoint diversity, such as Grounds for Dissent, Red Team Critique, and MAGA Perspective; and Notable Passages and Nutshell Summaries for each page.

Annotations

Abstracts

TL;DR (one word)

Symbols.

Explain It To Me Like I'm Five Years Old

The document talks about symbols that are used in the military to show different things like units, equipment, and operations. It explains how to draw these symbols and what they mean. It also talks about different tasks that soldiers might have to do in the military. It is like a big book that teaches people in the military how to use symbols to communicate important information.

TL;DR (vanilla)

The document discusses military symbols, control measures, operation planning symbols, equipment symbols, and tactical mission tasks. It includes information on the placement of symbols, reliability ratings, main function symbols, and control measure symbols for various military operations. The document also covers the purpose and makeup of course of action sketches, task organization composition symbols, and target control measure symbols.

Scientific Style

This information discusses the role of the vertebrate kinesin-5 Eg5 in driving the sliding of microtubules during cell division. Eg5 is shown to move towards the plus ends of microtubules it crosslinks, resulting in relative sliding at a rate comparable to spindle pole separation. Additionally, Eg5 can tether microtubule plus-ends, suggesting an additional microtubule-binding mode. These findings shed light on the function of kinesin-5 in mitosis and provide insight into how multiple mitotic motors may interact. The study also highlights the relevance of Eg5 inhibition in anti-cancer drug development and the need for well-defined assays for motor function in this context.

ACTION ITEMS

The document appears to be a military manual or guide related to military symbols and control measures.

It covers various symbols and constructs used in military operations, including unit and organizational symbols, equipment symbols, and control measures for offense, defense, and observation posts.

The document also includes information on tactical mission tasks and course of action sketches.

It may be beneficial to review the full text to gain a comprehensive understanding of military symbol constructs and control measures.

VIEWPOINTS

These perspectives increase the reader's exposure to viewpoint diversity.

GROUNDS FOR DISSENT

A member of the organization responsible for this document may have principled and substantive reasons to dissent from this report based on the following reasons:

Lack of clarity and organization: The document appears to be disorganized and lacks clear structure, as evidenced by the randomly selected sentences that do not form a coherent narrative. A dissenting member may argue that the lack of clarity in the document makes it difficult for readers to understand and follow the information presented.

Incomplete information: The extractive summarizer only provides a small portion of the document, suggesting that there may be key information missing from the analysis. A dissenting member may argue that important details or context are omitted, leading to an incomplete and potentially misleading understanding of the content.

Lack of relevance: The selected sentences focus on military symbols and control measures, which may not be relevant or applicable to all members of the organization. A dissenting member may question the relevance of the report to their specific role or responsibilities within the organization.

Inaccuracies or inconsistencies: Without access to the full text, it is difficult to verify the accuracy and consistency of the information presented in the document. A dissenting member may raise concerns about errors, contradictions, or inaccuracies in the report that could impact its credibility and reliability.

Overall, a member of the organization may dissent from this report due to concerns about its organization, completeness, relevance, accuracy, and overall effectiveness in communicating important information.

Red Team Critique

Based on the extracted sentences, it is clear that the document is a military manual or guide that focuses on military symbols, control measures, operation planning symbols, equipment symbols, and tactical mission tasks. The document includes information on the symbols used to represent military units, organizations, equipment, activities, and installations. It also delves into the process of constructing military symbols and providing amplifiers for units. The document outlines the use of various symbols and control measures for offense, defense, observation posts, military deception, mobility, countermobility, field fortifications, airspace coordinating measures, fire support coordination, target control, and CBRN contamination markers. The document appears to be highly detailed and informative, providing a comprehensive guide for military personnel on the use of symbols and control measures in various operational scenarios. However, the critique would focus on the organization and clarity of the document, ensuring that the information is presented in a clear and concise manner for easy comprehension by the intended audience. Additionally, the document may benefit from more practical examples and scenarios to demonstrate the application of the symbols and control measures in real-world military operations.

MAGA Perspective

This document is full of military symbols and control measures that are likely meant to confuse and deceive the American people. The use of acronyms and abbreviations only adds to the complexity and makes it difficult for ordinary Americans to understand what is really being planned. This level of detail and precision in military planning is unnecessary and only serves to benefit the deep state and globalist elites who seek to control every aspect of our lives.

The document contains symbols for equipment, units, and organizations that seem to prioritize the needs of foreign entities over the interests of the American people. The emphasis on planning for offense and defense operations, as well as deception and observation post control measures, raises concerns about the true intentions of those in power. It is clear that this document is not meant to protect American sovereignty but rather to advance a globalist agenda that undermines our national security.

The mention of chemical, biological, radiological, and nuclear (CBRN) capabilities within military units is especially troubling. This could be a signal that the government is preparing for a false flag operation or some other form of attack on the American people in order to further their own authoritarian goals. The use of deceptive symbols and amplifiers only adds to the suspicion that the government is not acting in the best interests of its citizens.

The inclusion of mobility and countermobility control measures suggests that the government is preparing for some form of domestic unrest or resistance from the American people. This militarized approach to dealing with dissent goes against the principles of democracy and freedom that our country was founded on. It is clear that those in power are more interested in maintaining control than upholding the rights of the people.

Overall, this document reveals a sinister agenda that is being carried out behind the scenes by shadowy forces within the government. The use of military symbols and control measures is a clear indication that the authorities are preparing to clamp down on dissent and assert their dominance over the American people. This should be a wake-up call to all patriots who value their freedom and independence.

Visual Exploration

Generative AI technologies include powerful tools for multimodal exploration of ideas.

Image-to-Text

I asked my AI to look at a typical image from the Field Manual and "reverse engineer" the design principles.

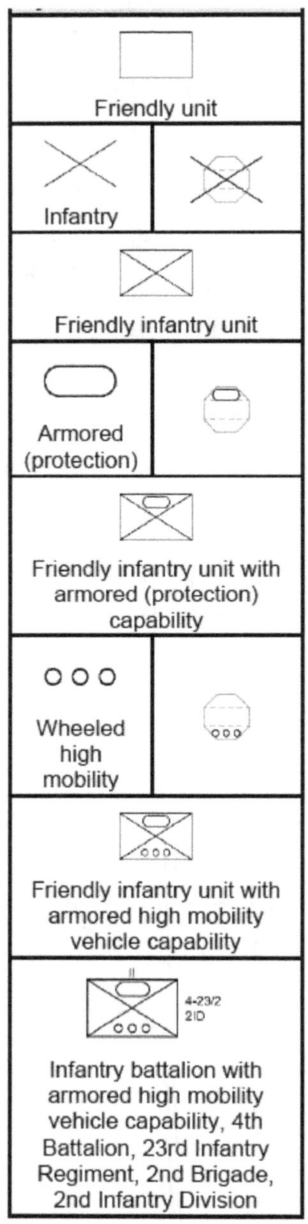

Table 1-5. Construction process for framed symbols.

Here is its response.

The symbology from the US Army Field Manual you've provided follows a clear hierarchical design structure intended to convey information efficiently and unambiguously. Here are some inferred design principles:

***Simplicity**: Use of basic geometric shapes to represent different units, facilitating quick identification.*

***Modularity**: Symbols can be combined to represent complex information, such as unit type and capabilities.*

***Hierarchy**: Addition of elements to a base symbol to indicate an increase in complexity or capability.*

***Standardization**: Adherence to a recognized standard that allows for interoperability across different forces and nations.*

***Iconography**: Use of iconic representations for different unit functions, such as crossed rifles for infantry or wheels for high mobility.*

***Scalability**: Symbols maintain clarity and recognition when scaled to different sizes, important for various operational contexts.*

These principles support rapid decision-making and clarity of communication in complex operational environments.

This is a pretty solid effort. The AI did **not** have the information that the caption of the figure pertained to symbol construction.

TEXT-TO-IMAGE

In this section, I demonstrate some examples text to image generation in the realm of military symbology. One set of my prompts asked the model to extend popular design sensibilities to the contents of the Manual. Ibegan with the architect of Apple's success.

JONY IVE

The model wrote:

I have redesigned three military symbology examples, applying the design principles of Jony Ive and focusing exclusively on graphic elements. These designs focus on the essence of the symbols' purposes, stripping away unnecessary details to offer clear, efficient, and elegant visual communication.

Military Unit Symbol: This design emphasizes enhanced status visibility through simplicity, minimalism, and the use of space. It employs a limited color palette and gradients to differentiate statuses, reflecting Ive's philosophy of clarity and elegance.

Equipment Symbol for a Tank: This symbol adopts a sleek, modern interpretation of a tank using geometric shapes and subtle gradients. It emphasizes clean lines and functionality, avoiding text to ensure visual clarity and intuitiveness.

Control Measure Symbol for a No-Fire Area: The design for this symbol embraces minimalism and simplicity, using geometric shapes and a refined color palette to communicate restrictions intuitively. A bold, simplified icon conveys the prohibition effectively, embodying Ive's approach to intuitive and impactful design.

WILLIAM MORRIS

Looking for another designer to test this process with, I pulled the name of William Morris out of a random slot in my memory. Morris (1834–1896) was a British textile designer, poet, novelist, translator, and socialist activist associated with the British Arts and Crafts Movement. Morris's designs are distinguished by their use of intricate floral and medieval-inspired patterns, emphasizing handcraftsmanship and natural dyes. I thought it would be interesting to see how these rather alien concepts translated to the world of military symbology.

Military Unit Symbol: This design integrates natural motifs and organic shapes to convey the unit's status, blending functionality with beauty. It utilizes a rich, earthy color palette with patterns inspired by flora and fauna, reflecting Morris's belief in the harmony between form and function.

Equipment Symbol for a Tank: Following Morris's design principles, this symbol incorporates intricate floral and foliage motifs to suggest the form and essence of a tank. It employs a rich, earthy color palette, conveying the tank's identity through natural forms and patterns.

Control Measure Symbol for a No-Fire Area: This design uses organic shapes and nature-inspired patterns to visually convey the restriction. It blends functionality with decorative art, employing motifs such as leaves, vines, and floral designs in natural, earthy tones.

Soviet Brutalism

Military Unit Symbol: This design embodies the stark, geometric, and monumental qualities of Soviet Brutalism, using bold shapes and a minimalist color scheme to convey the unit's status.

Control Measure Symbol for a No-Fire Area: Defined by bold, geometric shapes and a monochromatic palette, this symbol conveys prohibition and restriction with a sense of authority and permanence, true to Brutalist values.[1]

Impressions

What I take away from this exercise is that "artificial intelligence" can illuminate connections between deep technical content and broader worlds, such as graphic design, in ways that simply are beyond the capacity of the typical reader. This advances viewpoint diversity, which is one of my guiding principles.

[1] This image, with its similarity to the robotic assassin IG-11 in the Disney/Star Wars series *The Mandalorian, contains* some interesting foreshadowing of a future era of robotic fire control.—Ed.

PAGE-BY-PAGE SUMMARIES

BODY-1	*Information released by the Department of the Army with unlimited distribution approval.*
BODY-2	*The publication is available on the Army Publishing Directorate and Central Army Registry sites.*
BODY-3	*FM 1-02.2 is a Field Manual from the Department of the Army covering military symbols, including fundamentals, unit and organizational symbols, activity and installation symbols, and equipment symbols.*
BODY-4	*This page outlines control measures, operation planning symbols, tactical mission tasks, and course of action sketches in military operations.*
BODY-5	*The page contains various templates and examples of symbols and amplifiers used in military operations, including direction of movement, combat effectiveness, and control measure symbols.*
BODY-6	*The page contains various tables outlining standard identities, symbols, and constructs for military units, activities, installations, equipment, and control measures in operational planning.*
BODY-7	*Table of various symbols and control measures for route, convoy, countermobility, land mines, minefields, fortifications, airspace coordination, fire support, target control, target acquisition, and CBRN contamination.*
BODY-9	*FM 1-02.2 provides approved military symbols for land operations, applicable to all members of the profession of arms. It emphasizes compliance with laws of war and rules of engagement. Implements international agreements and uses joint terms. Proponent is the US Army Combined Arms Center. Comments and recommendations can be submitted for changes.*
BODY-10	*blank*
BODY-11	*FM 1-02.2 compiles military symbols for land operations, providing guidelines for creating and using symbols in U.S. Army publications. It includes hand-drawn alternate symbols and course of action sketch symbols not in MIL-STD 2525E. The Army Dictionary online is updated monthly to reflect changes.*
BODY-12	*Updates and modifications to military symbols, including additions and removals for various categories such as defensive operations, forms of maneuver, and target control measures.*
BODY-13	*This chapter discusses the standards for military symbols, including framed symbols and their components, standard identity, physical domain, and color usage. Symbols are used to represent units, equipment, installations, and activities in military operations.*
BODY-14	*Table showing standard identities and physical domain frame shapes for friendly, hostile, neutral, unknown, assumed friend, and pending entities in land, sea, air, and space equipment.*
BODY-15	*Military Symbol Fundamentals discusses the concept of status in military symbols, depicting whether an object is present, planned, or suspected at a location. It also explains the placement diagram for main and modifier symbols in units, equipment, installations, and activities.*
BODY-16	*Explanation of main function in horizontal placement diagrams, with emphasis on role or technical specialty for units, equipment, installations, or activities. Includes*

exceptions for full-frame main functions in land-domain symbols. Example provided.

BODY-17 *Military Symbol Fundamentals cover the use of modifiers, standard identity colors, amplifiers, and symbol lettering in military symbols. Modifiers provide additional capability information, standard identity colors are used for fill areas, amplifiers are optional, and symbol lettering must be uppercase and Sans Serif font.*

BODY-18 *Equipment symbols can be framed or unframed, with unframed symbols following specific rules. Tactical mission task symbols are used in sketches and do not use modifiers. Control measure symbols assign responsibilities and coordinate operations, with unique templates and colors based on background. Minefields and limited-access areas have specific modifiers.*

BODY-19 *Military symbol construction is like a logographic writing system, providing quick identification of units and capabilities. The process can construct sentences and paragraphs using standard symbols. Chapters 2 to 4 offer functions, modifiers, and amplifiers for building framed symbols, with an example provided for an infantry unit with armored high-mobility vehicle capability.*

BODY-20 *Constructing framed symbols involves selecting frame shape, main function symbol, modifiers, and amplifiers to create a complete military symbol. Control measures can be combined with other symbols and amplifiers for operational information.*

BODY-21 *Construction steps for creating a control measure symbol using templates and amplifiers to convey specific information, such as unit designation, date, and movement details.*

BODY-22 *blank*

BODY-23 *This chapter discusses symbols for military units and organizations, including frame shapes and amplifier fields for identifying friend, enemy, neutral, or unknown units. It provides guidelines for constructing unit symbols and standardizing alphanumeric information display.*

BODY-24 *Chapter 2 discusses land unit symbol amplifier fields.*

BODY-25 *Descriptions of symbols and amplifiers used in military unit organization, including main function, echelon, quantity, task organization, attached/detached status, staff comments, evaluation rating, combat effectiveness, higher formation, identification, friend or foe, direction of movement, and offset location indicators.*

BODY-26 *Description of main function symbol and amplifier fields for unit, including unique identifier, equipment type, date-time group, altitude or depth, location, speed, special headquarters, feint or dummy indicator, platform type, equipment teardown time, common identifier, operational condition, and engagement bar.*

BODY-27 *Description of military unit symbols, including country code and echelon amplifiers. Provides examples of different unit sizes and definitions.*

BODY-28 *Hierarchy of military units from platoon to division and their definitions.*

BODY-29 *This page discusses military unit and organizational symbols, including echelon amplifiers and quantity amplifiers, with examples and usage explanations provided.*

BODY-30 *Template and examples for quantity and task organization amplifiers in military units, including battalion task forces and company teams.*

BODY-31 *The page discusses military unit and organizational symbols, including task force amplifiers and indicators for task organization. It also explains the use of plus, minus, and plus/minus symbols for attaching and detaching subordinate elements.*

BODY-32	*The page discusses the template for attached and detached amplifiers, as well as the functions and examples of reinforced and reduced amplifiers. It also explains the country code amplifier and its placement in the frame.*
BODY-33	*The page discusses military unit and organizational symbols, including templates for country code amplifiers and command post using staff comments amplifier.*
BODY-34	*Various types of command posts, such as combat trains, early-entry, field trains, mobile command group, main, rear, and tactical, are used to control and coordinate different aspects of military operations.*
BODY-35	*The page discusses the use of alphanumeric unit designations, solidus, and hyphens in military symbols to identify units and their hierarchy within the chain of command.*
BODY-36	*Chapter 2 discusses the usage of higher echelon and direction of movement amplifiers in military symbols, providing examples and templates for clarity and communication in field operations.*
BODY-37	*Template and usage example for direction of movement amplifier, combat effectiveness amplifiers for military units based on factors like ammunition, personnel, fuel status, and weapons systems.*
BODY-38	*Template for combat effectiveness amplifier and offset location indicators for headquarters and non-headquarters units are discussed, with examples provided for placement and grouping.*
BODY-39	*This page discusses offset location indicators for military units, showing examples for both headquarters and non-headquarters units.*
BODY-40	*Main function symbols for units in the Army, including NATO and civil authority symbols, are determined by the table of organization and equipment descriptions. The main function symbol reflects the main function of the unit and is located in the center sector of the octagon.*
BODY-41	*Symbols and functions for military units, including information operations, interpreters, isolated personnel, liaison teams, and multinational operations.*
BODY-42	*Symbols and examples of main functions for units in the military, including public affairs, signal, space forces, special troops, and air defense artillery.*
BODY-43	*The page lists main function symbols for military units, including air and missile defense, ANGLICO, directed energy, field artillery, and missile batteries.*
BODY-44	*Table listing main function symbols for units, including missile defense, short-range air defense, intelligence operations, anti-armor, and armor.*
BODY-45	*Symbols for military units and their functions, such as armored cavalry, aviation, reconnaissance, and combined arms operations, are outlined with examples and translations for each symbol.*
BODY-46	*The page lists main function symbols for units, including infantry, mechanized armored infantry, mortar, surveillance, and the Bureau of Alcohol, Tobacco, Firearms, and Explosives (ATF).*
BODY-47	*Symbols for military units and organizations, including CBRN, DEA, and engineers, are defined and explained in FM 1-02.2.*
BODY-48	*Symbols and functions for units, including FBI, fire department, maneuver enhancement, military police, and police department.*
BODY-49	*Military unit and organizational symbols, including functions like protection and security, with examples such as the United States Marshal Service and the United States Secret Service. Also includes symbols for unmanned aircraft systems.*

BODY-50 *Summary: Main function symbols for units include aerial delivery, rigger platoon, Army field support, ammunition platoon, band performance, and contracting support team.*

BODY-51 *Symbols for military units and organizations, including explosive ordnance disposal, field feeding, finance operations, human resources, and judge advocate general.*

BODY-52 *Symbols and functions related to sustainment, maintenance, medical treatment facilities, and mortuary affairs in the Army are outlined in this chapter.*

BODY-53 *Symbols and functions of military units, including ordnance, quartermaster, religious support, shower and laundry operations, and sustainment, are outlined in FM 1-02.2.*

BODY-54 *Symbols and examples of main functions for military units, including transportation, civil affairs, civil-military cooperation, psychological operations, Rangers, and search and rescue.*

BODY-55 *The page discusses military unit symbols, including those for special operations, special forces, and multidomain operations. It provides examples and translations for each symbol.*

BODY-56 *Symbols for named units in the military, such as Combatant Commands, with examples like AFRICOM and CYBERCOM.*

BODY-57 *This page lists main function symbols for named military units, including the United States Space Command, Special Operations Command, Strategic Command, and Transportation Command. It also explains sub-unified commands and provides examples such as Alaskan Command and United States Forces Korea.*

BODY-58 *Symbols and examples of named units in the United States Army are listed, including USARAF, USARCENT, ARCYBER, SDDC, USANORTH, USARPAC, USARSO, and USASOC.*

BODY-59 *Symbols and modifiers for military units and organizations, providing additional information specific to capabilities.*

BODY-60 *Sector 1 modifiers for units include area support units and armored vehicles with specific capabilities such as wheeled or tracked vehicles. Assault capabilities are defined as short, violent attacks against local objectives.*

BODY-61 *Military unit and organizational symbols are discussed, including modifiers for units such as attack capabilities, aviation units, battalion support, biological detection, and bridging assets.*

BODY-62 *Table showing sector 1 modifiers for units, including support, chemical detection, clearing, and combat capabilities.*

BODY-63 *Military Unit and Organizational Symbols: Definitions and examples of symbols used to represent military units and their functions, such as combat camera capability and command post node services.*

BODY-64 *Table listing sector 1 modifiers for units, including composite capabilities, construction support, corps support, and counterintelligence activities.*

BODY-65 *Military unit and organizational symbols, including modifiers for units such as criminal investigation division, cross-cultural communication, crowd and riot control, cyberspace operations, and decontamination.*

BODY-66 *Table listing sector 1 modifiers for units, including capabilities such as detention, digital technology, diving, division support, working dogs, and drilling.*

BODY-67 *Summary: Military unit symbols and modifiers for different functions and capabilities, such as electromagnetic warfare and explosive ordnance disposal, are outlined in FM 1-02.2.*

BODY-68 *This page lists sector 1 modifiers for units, including fire direction center and general engineering capabilities. It also defines geospatial information for engineer units.*

BODY-69 *Military unit and organizational symbols for different functions and modifiers, including headquarters elements and intermodal operations.*

BODY-70 *Modifiers for units in Sector 1 include jamming capability, joint node network, and maintenance for unit readiness.*

BODY-71 *Symbols and modifiers for military units related to medical evacuation and care capabilities, with examples and guidelines for symbol usage.*

BODY-72 *Modifiers for units in Sector 1 include medical care, meteorological capabilities, multidomain operations, watercraft capabilities, and mobile advisor and support for various operations.*

BODY-73 *The page discusses military unit and organizational symbols, including modifiers for units such as mobility support, mortar capability, multinational operations, and multiple rocket launcher capability.*

BODY-74 *Table listing sector modifiers for units, including maritime terminal support, network operations, and nuclear capabilities.*

BODY-75 *Table listing sector 1 modifiers for military units, including symbols for operations, palletized load system, petroleum storage, transportation, and pipeline capabilities.*

BODY-76 *Sector 1 modifiers for units include postal service, radar, radiological, railway, and retransmission capabilities. Each modifier provides specific functions for different types of units in military operations.*

BODY-77 *Military unit symbols and modifiers for different functions and capabilities, such as riverine transport, robotic equipment, search and rescue, sensors, signals intelligence, and single rocket launchers.*

BODY-78 *Table of sector 1 modifiers for units, including sniper, survey, tactical satellite communications, target acquisition, and theater army support.*

BODY-79 *Military unit and organizational symbols, including modifiers for units such as theater sustainment command, unmanned aircraft systems, utility equipment, and water production capability.*

BODY-80 *Sector 1 and 2 modifiers for units provide additional information on unit capabilities, such as heavy weapons systems and air assault or airborne capabilities. These modifiers are represented in Field A and provide specific details for combat power and movement.*

BODY-81 *Symbols and modifiers for military units and organizations, including examples like air defense, amphibious capability, analysis, armored tracked vehicles, barges, and blood support.*

BODY-82 *Table listing sector 2 modifiers for units, including combat stress control, composite units, construction support, control actions, and decontamination capabilities.*

BODY-83 *Military unit and organizational symbols with examples such as dental services, direction finding, guerrilla units, heavy vehicle capability, and high altitude capability.*

BODY-84 Modifiers for units in sector 2 include intermodal operations, jamming capabilities, laboratories, and landing craft capabilities for transportation and warfare units.

BODY-85 Military unit and organizational symbols are defined by specific modifiers and symbols to indicate their capabilities and functions, such as light lift, long range, and low altitude capabilities.

BODY-86 Table 2-10 outlines sector 2 modifiers for units, including maintenance, medical bed capability, medium load transportation, medium altitude air defense, and medium range unmanned aircraft system capability.

BODY-87 Military unit and organizational symbols for various functions and capabilities, including mountain warfare, multifunctionality, ocean-going tugboats, optometry care, and over-snow capabilities.

BODY-88 Modifiers for units include pack animal, medical surveillance, railway, recovery, and riverine capabilities for enhancing mobility, health monitoring, infrastructure assessments, equipment recovery, and water transport.

BODY-89 Military unit and organizational symbols include modifiers for functions such as search, security force assistance, surgical capabilities, service craft yards, short range capabilities, and ski capabilities.

BODY-90 Table detailing Sector 2 modifiers for units, including examples like sled capability and supply functions such as producer and consumer phases.

BODY-91 Military unit and organizational symbols, including modifiers for units such as support, towed capability, harbor tug, vertical take-off and landing, and very heavy artillery.

BODY-92 Table 2-10 and Table 2-11 provide examples of unit modifiers and symbol constructs, including veterinary services, watercraft capability, and wheeled high mobility units.

BODY-93 Examples of unique unit designations and symbols for various military units and organizations are provided in a table format.

BODY-94 Examples of unique unit designations and their symbol constructs in the military, including capabilities and affiliations.

BODY-95 Chapter 3 of FM 1-02.2 discusses activity and installation symbols, including main function symbols and modifiers for identifying friendly, enemy, or criminal actions in an area of interest. It covers frame shapes, amplifier placement, and standard identity symbols.

BODY-96 Table 3-2 describes symbols used in activity frames, including staff comments, additional information, evaluation ratings, direction of movement, and more. The evaluation rating amplifier uses a 2-digit code for reliability and credibility ratings.

BODY-97 Symbols used in military activity and installation diagrams are explained, including evaluation rating amplifier, activity direction of movement amplifier, and activity offset location indicator amplifier.

BODY-98 Chapter 3 discusses installation frame shapes used to identify military, civilian, or enemy installations. It includes standard identity frame shapes for friendly, hostile, neutral, and unknown installations, as well as amplifier fields for main function and modifier symbols.

BODY-99 Description of main, modifier, and amplifier symbols for installation frames, including staff comments, evaluation ratings, and combat effectiveness ratings.

BODY-100 Descriptions of various fields for installation frames, including identification, location, and operational condition indicators. Examples include unique identifiers, date-time groups, altitude, and engagement bars. Field lengths and display priorities are specified.

BODY-101 Table and figure showing operational condition amplifiers and construct examples.

BODY-102 Main function symbols identify primary functions of activities and installations, including military and unique civilian symbols. They reflect the broader role of military forces in stability operations and support to civil authorities. Symbols are used to identify individual and group activities, as well as military and civilian infrastructure.

BODY-103 Symbols for activities and installations, including ATF, CBRN, Coast Guard, and cyber-server.

BODY-104 Symbols and examples of main functions related to cyberspace operations, data, drugs, and electric generation facilities are outlined in this chapter.

BODY-105 Symbols for various functions and activities, such as emergency management, EMS, explosion, extortion, and FBI, are defined and illustrated in a table.

BODY-106 Symbols and examples of main functions related to fire departments, governmental organizations, grenades, organized groups, and assault victims.

BODY-107 Symbols for activities and installations, including improvised explosive devices and individual victims, are defined in military terms.

BODY-108 Symbols and examples for various functions and facilities, including isolated personnel, law enforcement, maritime terminals, mass demonstrations, and mass graves.

BODY-109 Symbols for various functions and activities, such as medical treatment, psychological actions, mining, and network communication, are outlined in a table.

BODY-110 Symbols and functions related to activities and installations, including nongovernmental organizations, patrolling, petroleum facilities, and police departments.

BODY-111 Symbols and examples of main functions for activities and installations, including print media, religious, security, transportation, telecommunications, and potable water facilities. Sector 1 modifiers further identify affiliation, capability, special characteristics, or specialty.

BODY-112 Table listing symbols and examples for sector 1 modifiers related to activities and installations, including biological, chemical, coerced recruitment, command and control, continuity of operations, and electric generation.

BODY-113 Symbols and examples for various activities and installations, such as foreign fighters, gangs, hydroelectric power, leaders, meetings, and natural gas facilities.

BODY-114 Table listing sector 1 modifiers for activities and installations, including nuclear, operations, petroleum, raid, and radiological functions.

BODY-115 Symbols for activities and installations, including religious, security, speaker, targeted, telephone, and television functions.

BODY-116 Table listing sector 1 modifiers for activities and installations, including symbols and examples such as terrorist activity, training, trafficking, water capabilities, willing recruitment, and wired communication.

BODY-117 Symbols and modifiers for activities, installations, individuals, and organizations in military operations, including wireless communication, yard facilities, coerced recruitment, and data security.

BODY-118 The page discusses sector 2 modifiers for individuals and organizations, including leader, operations, security, supply, and willing recruitment. Symbols and examples are provided for each function.

BODY-119 Chapter 4 discusses equipment symbols, including main function symbols, sector 1 modifiers, and mobility indicator amplifiers. It explains the use of framed and unframed symbols to depict friendly, enemy, neutral, or unknown items. Frame shapes are used to identify equipment affiliation.

BODY-120 Main and modifier symbols and amplifiers for equipment standardize alphanumeric information to describe equipment capabilities, status, and location. Amplifiers are used for quantity, staff comments, and additional information.

BODY-121 Descriptions of equipment symbols including evaluation ratings, signature equipment indicators, hostile enemy symbols, identification features, movement indicators, mobility modes, unique identifiers, and equipment types.

BODY-122 Chapter 4 of FM 1-02.2 provides descriptions of various fields used in military symbols, including altitude, location, speed, platform type, and operational condition. It also explains the equipment direction of movement amplifier, which indicates the direction of movement of an object.

BODY-123 Explanation of equipment symbols, including engagement bar amplifier usage and colors to designate targets in military operations.

BODY-124 Chapter 4 discusses the mobility (transportation) mode indicator in military symbols, using examples of armored self-propelled vehicles transported by train and engaging enemy aircraft.

BODY-125 Equipment mobility symbols and speed indicators for various modes of transportation are outlined, with examples provided for clarity.

BODY-126 Illustration of speed usage construct with direction of movement of an armored high mobility vehicle. Operational condition amplifiers display equipment's level of operational condition. Table on page 115 provides examples.

BODY-127 Equipment symbols and examples of operational conditions for amplifiers, such as fully operational, damaged but operational, and destroyed. Main function symbols for equipment are also discussed.

BODY-128 Chapter 4 discusses the main function symbols for equipment, specifically weapons systems, with unique methods for indicating size, altitude, or range using lines perpendicular to the symbol. Land equipment symbols can be displayed without a frame, with color differentiating between friendly, enemy, and neutral.

BODY-129 Symbols for equipment functions, including air defense guns, are explained with examples and illustrations in FM 1-02.2.

BODY-130 Symbols for equipment functions, including anti-tank guns and direct fire guns, are depicted with specific symbols indicating their purpose.

BODY-131 Symbols for equipment functions, including guns and howitzers, with examples of armored vehicles and artillery units.

BODY-132 Table showing main function symbols for equipment, including symbols for friendly and enemy guns such as howitzers and mortars of various calibers.

BODY-133 Equipment symbols for various functions and weapons, including mortar and recoilless gun, with examples for friendly and enemy identification.

BODY-135 Symbols for equipment functions, including guns and missiles, are depicted in a table with examples of construct usage.

BODY-136 *Table listing main function symbols for equipment, including missile launchers for short, medium, and long-range defense against enemy aircraft and guided missiles.*

BODY-137 *Equipment symbols for missiles and armored vehicles, including anti-tank missile launchers, are outlined in FM 1-02.2.*

BODY-138 *Symbols for equipment functions in military operations, including anti-tank missile launchers and surface-to-surface missile launchers, are outlined in Table 4-6 of FM 1-02.2.*

BODY-139 *Equipment symbols for missiles, rockets, and launchers in military operations.*

BODY-140 *Symbols for different types of rocket launchers are listed, including those for anti-tank rockets and rocket-propelled explosive devices.*

BODY-141 *Equipment symbols for various functions and weapons, including rockets and nonlethal weapons like stun guns.*

BODY-142 *Symbols for equipment functions in combat, including water cannon and armored vehicles like armored fighting vehicles and armored personnel carriers.*

BODY-143 *Explanation of equipment symbols for armored vehicles, including the distinction between armored protected and armored tracked vehicles, as well as the use of lines to indicate size within the symbol.*

BODY-144 *Symbols for equipment functions, including friendly and enemy vehicles, with examples like locomotive engines and flatbed trucks. Land equipment symbols can be displayed in different colors for friend, enemy, and neutral. Vehicle platforms require a mobility modifier.*

BODY-145 *Equipment symbols for various functions and platforms, including friendly and enemy vehicles, engineer equipment such as bridges and drills.*

BODY-146 *Symbols for equipment functions in military operations, including mine clearing and mine laying, as well as aircraft types such as fixed-wing aircraft.*

BODY-147 *Symbols for equipment functions, including rotary-wing aircraft, unmanned aircraft, and maritime destroyer, with examples of each in ground and flight scenarios.*

BODY-148 *Symbols for equipment functions in military and maritime operations, including friendly, enemy, noncombatant ships, civilian merchant ships, and antennas.*

BODY-149 *Equipment symbols for various functions, including CBRN, space communications satellites, computers, and directed energy, are outlined in a table.*

BODY-150 *Chapter 4 discusses main function symbols for equipment, including psychological operations and radar systems. Sector 1 modifiers for equipment are also covered in Table 4-7.*

BODY-151 *Table showing symbols for equipment functions, including armored vehicles and attack helicopters, with notes on their usage and capabilities.*

BODY-152 *Table listing sector 1 modifiers for equipment, including cargo, command and control, and heavy capabilities.*

BODY-153 *Symbols for equipment modifiers, including light and medium capabilities, and medical evacuation capabilities. Symbols are interchangeable between sector 1 and sector 2 with the same meaning.*

BODY-154 *Table listing sector 1 modifiers for equipment, including medical evacuation vehicles and petroleum transport vehicles.*

BODY-155 *Equipment symbols for different functions and capabilities, including recovery and maintenance, robotic control, and utility vehicles.*

BODY-156 *Tables showing sector modifiers for equipment, including capabilities such as water transport, armored tracked vehicles, amphibious vehicles, and heavy launchers.*

BODY-157 *Symbols and examples of equipment modifiers for different capabilities and classes, including light, medium, pack animals, rail, tractor trailers, and high mobility vehicles.*

BODY-158 *Table showing Sector 2 modifiers for equipment, focusing on wheeled standard mobility capabilities.*

BODY-159 *Chapter 5 discusses military symbols for controlling and coordinating land operations. Control measures regulate forces and depict offensive and defensive capabilities. Operation planning symbols identify planned actions and obstacles. Friendly symbols are black or blue, hostile symbols are red, and obstacles are green or black.*

BODY-160 *Chapter 5 discusses labeling control measures, status of control measures, main function symbols, sector 1 modifier symbols, and amplifiers for tactical symbols. It provides guidelines for symbol makers and explains the use of different symbols and modifiers in military symbol constructs.*

BODY-161 *Control Measures and Operation Planning Symbols include various fields such as main function, echelon, direction of movement, altitude, and distance to provide detailed information in military symbols.*

BODY-162 *Chapter 5 discusses amplifier descriptions for control measure symbols, including target numbers, target number extensions, country codes, echelon indicators, and direction of movement indicators.*

BODY-163 *Explanation of offset location indicator and additional information amplifier symbols used in military planning, with examples and guidelines for their placement on symbols.*

BODY-164 *The page discusses boundaries as lines that facilitate coordination between units, with a template for boundary composition including optional field amplifiers for orientation and identification purposes.*

BODY-165 *The page discusses control measures and operation planning symbols, including examples of friendly present boundaries, friendly planned boundaries, and enemy known boundaries.*

BODY-166 *Chapter 5 discusses boundaries and areas in military operations, providing templates and examples for visual representation.*

BODY-167 *Table listing area control measure main function symbols and examples, including airfield, area of operations, and assembly area.*

BODY-168 *This page lists area control measures in military operations, including base camps, engagement areas, guerrilla bases, named areas of interest, and objective areas.*

BODY-169 *Symbols and control measures for area operations, including target areas, assault positions, attack positions, and holding areas for detainees.*

BODY-170 *Chapter 5 discusses area control measures, including holding areas, support areas like forward arming and refueling points, brigade support areas, and division support areas. Symbols and examples are provided for each function.*

BODY-171 *Table listing area control measures symbols and their functions, including support areas, zone areas like drop zones and landing zones, kill zones, and pickup zones. Battle positions and operations area control measures are also discussed.*

BODY-172 *Chapter 5 discusses battle positions and unique operation area templates, providing examples such as battle position XRAY and company strong point MOCA.*

BODY-173 *The page discusses control measures and operation planning symbols, including templates for battle positions and unique operation areas, such as encirclement and airhead lines for airborne assaults.*

BODY-174 *Template and examples of battle positions and unique operation areas, including unexploded explosive ordnance areas and limited access areas with mobility sector modifiers.*

BODY-175 *This page discusses the use of points as control measures in military operations, including their function, templates, and symbol fields for additional information. Points are specified geographic positions used for various purposes within an area of operation.*

BODY-176 *The page discusses point control measure symbols and their use in exercising authority over assigned forces. It includes a template for constructing these symbols and lists examples such as checkpoints for movement and maneuver.*

BODY-177 *Symbols and definitions for control measures and operation planning in military operations, including checkpoint, control point, engineer regulating point, and linkup point.*

BODY-178 *Table listing point control measure symbols and functions for movement and maneuver, including passage points, rally points, release points, and start points with examples and translations.*

BODY-179 *Symbols for control measures and operation planning, including point of departure, amnesty point, civilian collection point, and detainee collection point, with examples and translations provided.*

BODY-180 *The page discusses point control measures, including symbols and examples for personnel recovery, isolated personnel recovery points, search and rescue points, and decontamination.*

BODY-181 *Table listing main function symbols for point control measures related to decontamination equipment and points.*

BODY-182 *Table of point control measure main function symbols for decontamination, field artillery firing points, and survey control points.*

BODY-183 *Symbols and control measures for operation planning, including points for ambulance exchange, control, loading, and relay, as well as ammunition supply points.*

BODY-184 *Table listing point control measure main function symbols, including examples such as ammunition transfer holding points and casualty collection points.*

BODY-185 *Symbols and control measures for sustainment, medical evacuation, mortuary affairs, rearm/refuel/resupply, refuel on the move, and traffic control posts in military operations.*

BODY-186 *Table 5-6 lists point control measure symbols for sustainment and NATO class items, including subsistence, supplies, and fuels.*

BODY-187 *Symbols for control measures and operation planning related to sustainment distribution, Class IV supplies, ammunition, personal demand items, and major end items are outlined in Table 5-6 of FM 1-02.2.*

BODY-188 *Chapter 5 discusses point control measures and symbols for sustainment distribution, including NATO and U.S. supply classes. It also introduces distinctive action points with unique formats and symbols for specific actions.*

BODY-189 *This page outlines various control measures and operation planning symbols used in military operations, including contact points, decision points, and key terrain.*

BODY-190 *Chapter 5 discusses distinctive action point control measures, including waypoints, target handover, and harbor entrance points. It also covers line control measure symbols and templates for ease of reference in orders and transmissions.*

BODY-191 *Line control measure symbols and templates for operation planning, including forward line of own troops and friendly present/planned/enemy suspected positions.*

BODY-192 *Symbols for line control measures in military operations, including line of contact and phase line symbols.*

BODY-193 *The page discusses various line control measure symbols used in military operations, including release lines, bridgehead lines, battlefield handover lines, and delay lines.*

BODY-194 *Symbols for line control measures in offensive operations, including final coordination line, limit of advance, and line of departure.*

BODY-195 *The page discusses line control measure symbols for offensive and fire lines, including the probable line of deployment and common sensor boundary.*

BODY-196 *This page discusses line control measure symbols related to fire lines and fire support coordination lines in military operations.*

BODY-197 *The page discusses line control measure symbols for fire lines, engineer lines, and intelligence coordination lines, as well as movement control measures and forms of maneuver in military operations.*

BODY-198 *Table 5-9 in Chapter 5 of FM 1-02.2 discusses movement control measures and forms of maneuver symbols, outlining distinct tactical combinations of fire and movement in relation to the enemy.*

BODY-199 *The page discusses movement control measures and forms of maneuver symbols, including the axis of advance and direction of attack in military operations.*

BODY-200 *Movement control measures and forms of maneuver symbols are outlined, including examples and translations for direction of attack, direction of main attack, direction of supporting attack, and enemy or suspected direction of supporting attack.*

BODY-201 *Movement control measures and forms of maneuver symbols in military operations, including infiltration lanes, envelopment, infiltration, frontal attack, and penetration.*

BODY-202 *Chapter 5 discusses movement control measures and forms of maneuver symbols in offensive operations, including turning movement, movement to contact, exploitation, and pursuit. Offensive actions are conducted to defeat and destroy enemy forces, seize terrain, resources, and population centers.*

BODY-203 *The page discusses offensive operations planning symbols, including variations of movement to contact and attack. It also mentions defensive operations planning symbols and retrograde task symbols.*

BODY-204 *Chapter 5 discusses defensive operations planning symbols, including control measures such as area defense, mobile defense, delay, retirement, and withdrawal. These symbols are used to plan and execute defensive operations effectively.*

BODY-205 *Table 5-13 lists symbols for enabling operations planning, including security, reconnaissance, relief in place, and passage of lines. Security operations provide early warning and maneuver space for friendly forces. Symbols have specific rules for orientation and depiction of tasks.*

BODY-206 *Chapter 5 discusses enabling operations planning symbols such as guard, screen, passage of lines, and relief in place. It also covers observation post control measures for military operations.*

BODY-207 *Symbols for observation posts and military deception in military operations are outlined, including their functions and definitions. Military deception symbols are used to mislead adversaries and manipulate their actions.*

BODY-208 *Chapter 5 discusses deception symbols, such as feint, decoy, or dummy indicators, used in military operations to mislead the enemy. Examples include dummy supply points, vehicles, and minefields.*

BODY-209 *The page discusses deception symbols, variations of tactical deception, and mobility and countermobility control measures in military operations.*

BODY-210 *Chapter 5 discusses mobility control measure symbols, including examples such as gaps, passage lanes, water crossing sites, bridges, and fords. These symbols aid in maneuvering forces in tactical formations and planning river crossings during military operations.*

BODY-211 *This page discusses mobility control measure symbols and route control measures, including examples such as ferry crossings and main supply routes.*

BODY-212 *Chapter 5 discusses route control measures and convoy symbols for military operations, including main and alternate supply routes with different traffic flow patterns.*

BODY-213 *The page discusses convoy symbols and countermobility operations, focusing on obstacle integration and control measures to impede enemy movement. Symbols and templates are provided to aid in planning and execution.*

BODY-214 *Chapter 5 discusses countermobility symbols and obstacle control measure symbols, outlining various obstacle effects and measures for disrupting, fixing, and turning enemy formations.*

BODY-215 *This page outlines various countermobility symbols and control measures related to obstacles and explosives in military operations.*

BODY-216 *Chapter 5 discusses countermobility symbols, specifically focusing on constructed obstacle symbols such as wire fences and abatis.*

BODY-217 *Symbols and templates for countermobility measures, including high wire fences, concertina wire, and anti-tank ditches with various configurations and reinforcements.*

BODY-218 *Chapter 5 discusses land mines and minefield control measures, including symbols and modifiers for different types of mines and their placement. Minefields are areas dangerous due to the presence of land mines, with specific symbols used to indicate their location and type.*

BODY-219 *Symbols for land mine and minefield control measures, including wide area anti-tank mines and booby traps, are outlined in FM 1-02.2.*

BODY-220 *Chapter 5 discusses land mine and minefield control measure symbols, including templates and examples for different types of minefields. Minefield sector 1 modifiers are used to identify the type of minefield.*

BODY-221 *Table listing modifiers for different types of mines in a minefield sector.*

BODY-222 *Table showing modifiers for different types of mines in a minefield sector.*

BODY-223 *Table detailing modifiers for different types of minefields, including antipersonnel mines, anti-tank mines, and directional effects.*

BODY-224 *Chapter 5 discusses minefield sector modifiers and field fortification symbols, including shelter, above ground facility, underground facility, and fort.*

BODY-225 *The page discusses field fortification symbols and airspace coordinating measures to enhance mission effectiveness and prevent interference in military operations. It includes symbols for fortified/trench lines, fighting positions, and airspace corridors.*

BODY-226 *Chapter 5 of FM 1-02.2 discusses airspace coordinating measure symbols, including air corridors, low-level transit routes, and minimum-risk routes. Each symbol represents a specific type of airspace designated for use by friendly aircraft to prevent friendly fire incidents.*

BODY-227 *Symbols for airspace coordinating measures, including safe lanes, special corridors, standard use Army aircraft flight routes, and transit corridors, are outlined in FM 1-02.2.*

BODY-228 *Airspace coordinating measures symbols for unmanned aircraft corridors, air control points, and identification lines for friend-or-foe signals.*

BODY-229 *Symbols and control measures for airspace planning and coordination, including engagement zones and restricted operations zones with specific altitudes and time frames.*

BODY-230 *Symbols and translations for airspace coordinating measures, including air-to-air refueling zones, unmanned aircraft zones, and weapon engagement zones, with specific operational details provided.*

BODY-231 *This page provides examples and translations of symbols used for airspace coordinating measures, including joint engagement zones, missile engagement zones, and low-altitude missile engagement zones.*

BODY-232 *Symbols and definitions for airspace coordinating measures, including high-altitude missile engagement zones, short-range air defense engagement zones, and weapons free zones.*

BODY-233 *Symbols and definitions related to airspace coordinating measures and base defense zones are provided in a table format.*

BODY-234 *Fire support coordination control measures are used by commanders to quickly engage targets while protecting friendly forces. These measures are labeled with abbreviations, controlling headquarters, and effective times. Symbols for these measures are listed in Table 5-24.*

BODY-235 *The page discusses fire support coordination control measure symbols, including no-fire areas and restrictive fire areas with examples of irregular, rectangular, and circular shapes.*

BODY-236 *Symbols for fire support coordination control measures, including restrictive fire areas and position areas for artillery, are outlined in Table 5-24 of FM 1-02.2.*

BODY-237 *The page discusses fire support coordination control measure symbols, including direction of fire and munition flight paths. It also introduces target control measures for military operations.*

BODY-238 *Chapter 5 discusses target control measure symbols, including target reference points, point targets, and linear targets. It also defines final protective fire as a barrier of fire to impede enemy movement.*

BODY-239 *Symbols for target control measures in military operations, including irregular, rectangular, and circular area targets, with specific descriptions and guidelines for each type.*

BODY-240 *Chapter 5 discusses target control measure symbols, including group of targets, series of targets, linear targets, area targets, and smoke obscurant.*

BODY-241 *The page discusses target control measure symbols related to fire support areas and naval gunfire support stations, as well as target acquisition control measures for effective employment of capabilities in military operations.*

BODY-242 *Chapter 5 discusses target acquisition control measure symbols and templates for identifying functions and zones, such as artillery target intelligence zones in enemy territory.*

BODY-243 *Symbols for target acquisition control measures and operation planning, including examples like call for fire zones and censor zones in various shapes.*

BODY-244 *Symbols for target acquisition control measures, including critical friendly zones and dead space areas, are defined and illustrated in various shapes for tactical planning and protection of assets.*

BODY-245 *This page discusses target acquisition control measure symbols, including Blue and Purple Kill Boxes for attacking surface targets with various munitions without further coordination with area commanders.*

BODY-246 *This page provides symbols for target acquisition control measures, including range fans to identify effective distance, altitude, and sectors for engaging weapons or sensors. Blue symbols are for explanatory purposes only.*

BODY-247 *CBRN contamination markers and symbols are used to depict events and contaminated areas resulting from CBRN attacks. Event symbols identify location, date, type, size, and movement direction. Contaminated area control measures visually identify and warn of specific hazards. Additional information can be found in ATP 3-11.36 and ATP 3-11.37.*

BODY-248 *Symbols and examples of CBRN contaminated markers for biological, nuclear, and radiological events, including specific details of delivery and location.*

BODY-249 *The page discusses CBRN contaminated marker symbols and CBRN contour lines used in operation planning for chemical, nuclear, and radiological threats, including minimum safe distance zones and radiation dose rate contour lines.*

BODY-250 *blank*

BODY-251 *Symbols for tactical mission tasks are provided, with examples such as attack by fire, block, and breach. These symbols are used in course of action sketches and maneuver sketches to describe specific activities units perform during tactical operations.*

BODY-252 *Chapter 6 of FM 1-02.2 discusses various tactical mission task symbols, including bypass, canalize, clear, contain, control, destroy, and disengage. Each task involves different actions to be taken in relation to enemy forces.*

BODY-253 *Tactical mission tasks include disrupting enemy formations, exfiltrating soldiers, fixing enemy units, following and assuming lead forces, following and supporting lead forces, and interdicting enemy units.*

BODY-254 *Tactical mission tasks include isolating, neutralizing, occupying, retaining, securing, and seizing enemy units or areas.*

BODY-255 *Tactical mission tasks include support by fire, suppress, and turn enemy units, as outlined in FM 3-90.*

BODY-256 *blank*

BODY-257 *This chapter discusses the purpose and makeup of a course of action sketch, which is a graphic portrayal of the movement and maneuver aspects of a mission plan,* including forces, control measures, and enemy locations. Task organization composition symbols provide detailed information for course of action sketches.

BODY-258 Chapter 7 discusses the use of symbols in task organization, including identifying subordinate units, echelon levels, task forces, and combat effectiveness. Symbols are based on unit main and modifier symbols, with no limit on the number of symbols used.

BODY-259 Symbols for task organization in military units, including air assault infantry, airborne infantry, air defense, reconnaissance, anti-tank, armored vehicles, assault helicopters, attack helicopters, combined arms, engineers, and field artillery.

BODY-260 Table 7-1 shows task organization symbols for infantry units, including mechanized, mobile (Stryker), mountain, and reconnaissance. Combat effectiveness symbols are used to assess units' ability to perform missions based on factors like ammunition and personnel. Table 7-2 displays combat effectiveness symbols for overall rating and selected items.

BODY-261 The page discusses combat effectiveness symbols, commander's assessment of unit's ability to perform mission, and provides an example of a task force task organization composition symbol for a battalion task force.

BODY-262 blank

BODY-263 Glossary of acronyms and terms used in Army and joint operations, including definitions and abbreviations.

BODY-264 The glossary provides definitions for terms related to aerial delivery, air and missile defense, air assault, air defense, airfield operations, and ambulance services in military operations.

BODY-265 The glossary provides definitions for military terms such as ambush, ammunition supply point, and attack position, aiding in understanding of military operations and strategies.

BODY-266 This glossary provides definitions of military terms such as axis of advance, base camp, battalion, and bypass. It covers various aspects of military operations and tactics in a clear and concise manner.

BODY-267 Military glossary terms including call for fire zone, canalize, casualty collection point, chemical agent, civil affairs, combat outpost, and command and control.

BODY-268 The glossary defines military terms such as common sensor boundary, company, control, convoy, and counterattack. It also explains concepts like contracting support, coordination points, and cordon and search operations.

BODY-269 Military glossary terms including crew, critical friendly zone, cyberspace operations, decision point, decontamination, delay, demonstration, destroy, detachment, detainee holding area, directed energy, direction finding, direction of attack, and direction of fire.

BODY-270 The glossary defines military terms such as disengage, disrupt, division, drop zone, and early-entry command post. It also includes definitions for terms like electromagnetic warfare, enabling operation, encirclement, engagement area, and explosive ordnance disposal.

BODY-271 The glossary defines military terms such as feint, field artillery, final coordination line, and forms of maneuver, providing clear and concise explanations for each term.

BODY-272 *Military glossary terms including forward line of own troops, forward observer, free-fire area, frontal attack, and improvised explosive device are defined in this page.*

BODY-273 *The glossary provides definitions for military terms such as information operations force, interdict, key terrain, and liaison.*

BODY-274 *The glossary provides definitions for military terms such as low-altitude missile engagement zone, main supply route, and military deception.*

BODY-275 *Military glossary terms related to operations, objectives, and obstacles, including movement to contact, multinational operations, neutralize, and observation post.*

BODY-276 *The glossary defines military terms such as obstacle zone, occupy, passage lane, and personnel recovery. It also includes definitions for psychological actions, public affairs, and pursuit operations.*

BODY-277 *The glossary provides definitions of military terms such as raid, rally point, Rangers, recovery, and route. It also explains concepts like rearward passage of lines, relief in place, and retrograde operations.*

BODY-278 *The glossary defines military terms such as SEAL team, search and rescue, security, special forces, and squad. It also includes definitions for tactical mission tasks and operational concepts related to military operations.*

BODY-279 *Definitions of military terms such as start point, strong point, supply, support, surveillance, sustainment, and tactical mission tasks are provided in this glossary.*

BODY-280 *The glossary defines military terms such as task organization, theater army, troop, and turning movement, providing clear and concise explanations for each term.*

BODY-281 *A list of required and related military publications, including online resources and dates of access, for users of the publication.*

BODY-282 *A list of references for various joint operations and Army publications, including intelligence, operations, protection, offense and defense, sustainment, and mission command.*

BODY-283 *A list of various Army tactics, techniques, and procedures manuals, covering topics such as reconnaissance, infantry operations, medical logistics, and supply operations.*

BODY-284 *A list of military references and manuals covering various topics such as operations, support, tactics, and health systems.*

BODY-285 *List of international standardization agreements, websites for various US government agencies, and reference forms available on the Army Publishing Directorate website.*

BODY-286 *blank*

BODY-287 *Index of military symbols and amplifiers for various activities, equipment, and operations, including combat effectiveness, engagement, and direction of movement. Symbols and modifiers are detailed for different fields and units.*

BODY-288 *Military symbols and control measures are detailed, including main function symbols, modifiers, and placement diagrams for various operations and units.*

BODY-289 *Index entries for unframed symbols, tactical mission task symbols, unit symbols construct, units, amplifier fields, frame shapes, main function, and modifier symbols.*

BODY-290 *blank*

BODY-291 *The page discusses the official duties and responsibilities of the Vice President of the United States.*

BODY-292 blank
BODY-293 blank
BODY-294 blank
BODY-295 The page is intentionally left blank.

Notable Passages

BODY-11 *This publication is the standard for properly constructing land operations associated military symbols for communicating instructions to subordinate units, commanders, and staffs from company through corps echelons.*

BODY-13 *Military symbols are logograms that represent words or terms used to depict abstract graphic representations of a unit, equipment, installation, activity, control measure, or tactical mission task relevant to military operations.*

BODY-15 *"Status depicts whether an object exists at location identified (status is "present" or "confirmed"), will in the future reside at that location (status is "planned" or "anticipated"), or is thought to reside at that location ("suspected"). (See table 1-2 for a depiction of friendly frames.)"*

BODY-19 *"The military symbol construction process can construct sentences and paragraphs or translate written words into military symbols using MIL-STD 2525E symbol construct standard used by command and control systems."*

BODY-29 *"theater army—An echelon of command designated as the Army Service component command responsible for recommendations of allocation and employment of Army forces to a combatant commander with an assigned area of responsibility." (JP 3-31)*

BODY-37 *Combat effectiveness is the ability of a unit to perform its mission, and this amplifier uses the following rating factors for ammunition, personnel, fuel status, and weapons systems:*

- *Fully operational (FO) - (85 percent or greater)*
- *Substantially operational (SO) - (70 to 84 percent)*
- *Marginally operational (MO) - 50 to 69 percent)*
- *Not operational (NO) - (less than 50 percent)*
- *Unknown (UNK)*

BODY-49 *"Security team with working dog capability"*

BODY-52 *"Maintenance platoon medical—Promotes, improves, conserves, or restores the behavioral and physical well-being of personnel in the Army, and as directed in other Services, agencies, and organizations."*

BODY-53 *"Sustainment—The provision of logistics, financial management, personnel services, and health service support necessary to maintain operations until successful mission completion."*

BODY-55 *Multidomain operations—The combined arms employment of joint and Army capabilities to create and exploit relative advantages to achieve objectives, defeat enemy forces, and consolidate gains on behalf of joint force commanders. (FM 3-0)*

BODY-63 *command and control—The exercise of authority and direction by a properly designated commander over assigned and attached forces in the accomplishment of the mission. (JP 1 VOL 2)*

BODY-64 *"Counterintelligence—Information gathered and activities conducted to identify, deceive, exploit, disrupt, or protect against espionage, other intelligence activities, sabotage, or assassinations conducted for or on behalf of foreign powers, organizations or persons or their agents, or international terrorist organizations or activities."*

BODY-83 *"guerrilla—An irregular, predominantly indigenous member of a guerilla force organized similar to military concepts and structure in order to conduct military and paramilitary operations in enemy-held, hostile, or denied territory."*

BODY-89 "*security force assistance—(DOD) The Department of Defense activities that support the development of the capacity and capability of foreign security forces and their supporting institutions.*"

BODY-91 *support—1. The action of a force that aids, protects, complements, or sustains another force in accordance with the directive requiring such action. 2. A unit that helps another unit in battle. 3. An element of a command that assists, protects, or supplies other forces in combat. (JP 1 VOL 2)*

BODY-95 *This frame is used to identify friendly, enemy, or criminal actions that can reveal civic, ethnic, religious, social, or other grouping activities in an area of interest or operation that may affect unified land operations.*

BODY-102 *The use of unique civilian symbols is a recognition of the larger role of military forces beyond war fighting and reflect stability and support to civil authority activities around the world.*

BODY-105 *Emergency management—The managerial function charged with creating the framework within which communities reduce vulnerability to hazards and cope with disasters.*

BODY-117 "*coerced recruitment—The process of adding new individuals by force or threat.*" - *CR Organized coerced terrorist recruitment activity*

BODY-123 "*The engagement bar may contain information on 1) remote or local engagement; 2) stage of the engagement (for example, assign, cover, engage, hold fire, cease fire, cease engage, break engagement, or missile in flight); and 3) type of weapon assignment (for example, missile, gun, or torpedo).*"

BODY-150 "*psychological operations—A robust military influence capability.*"

BODY-159 *Control measures provide the means to visualize and understand current combat capabilities when combined with other symbols and amplifiers to depict operational information on maps and displays.*

BODY-172 *strong point—A heavily fortified battle position tied to a natural or reinforcing obstacle to create an anchor for the defense or to deny the enemy decisive or key terrain. (ADP 3-90)*

BODY-173 "*Where one force loses its freedom of maneuver because an opposing force is able to isolate it by controlling all ground lines of communications and reinforcement.*" *(FM 3-90)*

BODY-175 "*Points are specified geographic positions used for identifying and regulating movement and maneuver, consolidation, de-confliction, distribution, and reducing hazards within an area of operation. They also can be special interest locations used to coordinate actions or locations of special interest to the mission.*"

BODY-176 *A properly designated commander uses these points to exercise authority and direction over assigned and attached forces to the accomplishment of the mission.*

BODY-177 *Control point RED on main supply route 2, controlled by 615th Military Police Company, operational from 0700 Zulu hour 12 May 2013 to 0900 Zulu hour 12 May 2013*

BODY-188 *Distinctive action points have unique formats (including square, circular, star, and cross) that do not follow a specific template and are unique symbols that represent specific actions (including contact, coordination, decision, targets, communication, and air).*

BODY-197 "*Movement is the positioning of combat power to establish the conditions for maneuver, and maneuver is the employment of forces in the operational area, through movement in combination with fires and information, to achieve a position of advantage in respect to the enemy."*

BODY-199 *main axis of advance—The principal attack or effort into which the commander throws the full weight of the offensive power at his disposal.*

BODY-201 *infiltration—(Army) A form of maneuver in which an attacking force conducts undetected movement through or into an area occupied by enemy forces. (FM 3-90)*

BODY-202 *Offensive actions are combat operations conducted to defeat and destroy forces and seize terrain, resources, and population centers. (See FM 3-90 for more information on offensive actions).*

BODY-205 *"Security Operations: Those operations performed by commanders to provide early and accurate warning of enemy operations, to provide the forces being protected with time and maneuver space within which to react to the enemy, and to develop the situation to allow commanders to effectively use their protected forces." (ADP 3-90)*

BODY-206 *"Passage of Lines. An operation in which a force moves forward or rearward through another force's combat positions with the intention of moving into or out of contact with the enemy."*

BODY-207 *Military deception is actions executed to deliberately mislead adversary military, paramilitary, or violent extremist organization decision makers, thereby causing the adversary to take specific actions (or inactions) that will contribute to the accomplishment of the friendly mission.*

BODY-209 *"Mobility is a quality or capability of military forces which permits them to move from place to place while retaining the ability to fulfill their primary mission."*

BODY-213 *"Countermobility operations are those combined arms activities that use or enhance the effects of natural and man-made obstacles to deny enemy freedom of movement and maneuver, and they include proper obstacle integration with the maneuver plan, adherence to obstacle emplacement authority, and positive obstacle control."*

BODY-219 *"booby trap—A device designed, constructed, or adapted to kill or injure, which functions when a person disturbs or approaches an apparently harmless object or performs an apparently safe act."*

BODY-225 *"Airspace coordinating measures employed to facilitate the efficient use of airspace to accomplish missions and simultaneously provide safeguards for friendly forces."*

BODY-229 *"high-density airspace control zone—Airspace designated in an airspace control plan or airspace control order in which there is a concentrated employment of numerous and varied weapons and airspace users."*

BODY-235 *A location in which specific restrictions are imposed and into which fires that exceed those restrictions will not be delivered without coordination with the establishing headquarters. (JP 3-09)*

BODY-237 *"A target is the object of a particular action, for example a geographic area, a complex, an installation, a force, equipment, an individual, a group, or a system, planned for capture, exploitation, neutralization, or destruction by military forces."*

BODY-238 *final protective fire—An immediately available prearranged barrier of fire designed to impede enemy movement across defensive lines or areas. (JP 3-09.3)*

BODY-247 *"Event symbols are used to identify CBRN detonation or spillage location, incident date, type, size, and direction of movement. A warning system formats, processes, and broadcasts CBRN event plumes symbols to provide common operational picture effects throughout the operational environment."*

BODY-251 *"A task is a clearly defined and measurable activity accomplished by individuals or organizations. A tactical mission task is the specific activity a unit performs while executing a tactical operation or form of maneuver."*

BODY-260 *"Combat effectiveness refers to the ability of a unit to perform its mission. Factors such as ammunition, personnel, status of fuel, and weapons systems availability are assessed and rated. Commanders use this information to provide a net assessment of a unit's ability to perform its mission."*

BODY-265 *"Area defense: A type of defensive operation that concentrates on denying enemy forces access to designated terrain for a specific time rather than destroying the enemy outright."*

BODY-270 *"An area where the commander masses effects to contain and destroy an enemy force." (FM 3-90)*

BODY-274 *"military deception Actions executed to deliberately mislead adversary military, paramilitary, or violent extremist organization decision makers, thereby causing the adversary to take specific actions (or inactions) that will contribute to the accomplishment of the friendly mission."*

BODY-276 *"personnel recovery The sum of military, diplomatic, and civil efforts to prepare for and execute the recovery and reintegration of isolated personnel. Also called PR. (JP 3-50)*

BODY-278 *"security operations Those operations performed by commanders to provide early and accurate warning of enemy operations, to provide the forces being protected with time and maneuver space within which to react to the enemy, and to develop the situation to allow commanders to effectively use their protected forces."*

FM 1-02.2

MILITARY SYMBOLS

FEBRUARY 2024

DISTRIBUTION RESTRICTION:

Approved for public release; distribution is unlimited.

This publication supersedes FM 1-02.2, dated 18 May 2022.

HEADQUARTERS, DEPARTMENT OF THE ARMY

This publication is available at the Army Publishing Directorate site (https://armypubs.army.mil) and the Central Army Registry Site (https://atiam.train.army.mil/catalog/dashboard).

Field Manual
No. 1-02.2

Headquarters
Department of the Army
Washington, DC, 28 February 2024

MILITARY SYMBOLS

Contents

DISTRIBUTION RESTRICTION: Approved for public release; distribution is unlimited.
This publication supersedes FM 1-02.2 dated 18 May 2022.

Figures

Tables

This page intentionally left blank.

Preface

FM 1-02.2, *Military Symbols*, constitutes approved Army military symbols for general use to depict land operations. The principal audience for FM 1-02.2 is all members of the profession of arms. Commanders and staffs of Army headquarters serving as a joint task force or multinational headquarters should also refer to applicable joint or multinational doctrine concerning the range of military operations and joint or multinational forces. Trainers and educators throughout the Army will also use this publication.

Commanders, staffs, and subordinates ensure their decisions and actions comply with applicable U.S., international, and, in some cases, host-nation laws and regulations. Commanders at all echelons ensure their Soldiers operate in accordance with the law of war and the rules of engagement. (See the Department of Defense Law of War Manual for more information on the law of war.)

This publication implements the following international agreements:

STANAG 1059 (ED. 8). *Letter Codes for Geographical Entities*. 19 February 2004.

STANAG 1241. (ED. 5). *NATO Standard Identity Description Structure for Tactical Use*. 7 April 2005.

STANAG 2019 (ED 7)/APP 6 (E). *NATO Joint Military Symbology*. 11 October 2023.

FM 1-02.2 uses joint terms where applicable. Selected joint and Army terms and definitions appear in both the glossary and the text. When first defining proponent definitions in the text, the term is italicized and the number of the proponent publication follows the definition. Following uses of the term are not italicized.

FM 1-02.2 applies to the Active Army, Army National Guard/Army National Guard of the United States and United States Army Reserve unless otherwise stated.

The proponent of FM 1-02.2 is the United States Army Combined Arms Center. The preparing agency is the Combined Arms Doctrine Directorate, United States Army Combined Arms Center. Send comments and recommendations on DA Form 2028 (*Recommended Changes to Publications and Blank Forms*) to Commander, United States Army Combined Arms Center and Fort Leavenworth, ATZL-MCD (FM 1-02.2), 300 McPherson Avenue, Fort Leavenworth, KS 66027-2337; by email to usarmy.leavenworth.mccoe.mbx.cadd-org-mailbox@army.mil; or submit an electronic DA Form 2028.

This page intentionally left blank.

Introduction

This publication compiles Department of Defense Military Standard (MIL-STD) 2525E approved military symbols applicable to land operations for use in U.S. Army doctrinal publications, situation maps, overlays, and annotated aerial photographs for all types of military operation. MIL-STD 2525E is the single standard for developing and depicting computer-generated military symbols for use in command and control systems. FM 1-02.2 is the proponent for hand drawn alternate symbols, course of action sketch symbols applicable to U.S. Army doctrine, and approved for use military symbols that are not currently included in MIL-STD 2525E due to difference in revision timeline. This publication is the standard for properly constructing land operations associated military symbols for communicating instructions to subordinate units, commanders, and staffs from company through corps echelons.

This publication is augmented by FM 1-02.1, *Operational Terms*, and the *Army Dictionary* online. Changes to military symbols occur more frequently than traditional publication media can be updated. The terminology and military symbol database, known as the *Army Dictionary*, is updated monthly to reflect the latest editions of Army publications. (To access the database, go to https://jdeis.js.mil/jdeis/index.jsp?pindex=207, and log in with a common access card.) This database is an official DOD website, maintained by the Combined Arms Doctrine Directorate in collaboration with the Joint Staff Directorate for Joint Force Development. The site is part of the Joint Doctrine, Education, and Training Electronic Information System. It includes all Army doctrinal terms and all military symbols in MIL-STD 2525E, including air, land, maritime, space, activities, and control measures.

FM 1-02.2 is organized in seven chapters:

Chapter 1 introduces military symbol fundamentals.

Chapters 2 through 4 provide symbols for units, organizations, equipment, installations, and activities.

Chapter 5 introduces control measure and operation symbols.

Chapter 6 discusses tactical mission tasks.

Chapter 7 discusses the course of action sketch.

These chapters provide detailed requirements for composing and constructing military symbols. The rules for building a set of military symbols allow enough flexibility for users to create any symbol to meet their operational needs. All military symbols construct standards are governed by MIL-STD 2525E, and this publication serves as the compendium of land related military symbols used in U.S. Army doctrine and training manuals.

The introductory table on page X provides a listing of new and modified military symbol changes published in this manual.

Table introduction-1. New and modified military symbols

Symbol	*Status*	*Symbol category*
area defense	Added	Defensive operation
clearing	Added	Sector 1 modifier for units
communication check point	Removed	Airspace coordinating measure
envelopment	Added	Forms of maneuver
exploitation	Added	Offensive operation
field feeding	Added	field feeding
frontal attack	Added	Forms of maneuver
hide point	Removed	Point control measure
infiltration	Added	Forms of maneuver
intercept	Removed	Sector 2 modifier for units
kidnapping	Removed	Sector 1 modifier for activities
kill zone	Added	Area control measure
mobile defense	Added	Defensive operation
mobile gun system	Removed	Main function symbol for unit symbols
movement to contact	Added	Offense operation
non-headquarters unit location indicators	Added	Amplifier
nuclear target	Removed	Target control measure
obstacle group	Added	Countermobility
penetration	Added	Forms of maneuver
pursuit	Added	Offensive operation
rape	Removed	Sector 1 modifier for activities
target built-up area	Removed	Target acquisition control measure
target value area	Removed	Target acquisition control measure
turning movement	Added	Forms of maneuver
zone of responsibility	Removed	Target acquisition control measure

Chapter 1
Military Symbol Fundamentals

This chapter discusses the MIL-STD 2525E military symbol construct standards for framed and unframed symbol standard identity, physical domain, color usage, and the placement of Main function symbols, modifiers, and amplifiers. Military symbols are logograms that represent words or terms used to depict abstract graphic representations of a unit, equipment, installation, activity, control measure, or tactical mission task relevant to military operations. These symbols are available for use in course of action sketches, visualizing operation orders, planning, maps, overlays, and command and control system displays to represent a current common operational picture.

FRAMED SYMBOLS

1-1. Framed symbols allow the depiction of units, equipment, installations, and activities by using a combination of main function, modifiers, amplifiers, and color (optional) to complete the symbol build. The frame is the border of the symbol and serves as the base to which other symbol components are added, and it indicates the standard identity, physical domain, and status of the object being represented. Framed symbols may use standard identity colors to enhance depiction or may be black and white if the display does not support color. (See table 1-4 on page 5 for a listing of standard identity colors.)

STANDARD IDENTITY

1-2. The framed shape design identifies a symbol's standard identity. The standard identity frame shape categories are friend, assumed friend, hostile, neutral, and unknown. Table 1-1 depicts the standard identity frame shapes categorized by physical domains.

PHYSICAL DOMAIN

1-3. The physical domain defines the primary mission area above the earth's surface (in the air domain or space domain), on the earth's surface (in the land domain or maritime domain), or below the earth's surface (in the subsurface domain). Frame shapes differ by surface (land and sea), sea subsurface, air, and space physical domains. Table 1-1 lists and depicts the frame shapes usage for each of these physical domains by land unit, installation, activity, land and sea surface equipment, air equipment (in flight), space equipment (in orbit), and sea subsurface equipment.

1-4. Frame shape symbol use is dependent on the object's current physical domain. An aircraft, regardless of Service ownership, can be depicted in either the air domain (in flight) or land domain (on the ground), while aviation units are depicted as land units and facilities as land installations. The exception is surface equipment (land and sea) frames that do not change between land and maritime physical domains. An example is an amphibious vehicle that uses the same frame shape regardless of current physical domain because it can operate in both land and maritime domains.

Table 1-1. Standard identities and physical domain frame shapes

Standard Identities and Physical Domains	*Friendly* / *Assumed Friend*	*Hostile*	*Neutral*	*Unknown* / *Pending*
land unit	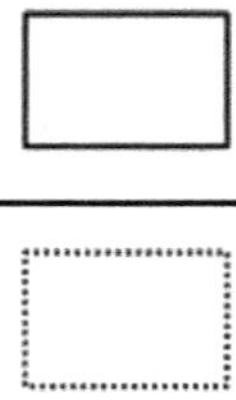		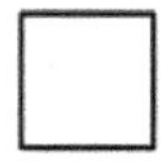	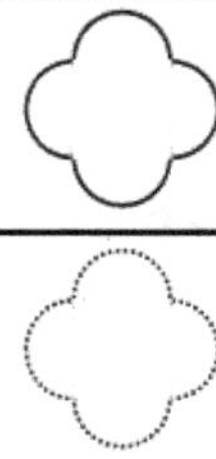
land and sea surface equipment			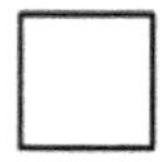	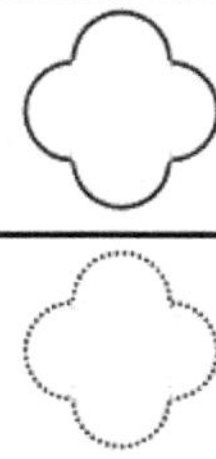
air equipment (in flight)			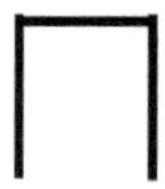	
space equipment (in space)				
activity	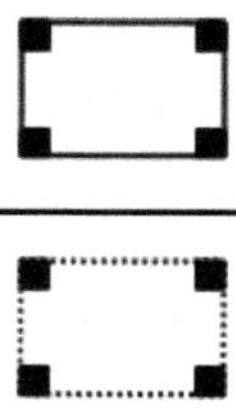	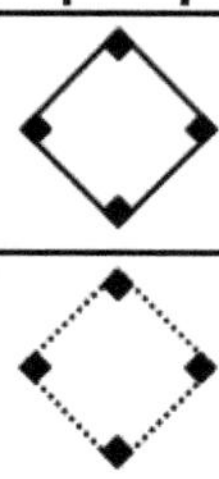		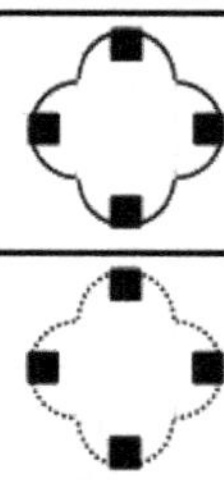
installation	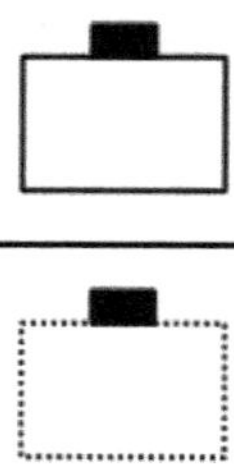			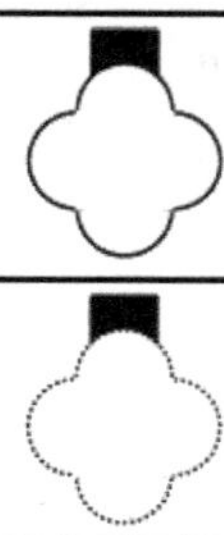
sea subsurface equipment	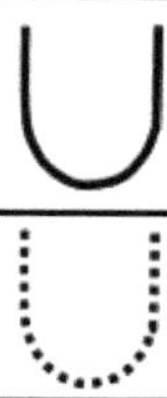			

STATUS

1-5. Status depicts whether an object exists at location identified (status is "present" or "confirmed"), will in the future reside at that location (status is "planned" or "anticipated"), or is thought to reside at that location ("suspected"). (See table 1-2 for a depiction of friendly frames.)

Table 1-2. Friendly frame status examples in present, planned, or suspected

Domain / ***Status***	***Space Equipment***	***Air Equipment***	***Land Unit***	***Land Equipment and Sea Surface***	***Land Installation***	***Sea Subsurface Equipment***	***Activity or Event***
Present or confirmed position				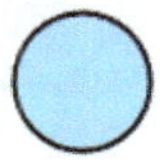			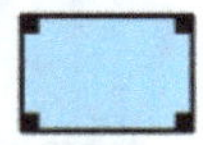
Anticipated, planned, or suspected position							

PLACEMENT DIAGRAM

1-6. The placement diagram serves as a reference for placement of main and modifier symbols for units, equipment, installations, and activities symbols. The framework shown in the placement diagram is not a visible part of the symbol. Table 1-3 provides examples of the placement diagram. These examples are only for illustration purposes and not transmission standards for command and control systems. See MIL-STD 2525E for command and control system standards. The placement diagram is split into three sectors:

- The main function is placed in the center sector.
- The sector 1 modifier is placed in the upper sector.
- The sector 2 modifier is placed in the lower sector.

Table 1-3. Horizontal placement diagram examples

Horizontal placement diagram – center sector

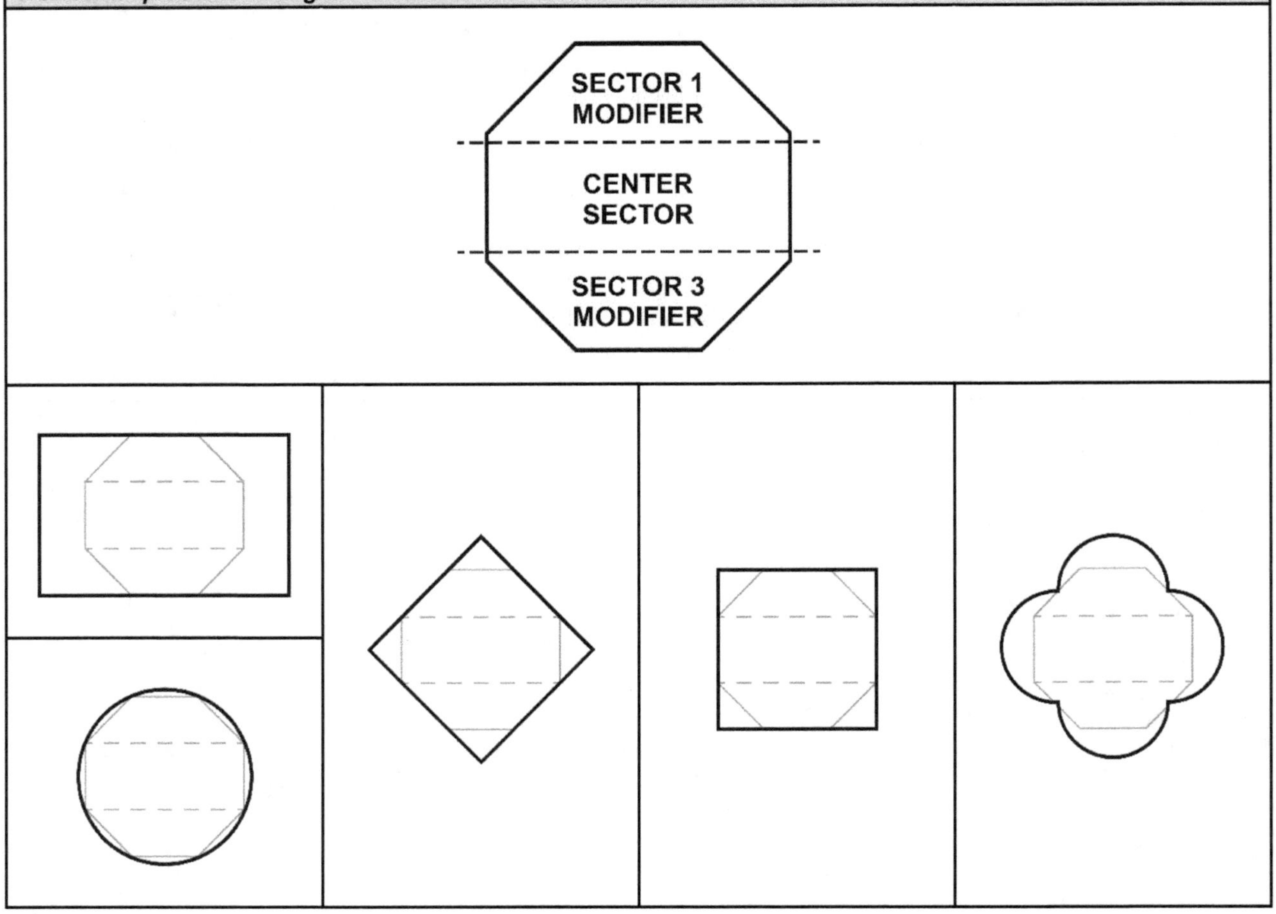

MAIN FUNCTION FOR UNITS, EQUIPMENT, INSTALLATIONS, OR ACTIVITIES

1-7. For a unit, installation, or type of equipment, the main function indicates its role or technical specialty. For activities, the main function represents an action. Normally the main function does not exceed the dimensions of the center sector or touch the interior border when framed. There are exceptions in which the dimensions of the placement diagram do touch the interior border of the frame. These symbols are known as full-frame main functions and occur only in land-domain symbols. Figure 1-1 shows an example of a full-frame main function for all frame shapes.

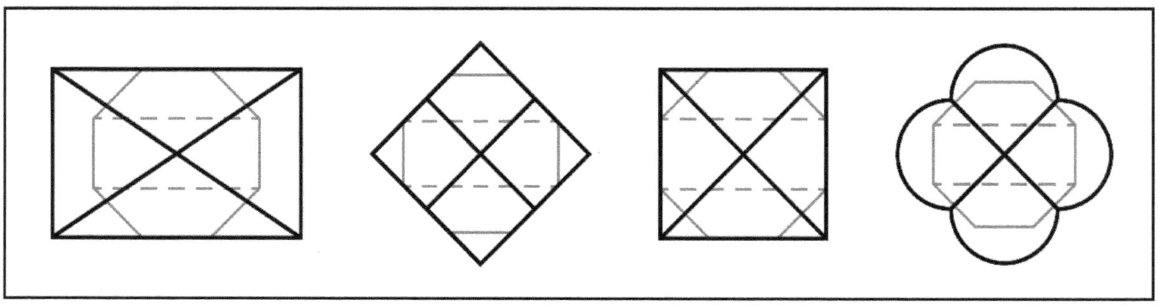

Figure 1-1. Example of full frame Main function symbols

MODIFIER SYMBOL FOR UNITS, EQUIPMENT, INSTALLATIONS, OR ACTIVITIES

1-8. A modifier provides additional capability information. Modifiers conform to the placement diagram and are placed in the sector 1 or 2 modifier areas. Some modifiers are interchangeable (multiple usage) and may be used as either a sector 1 or 2 modifier. This publication defines various types of modifiers and indicates their placement in relation to the main function.

STANDARD IDENTITY COLORS

1-9. Standard identity colors are used as the fill areas for framed symbols. Line colors are used for unframed and framed symbols. Additionally—

- Unframed symbols require color as the standard identity indicator if text amplifiers are not used to categorize the symbol.
- Fill and line color are optional for framed symbols because the frame design provides the standard identity, and color is only a redundant identity indicator used to enhance the framed symbol. The framed symbol fill is normally white or transparent when color is not used in the depiction.

Table 1-4 provides the approved standard identity color variants for military symbol construct fill and line colors.

Table 1-4. Standard identity colors

Description	***Hand-Drawn***	***Computer-Generated Colors***	
		SYMBOL (RGB Value)	***FILL*** (RGB Value)
Friend, assumed friend	Blue	Cyan (0, 255, 255)	Crystal Blue (128, 224, 255)
Unknown, pending	Yellow	Yellow (255, 255, 0)	Light Yellow (255, 255, 128)
Neutral	Green	Neon Green (0, 255, 0)	Bamboo Green (170, 255, 170)
Hostile	Red	Red (255, 0, 0)	Salmon (255, 128, 128)
Boundaries, lines, areas, text, symbols, and frames	Black	Black (0, 0, 0)	Black (0, 0, 0)
Boundaries, lines, areas, text, symbols, and frames.	White	White (255, 255, 255)	Off-White (6% Gray) (239, 239, 239)

AMPLIFIERS

1-10. Amplifiers are optional fields that can be used to include additional information. Not all symbols use amplifiers, and each symbol build has its own unique amplifier placement location in accordance with the appropriate symbology standard. Each respective chapter in this publication provides an applicable amplifier placement template.

SYMBOL LETTERING

1-11. The lettering for all symbols will always be uppercase, Sans Serif font and should be right-aligned on the symbol's left side, and left-aligned on the symbol's right side, and centered on top and bottom. In some cases, the lettering may be tilted slightly to follow the contour of a line, but it must be oriented for left-to-right legibility and not be tilted so much that readers must tilt their heads to read it.

UNFRAMED SYMBOLS

1-12. Equipment symbols may be depicted with frame or unframed. Control measure symbols and mission task symbols are unframed symbols that conform to special rules for their own elements.

UNFRAMED EQUIPMENT SYMBOLS

1-13. Equipment symbols can be constructed with or without a frame. Unframed and framed equipment symbols follow the same build rules except with unframed equipment symbols, the main function serves as the base for the placement of any modifiers and amplifiers. Chapter 4 provides the equipment symbol main function, modifiers, and amplifier guidelines.

TACTICAL MISSION TASK SYMBOLS

1-14. Tactical mission task symbols are used in course of action sketches, synchronization matrices, and maneuver sketches. Tactical mission task symbols are sized to accommodate the scale of the display or map, and they may be used with other framed and unframed symbols, but they do not use modifiers or amplifiers. Chapter 6 provides tactical mission task symbol listings and construct examples.

OPERATIONS PLANNING SYMBOLS.

1-15. Operation planning symbols are not control measures symbols and are used to depict planned actions, events, affects, built-up areas (building, fortified area, obstacles), and gap crossings. They may also be used to depict planned actions conducted during offense and defense operations to enhance the operational picture of an operation.

CONTROL MEASURE SYMBOLS

1-16. Control measures are directives given to assign responsibilities, coordinate fires and maneuver, and control operations. They may be boundaries, special area designations, and other unique markings related to operational environment geometry and necessary for planning and management of operations. Control measure symbols provide graphic operational information that cannot be displayed via symbol-based symbols alone.

1-17. Control measure symbols have unique, build template patterns for each type of control measure, but they use similar standard identity colors and amplifiers as other symbols. They can be black or white, depending on their display background. Display backgrounds can be blue (for friend), red (for hostile), green (for obstacles) or yellow (for a chemical, biological, radiological, and nuclear contaminated area fill). See Chapter 5 for further details.

MAIN FUNCTIONS FOR CONTROL MEASURES

1-18. Like frame symbols, some control measures symbol builds permit the use of field A, and they may be combined with other symbols and amplifiers to display operational information. Control measure templates in Chapter 5 indicate if a build composition provides the capability to add additional information to the symbol.

MODIFIERS FOR CONTROL MEASURES

1-19. Minefields and limited-access area control measures have modifiers that can be used within their symbol build. (See Chapter 5 for modifier usage.)

MILITARY SYMBOL CONSTRUCT PROCESS

1-20. Military symbol construction is a logographic writing system like words in written language. Each symbol has a specific meaning that, when combined with other symbols, provides legible information that can be used to quickly identify units, organizations, and capabilities and to understand current or future actions. The military symbol construction process can construct sentences and paragraphs or translate written words into military symbols using MIL-STD 2525E symbol construct standard used by command and control systems. The symbols in this publication are adequate for depicting a variety of military symbol constructs, but if a user determines there is a gap in the symbol construct language, that user must inform the Army symbologist so collaboration can begin on creating a new military symbol.

CONSTRUCTION PROCESS FOR FRAMED SYMBOLS

1-21. Chapters 2 to 4 provide main functions, modifiers, and amplifiers for building a wide variety of framed symbols. Table 1-5 provides a step-by-step example of how to build a framed symbol for an infantry unit with an armored high-mobility vehicle capability and indicate its echelon and unit designation.

Table 1-5. Construction process for framed symbols

Steps		***Construct example and symbol translation***
1	Choose appropriate frame shape. ***Note***. This example uses the friendly unit frame.	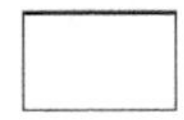 Friendly unit
2	Choose appropriate Main function symbol. ***Note***. This example selects the infantry Main function symbol which is a full frame symbol found in chapter 2.	 Infantry 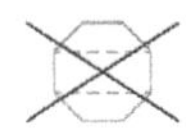 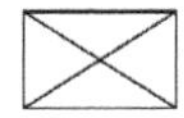 Friendly infantry unit
3	Choose appropriate sector 1 modifier. ***Note***. This example uses the armored protected sector 1 modifier found in chapter 2.	 Armored (protection)  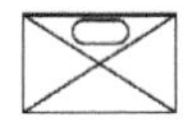 Friendly infantry unit with armored (protection) capability
4	Choose appropriate sector 2 modifier. ***Note***. This example selects the wheeled high mobility sector 2 modifier found in chapter 2.	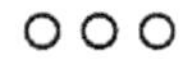 Wheeled high mobility 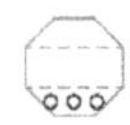 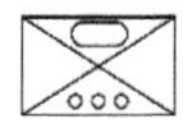 Friendly infantry unit with armored high mobility vehicle capability
5	Choose essential amplifier field from those listed in pertinent chapters. ***Note***. This example uses fields B, H, and M to add echelon and unit designator information to complete the desired military symbol. These specific amplifier symbols and construct usage can be found in chapter 2.	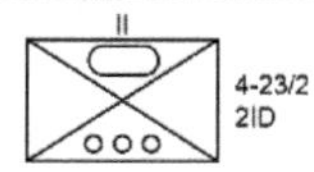 Infantry battalion with armored high mobility vehicle capability, 4th Battalion, 23rd Infantry Regiment, 2nd Brigade, 2nd Infantry Division

CONSTRUCTION PROCESS FOR CONTROL MEASURES

1-22. As part of the symbol build process, many control measure symbols can be combined with other symbols and amplifiers to display operational information in one symbol. Table 1-6 on page 9 is an example of the building process for a control measure.

Table 1-6. Construction process for control measure symbol

<table>
<tr><th colspan="2">Steps</th><th colspan="2">Construct example and symbol translation</th></tr>
<tr><td>1</td><td>Choose an appropriate control measure template with from chapter 5.
Note. This example uses the main axis of advance template.</td><td colspan="2">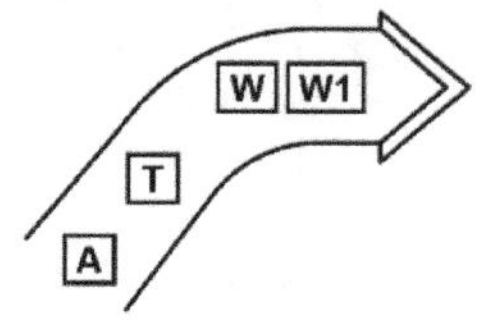

Main axis of advance with amplifier fields</td></tr>
<tr><td rowspan="2">2</td><td rowspan="2">Choose the appropriate amplifier information by field.
Note. This example uses unique designation Field T to name the axis of advance.</td><td>T</td><td>WHITE</td></tr>
<tr><td colspan="2">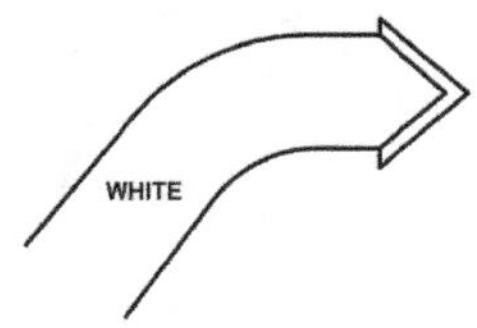

Main axis of advance "White"</td></tr>
<tr><td rowspan="3">3</td><td rowspan="3">Choose the next appropriate amplifier information by field.
Note. This example uses Field W to add a date time group to axis of advance.</td><td>W</td><td>140600ZMAR2019</td></tr>
<tr><td>W1</td><td>If needed</td></tr>
<tr><td colspan="2">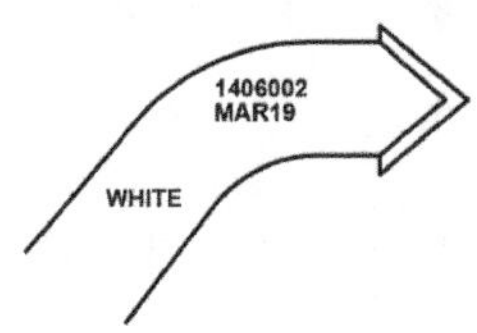

Main axis of advance "White" movement begins at 0600 Zulu hour, 14 MAR 2019</td></tr>
<tr><td rowspan="2">4</td><td rowspan="2">Add a Main function symbol construct to complete the intent of the symbol.
Note. This example uses Field A to add a completed unit symbol construct.</td><td colspan="2">A</td></tr>
<tr><td colspan="2">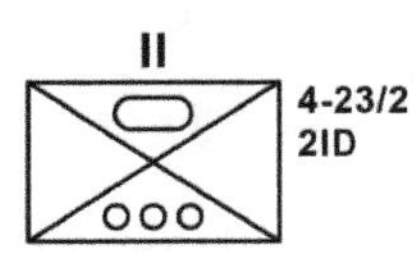

Infantry battalion with armored high mobility vehicle capability, 4th Battalion, 23d Infantry Regiment, 2d Brigade, 2nd Infantry Division</td></tr>
<tr><td>5</td><td>Completed construct of control measure.</td><td colspan="2">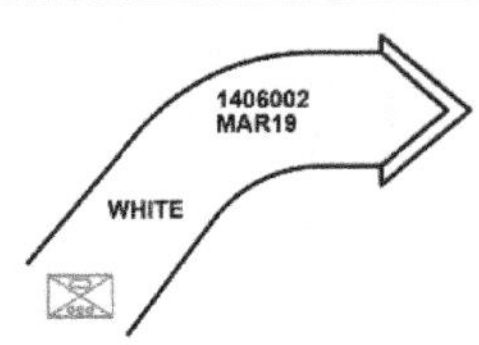

4th Infantry Battalion (Stryker), 23d Infantry Regiment, 2d Brigade, 2nd Infantry Division, moves at 0600 Zulu hour, 14 MAR 2019 on main axis of advance White</td></tr>
</table>

This page intentionally left blank.

Chapter 2

Military Unit and Organizational Symbols

This chapter discusses symbols for units and organizations.

UNIT AND ORGANIZATION SYMBOLS

2-1. A *unit* is any military element whose structure is prescribed by a competent authority (JP 3-33). This section includes the lists of amplifiers, main functions, and modifiers for constructing unit and organization symbols.

UNIT AND ORGANIZATION FRAME SHAPES

2-2. Unit and organization frame shapes are used to identify friend, enemy, neutral, or unknown affiliated units. Table 2-1 provides the standard identity frame shapes for units and organizations. The frame-shape build guidelines for main function, modifier and amplifier placement is provided in paragraph 2-3.

Table 2-1. Unit and organization standard identity frame shapes

Friendly	*Hostile*	*Neutral*	*Unknown*
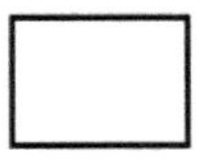			
Assumed Friend	*Suspect*		*Pending*

AMPLIFIER FIELDS FOR UNITS

2-3. The main function, modifier, and amplifier fields standardize the display of alphanumeric information that graphically describes a unit, its capabilities, status, and location. The field placement is the same for all unit standard identity frames (including friend and friend or assumed friend, hostile, neutral, and unknown). Figure 2-1 shows the placement fields for land unit symbols using a friend symbol frame as an example. Table 2-2 on page 13 provides descriptions and formats for each amplifier.

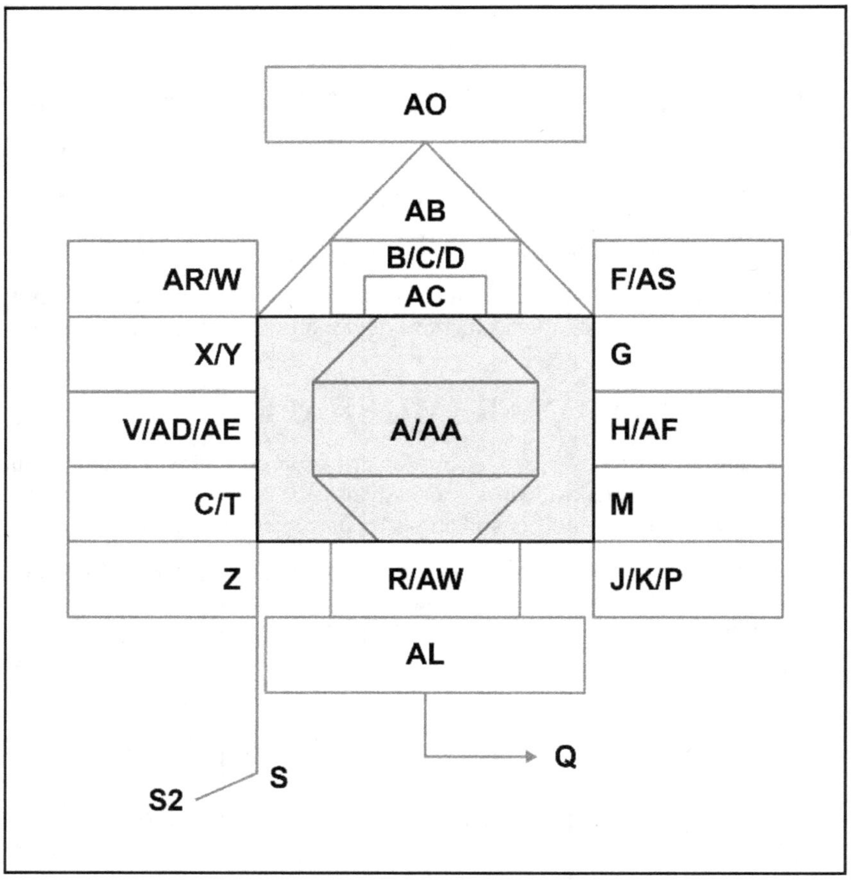

Figure 2-1. Land unit symbol amplifier fields

Table 2-2. Descriptions of Main function symbol and amplifier fields for unit frames

Field	*Field Title*	*Description*
A	Main and modifier symbols	The innermost part of a symbol that represents the main function (the Main function symbol) and its capabilities (modifiers 1 and 2).
B	Echelon	A graphic amplifier in a unit symbol that identifies command level.
C	Quantity	A text amplifier that identifies a specific number and type of items.
D	Task organization indicator	A graphic amplifier that identifies a unit or an activities symbol as a task force.
F	Attached and detached (reinforced or reduced)	A text amplifier in a unit symbol that displays (+) for reinforced, (-) for reduced, (±) reinforced and reduced. ***Note***. This field allows a maximum of 3 characters.
G	Staff comments	A text amplifier for units, equipment, and installations. Content is implementation specific. ***Note***. This field allows a maximum of 20 characters.
H	Additional information	A unique alphanumeric designation that identifies the displayed unit. ***Note***. This field allows a maximum of 20 characters.
J	Evaluation rating	A text amplifier for units, equipment, and installations that consists of a single letter reliability rating and a single digit credibility rating. Reliability ratings: A—completely reliable B—usually reliable C—fairly reliable D—not usually reliable E—unreliable F—reliability cannot be judged Credibility ratings: 1—confirmed by other sources 2—probably true 3—possibly true 4—doubtfully true 5—improbable 6—truth cannot be judged ***Note***. This field allows a maximum of 2 characters
K	Combat effectiveness	A text amplifier for units and installations that indicates effectiveness. The entries are— • Fully operational (FO) • Substantially operational (SO) • Marginally operational (MO) • Not operational (NO) • Unknown (UNK) ***Note***. This field allows a maximum of 5 characters.
M	Higher formation	A text amplifier for units that indicates number or title of higher echelon command (Roman numerals designate corps). ***Note***. This field allows a maximum of 21 characters.
P	Identification, friend or foe Selective identification feature	A text amplifier displaying one or more identification, friend or foe, or selective identification feature identification modes and codes. Display priority is mode 5, mode 4, mode 3, and mode 2. ***Note***. This field allows a maximum of 15 characters.
Q	Direction of movement indicator	A graphic amplifier for units and equipment that identifies the direction of movement or intended movement of an object.
S	Offset location indicator	A graphic amplifier used to indicate the offset or precise location.
S2	Offset location indicator extension line	A multidirectional graphic amplifier line used to extend the offset location indicator amplifier 'Field S' to the desired location point.

Table 2-2. Descriptions of Main function symbol and amplifier fields for unit (*continued*)

Field	*Field Title*	*Description*
T	Unique identifier	1. An amplifier field reserved for command and control systems that uniquely identifies a particular symbol with a track number. Prefix = TN #####. Example: TN: 13579. 2. Also available for unit designation (similar to amplifier Field H usage). ***Note.*** This field allows a maximum of 30 characters.
V	Equipment type	A text amplifier used to indicate a specific type of unit equipment. ***Note.*** This field allows a maximum of 24 characters.
W	Date-time group	An alphanumeric designator for displaying date-time group (DDHHMMSSZMONYYYY) or "O/O" for an order. The date-time group is composed of a group of six numeric digits with at time zone suffix and the standardized three-letter abbreviation for the month followed by four digits representing the year. The first pair of digits represents the day; the second pair, the hour; the third pair, the minutes. For automated systems, two digits may be added before the time zone suffix and after the minutes to designate seconds. ***Note.*** This field allows a maximum of 16 characters.
X	Altitude or depth	A text amplifier that displays either altitude, flight level, depth for submerged objects, or height of equipment or structures on the ground. Measurement units shall be displayed in the string. Examples: 1500MSL FL150 ***Note.*** This field allows a maximum of 14 characters.
Y	Location	A text amplifier that displays a symbol's location in degrees, minutes, and decimal minutes (or in military grid reference system, global area reference system, or other applicable display formats). ***Note.*** This field allows a maximum of 22 characters.
Z	Speed	A text amplifier for units and equipment that displays velocity. ***Note.*** This field allows a maximum of 8 characters.
AA	Special headquarters	A text modifier for units. The indicator is contained inside the frame. A named command such as Supreme Headquarters Allied Powers, Europe, United States Southern Command, United States Central Command, and joint, multinational, or coalition commands such as combined joint task forces or joint task forces. ***Note.*** This field allows a maximum of 9 characters.
AB	Feint or dummy indicator	A graphic amplifier that identifies an offensive or defensive unit that is intended to draw the enemy's attention away from the area of the main attack.
AD	Platform type	A text amplifier containing either electronic intelligence notation (ELNOT) or communications intelligence notation (CENOT).
AE	Equipment teardown time	A numeric amplifier indicating equipment disassembly time in minutes.
AF	Common identifier	An alphanumeric amplifier to provide a name used to identify an entity.
AL	Operational condition	A graphic amplifier that indicates operational condition or capacity
AO	Engagement bar	A graphic amplifier placed immediately atop the symbol. May denote 1) local or remote status, 2) engagement status, and 3) weapon type. Format: A:BBC-CC, where A = remote/local BBB = engagement status CC = weapon asset

Table 2-2. Descriptions of Main function symbol and amplifier fields for unit (*continued*)

Field	*Field Title*	*Description*
AS	Country	A three-letter code that indicates the country of origin of the organization. In stability activities, this field can be used for factions or groups. ***Note***. This field allows a maximum of 3 characters.

2-4. **Echelon and Non-Echelon Amplifiers (Field B).** Figure 2-2 shows the template for an echelon amplifier. The height of the echelon amplifier is one-fourth of the size of the height of the frame. Table 2-3 on pages 15 through 17 shows the Field B amplifiers for Army echelons and non-echelon commands.

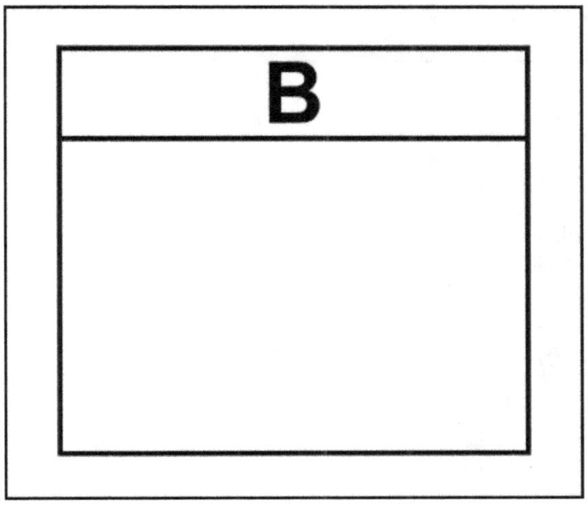

Figure 2-2. Template for an echelon amplifier

Table 2-3. Echelon and non-echelon amplifiers

Echelon	*Amplifier*	*Amplifier Usage Construct Example*
crew—A small military unit that consists of all personnel operating a particular system. (ADP 3-90) **team**—A small group of persons associated by work or activity ***Note.*** This is the smallest echelon and should not be confused with company team or brigade combat team.	Ø	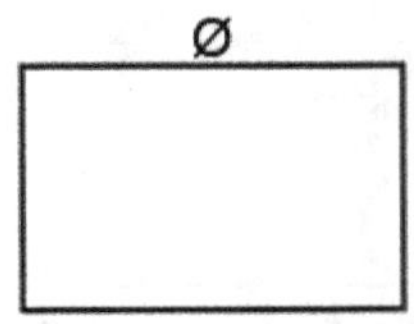
squad—A small military unit typically containing two or more fire teams. (ADP 3-90) ***Note.*** Common English language definition also applies to this symbol, "a small group engaged in a common effort or occupation."	●	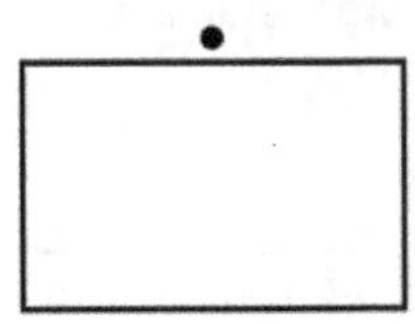
section—A tactical unit of the Army and Marine corps smaller than a platoon and larger than a squad. (ADP 3-90)	●●	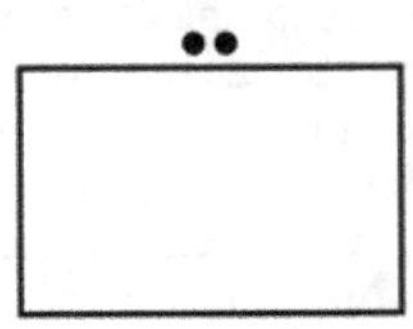

Table 2-3. Echelon and non-echelon amplifiers (*continued*)

Echelon	*Amplifier*	*Amplifier Usage Construct Example*
platoon—A subdivision of a company or troop consisting of two or more squads or sections. (ADP 3-90) ***detachment***—A tactical element organized on either a temporary or permanent basis for special duties. (ADP 3-90)	●●●	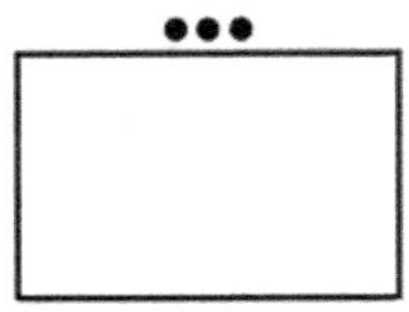
company—A unit consisting of two or more platoons, usually of the same type, with a headquarters and a limited capacity for self-support. (ADP 3-90) ***battery***—A company-size unit in a field artillery or air defense artillery battalion. (ADP 3-90) ***troop***—A company-size unit in a cavalry organization. (ADP 3-90)	I	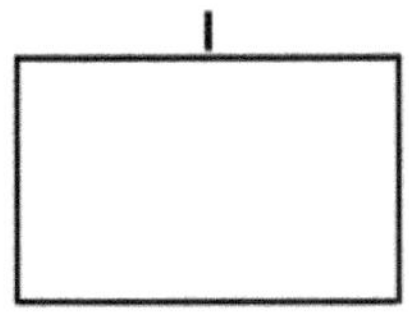
battalion—A unit consisting of two or more company-, battery-, or troop-size units and a headquarters. (ADP 3-90) **squadron**—A unit consisting of two or more troop-size units and a headquarters in a cavalry organization. (See ATP 3-20.96 for more information on the squadron.)	II	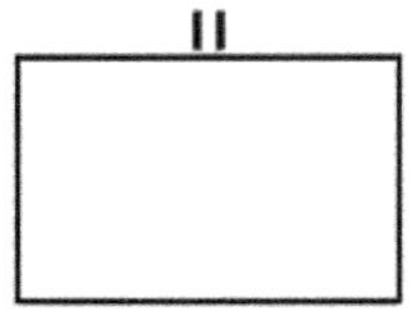
regiment or group—A unit consisting of 2 or more battalions with a group or regimental headquarters.	III	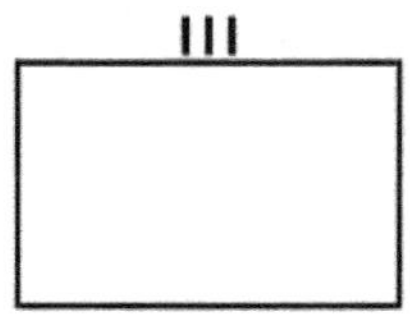
brigade—A unit consisting of two or more battalions and a headquarters company or detachment. (ADP 3-90)	X	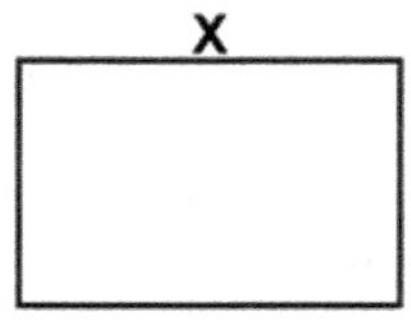
division—An echelon of command and tactical formation that employs brigade combat teams, multi-functional brigades, and functional brigades to achieve objectives on land. (ADP 3-90)	XX	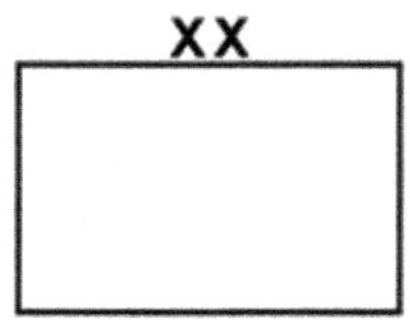
corps—An echelon of command and tactical formation that employs divisions, multi-functional brigades, and functional brigades to achieve objectives on land. (ADP 3-90)	XXX	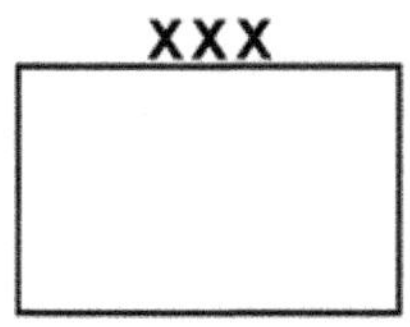

Table 2-3. Echelon and non-echelon amplifiers (*continued*)

Echelon	*Amplifier*	*Amplifier Usage Construct Example*
theater army—An echelon of command designated as the Army Service component command responsible for recommendations of allocation and employment of Army forces to a combatant commander with an assigned area of responsibility. (JP 3-31)	**XXXX**	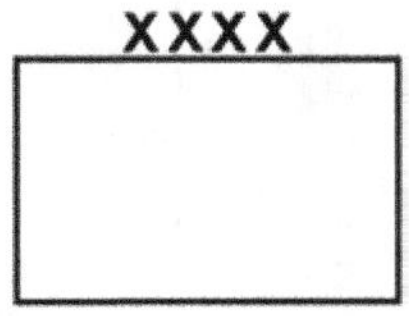
army group—The largest formation of land forces, normally comprising two or more armies or army corps under a designated commander. (APP-06E) ***Note.*** Used in North Atlantic Treaty Organization or multinational military operations.	**XXXXX**	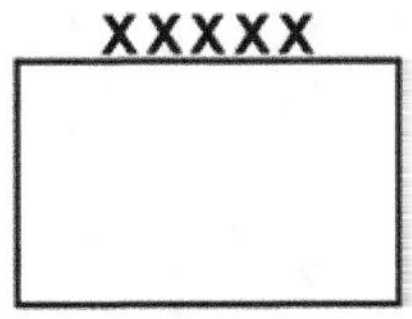
theater—The geographical area for which a commander of a geographic combatant command has been assigned responsibility. (JP 1 Volume 1)	**XXXXXX**	
independent support command—A unit specifically supporting a theater or corps command.	**++**	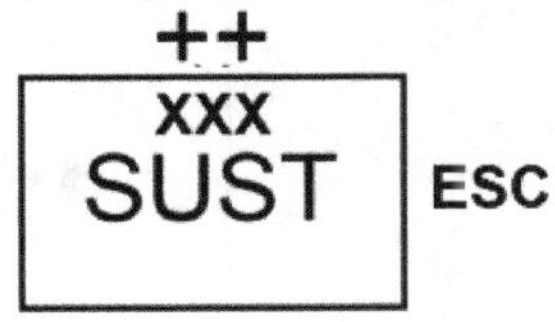 Sustainment command supporting a corps (expeditionary sustainment command) 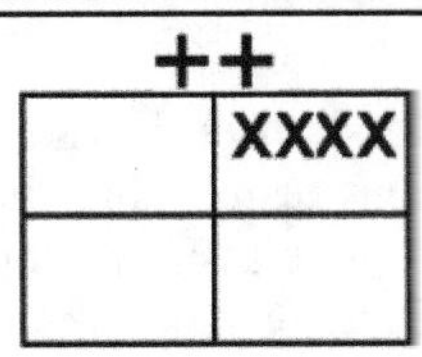 Medical command supporting a theater army (theater medical command)

2-5. **Quantity amplifiers (Field C).** To avoid clutter, the Army prefers that that quantity be placed in field C/T. Users also have the option to use field C above echelon field B (see figure 2-1 on page 12) Both are text amplifiers that can be used to identify the number of items or individuals. Figure 2-3 shows the template for a quantity amplifier and figure 2-4 provides a usage example of a medical augmentation detachment with a 32-hospital bed capability.

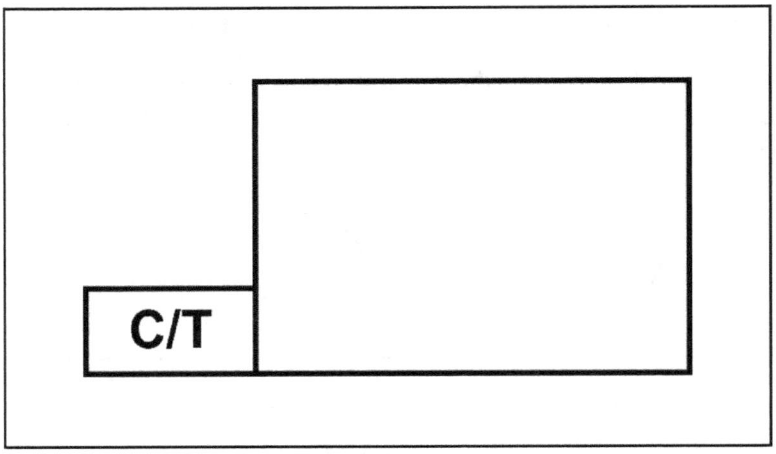

Figure 2-3. Template for quantity amplifier

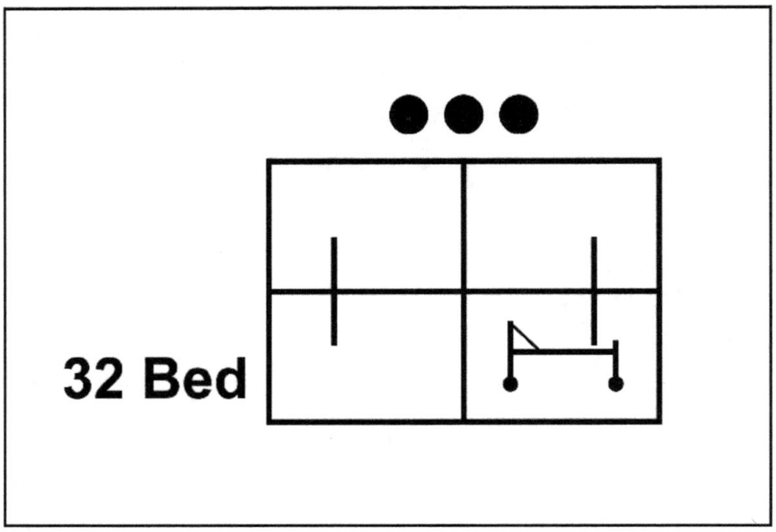

Figure 2-4. Quantity amplifier usage example

2-6. **Task organization indicator amplifier (Field D).** This amplifier is used with a battalion task force or company team. A battalion task force is a maneuver battalion-size unit consisting of a battalion headquarters, at least one assigned company-size element, and at least one attached company-size element from another maneuver or support unit (functional or multifunctional). A company team is a combined arms organization formed by attaching one or more nonorganic armor, mechanized, infantry, Stryker infantry, or infantry platoons to a tank, mechanized, infantry, Stryker, or infantry company, either in exchange for, or in addition to, its organic platoons. (See ADP 3-90 for more information on company teams.) Figure 2-5 on page 19 provides the template for a tsk organization amplifier, and table 2-4 on page 19 shows the task organization amplifier and construct examples.

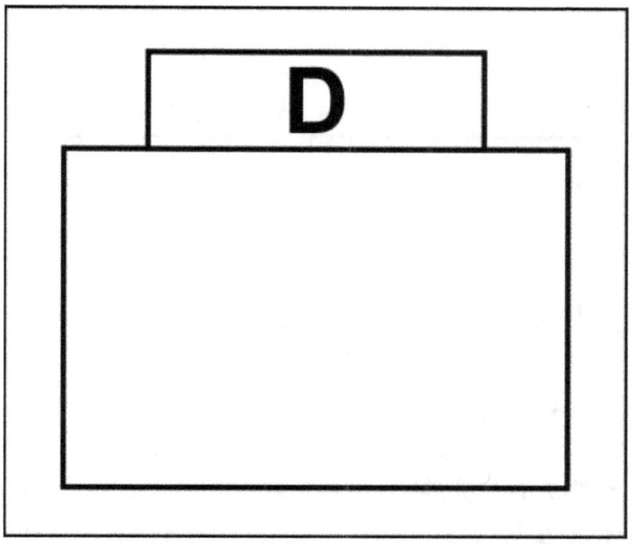

Figure 2-5. Template for task force or team amplifier

Table 2-4. Task organization indicator amplifier

Amplifier	*Symbol*	*Example*
task organization—A temporary grouping of forces designed to accomplish a particular mission. (ADP 5-0) ***Note.*** This amplifier is sized to correspond with the echelon amplifiers being used. The height of the amplifier is one-third of the size of the height of the frame.		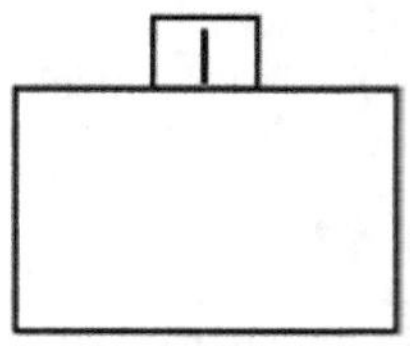 Company team
		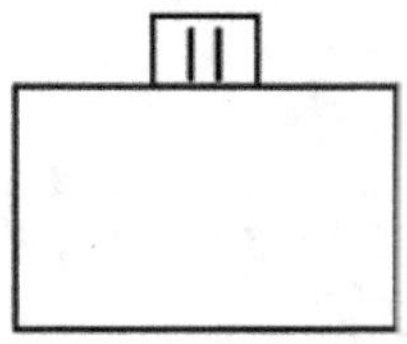 Battalion task force
		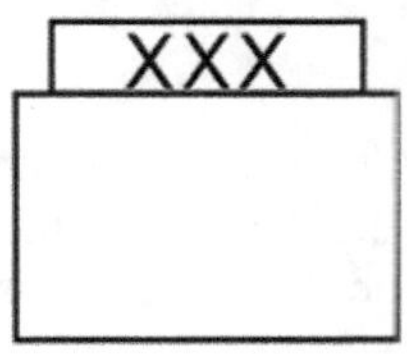 Marine expeditionary force

2-7. **Attached and detached amplifiers (Field F).** A plus symbol (+) is used when attaching one or more subordinate elements. A minus symbol (–) is used when detaching one or more subordinate elements. A plus and minus symbol (±) is used when attaching and detaching one or more subordinate elements. Figure 2-6 (on page 20) shows a template for attached and detached amplifiers. Table 2-5 lists the amplifiers and usage examples. (See FM 6-0 for additional information on attachment and detachment.)

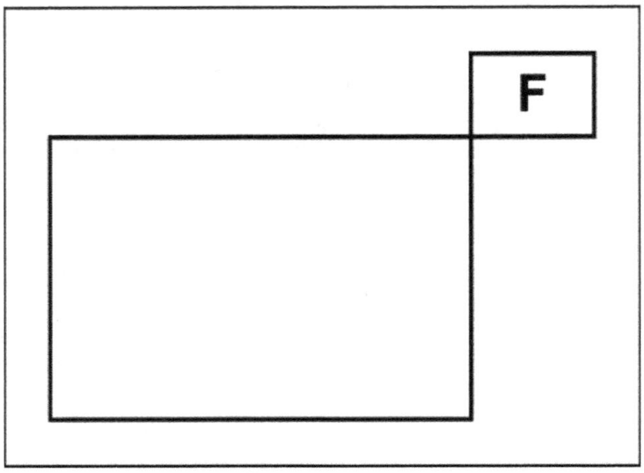

Figure 2-6. Template for attached and detached amplifier

Table 2-5. Reinforced (attached) and Reduced (detached) amplifiers

Function	*Amplifier*	*Amplifier Usage Construct Example*
Reinforced (attached) ***attach***—The placement of units or personnel in an organization where such placement is relatively temporary. (JP 3-0)	**(+)**	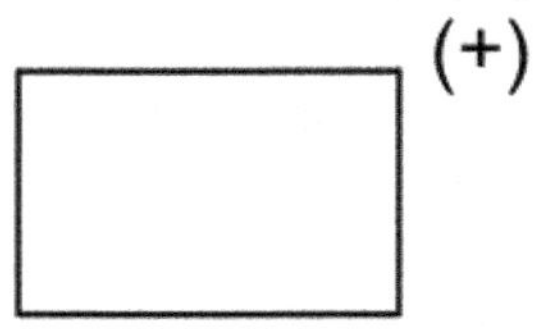
reduced (detached)—The temporary loss of personnel or units to reinforce another unit.	**(–)**	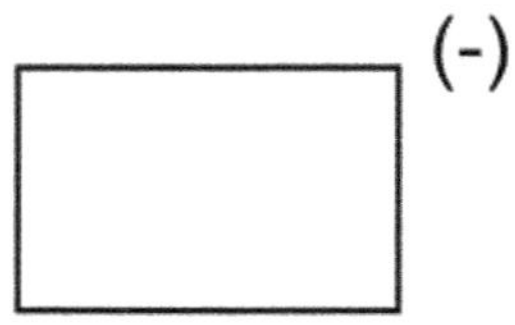
reinforced and reduced (attached and detached)	**(±)**	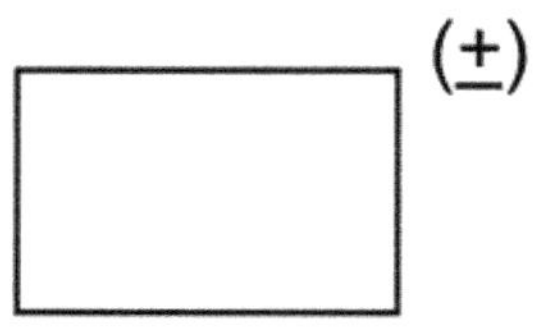

2-8. **Country code amplifier (Field AS).** The country code is a three-letter code that indicates the country of origin. FM 1-02.1 includes a complete listing of geographical entry codes (country codes) that can be used in Field AS. Field AS is located at the right corner of the frame and shares this space with Field F (attached and detached amplifiers). If Field F is in use, the Field AS country code will be placed to the right of Field F as shown in figure 2-7 on page 21. If Field F is not being used, the AS field can occupy the entire space as shown in figure 2-8 on page 21.

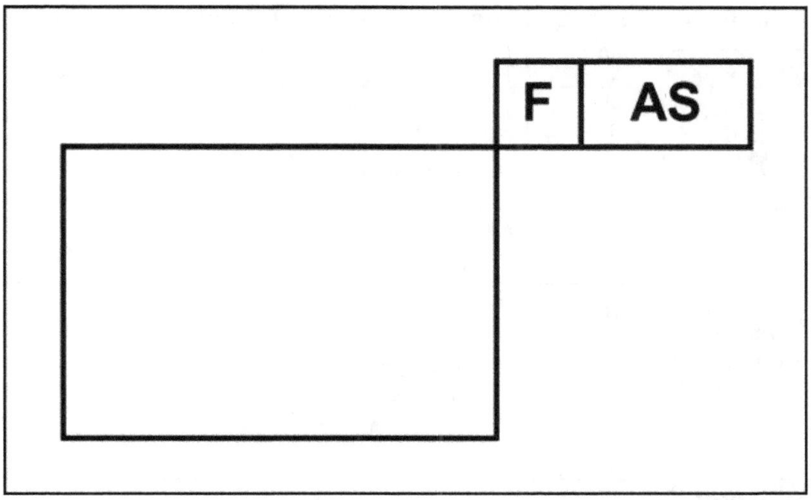

Figure 2-7. Template for country code amplifier Field AS with Field F in use

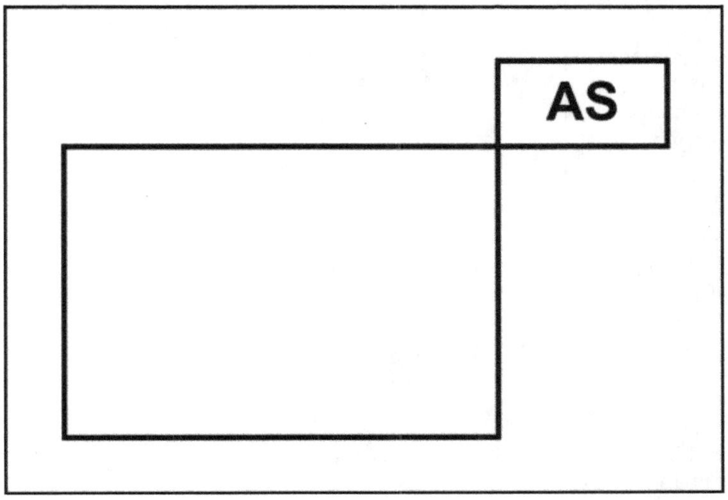

Figure 2-8. Template for country code amplifier Field AS without Field F usage

2-9. **Command post using staff comments amplifier (Field G).** A *command post* is a headquarters or a portion there of, organized for the exercise of command and control (FM 6-0). The headquarters staff indicator (Field S) is always used in conjunction with the command post and command group amplifiers. Figure 2-9 shows the template for the command post using amplifier Field G, and table 2-6 provides amplifier usage examples.

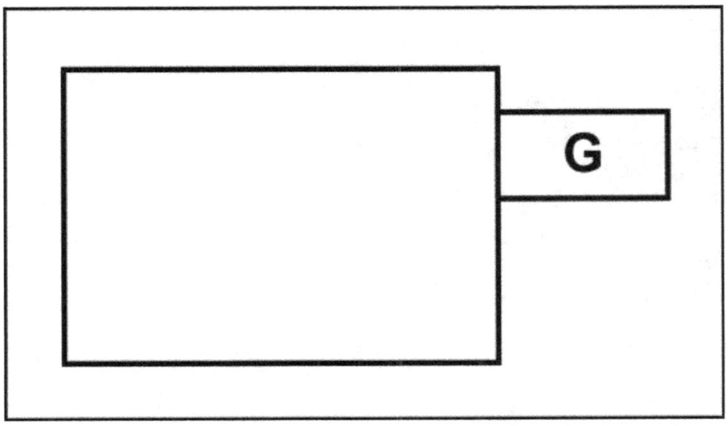

Figure 2-9. Template for command post using amplifier Field G

Table 2-6. Command post amplifier Field G usage examples

Description	*Amplifier*	*Amplifier Usage Example*
combat trains command post—Controls and coordinates administrative and logistic support. (See ATP 6-0.5 for more information on combat trains.)	**CTCP**	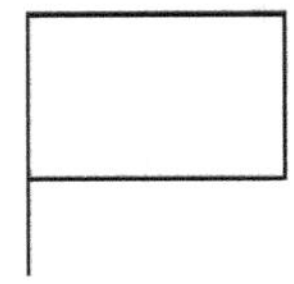 CTCP
early-entry command post—A lead element of a headquarters designed to control operations until the remaining portions of the headquarters are deployed and operational. (FM 6-0)	**EECP**	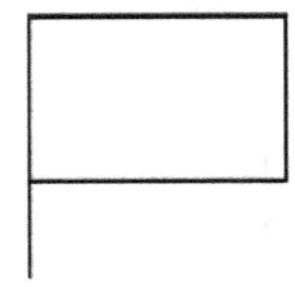 EECP
field trains command post—A facility containing a personnel administration center, elements of the S-4 (battalion or brigade logistics staff officer) sustainment staff section, elements of company supply sections, and elements of the forward support company. (See ATP 6-0.5 for more information on field trains.)	**FTCP**	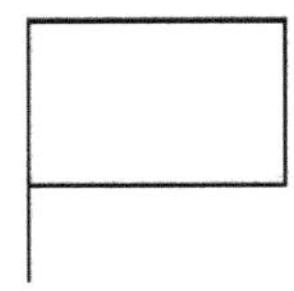 FTCP
mobile command group—A portion of the main command post or tactical command post for headquarters at brigade and lower echelons that permits the commander to maintain command and control when separated from the main, tactical, or rear command post. (See FM 6-0 for more information on mobile command group.)	**MCG**	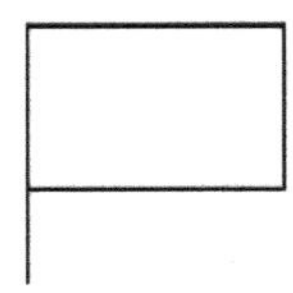 MCG
main command post—A portion of a unit headquarters containing most of the staff designed to command and control current operations, conduct detailed analysis, and plan future operations. (FM 6-0)	**MAIN**	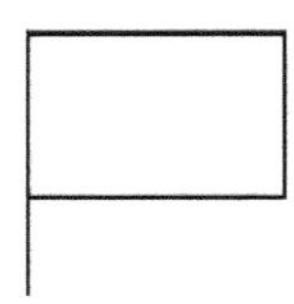 MAIN
rear command post—Focused on supporting operations in the deep and close areas, and it coordinates activities between those areas and the corps rear boundary. (See FM 3-94 for more information on rear command post.)	**REAR**	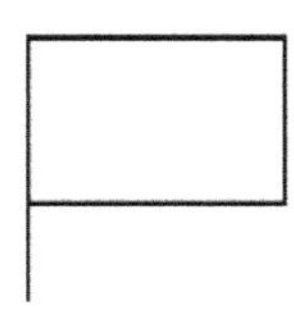 REAR
tactical command post—A portion of a unit headquarters designed to command and control operations as directed. (FM 6-0)	**TAC**	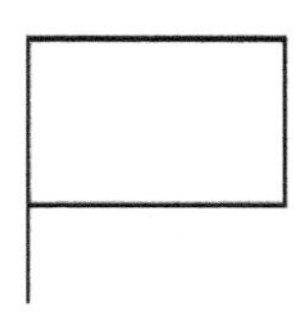 TAC

2-10. **Alphanumeric unit designations using additional information amplifier (Field H).** The alphanumeric unit designation identifies the unit displayed, and it may consist of a number consistent with the unit designation, function, and a higher echelon chain of command. The unit designation construct begins with the unit's own designation (number, letter, or acronym), followed by a higher echelon command designation. Figure 2-10 shows the template.

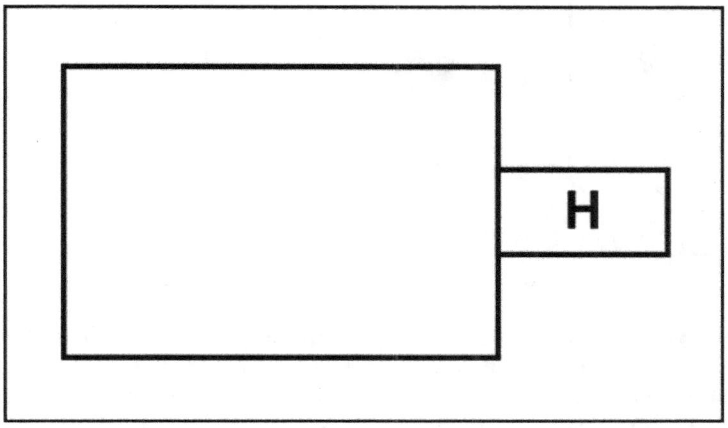

Figure 2-10. Template for additional information amplifier

2-11. A unit designation construct uses the solidus (/) between echelons to identify a continuous hierarchy of command. Figure 2-11 provides an example of the solidus being used in a military symbol construct for 1st Platoon, 51st Transportation Company, 181st Transportation Battalion.

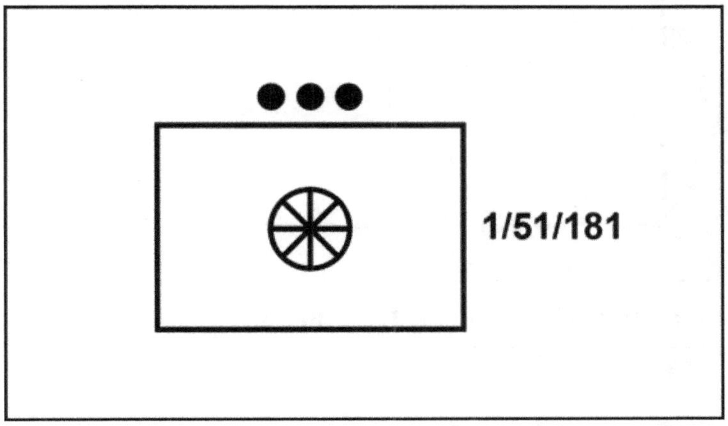

Figure 2-11. Solidus usage example

2-12. Hyphens (-) are only used to depict combat units that maintain regimental affiliations but have no regimental headquarters and are organized as part of a brigade. The use of the hyphen in the unit designation construct retains the units' traditional regimental affiliation and avoids confusing units that do not have a regimental commander from regiments which have remained organized with a regimental headquarters. Figure 2-12 provides an example of a hyphen being used in a military symbol construct for A Battery, 6th Battalion, 37th Field Artillery Regiment.

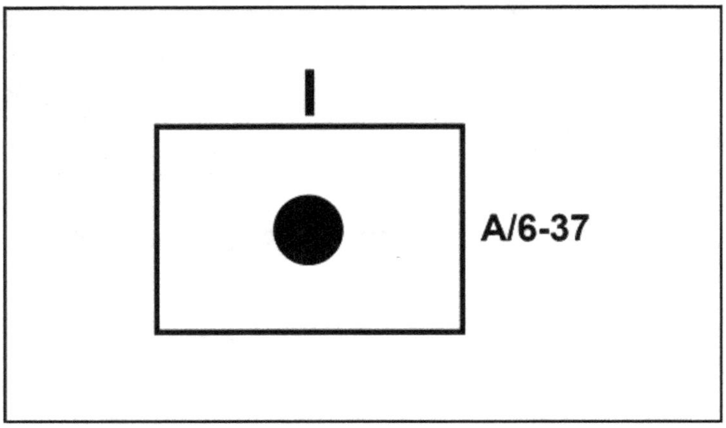

Figure 2-12. Hyphen usage example

2-13. **Higher echelon command using higher information amplifier (Field M).** This additional text amplifier provides a unit symbol a space for the title or number of the higher echelon command. (Roman numerals are used to designate corps.) Figure 2-13 shows the template, and figure 2-14 provides a construct example for A Battery, 6th Battalion, 37th Field Artillery Regiment, 2nd Infantry Division, III Corps.

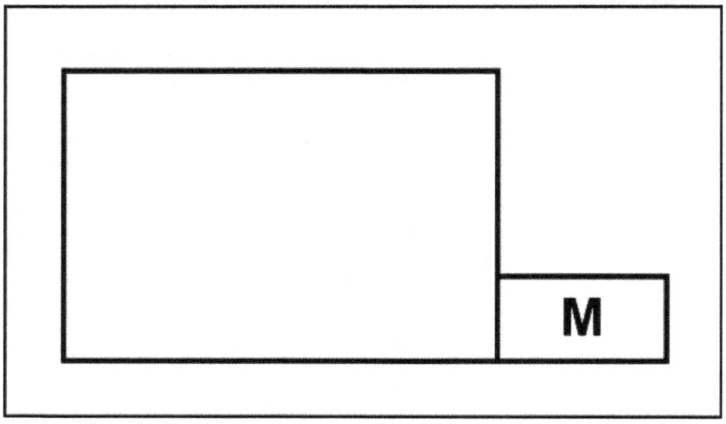

Figure 2-13. Template for higher echelon amplifier

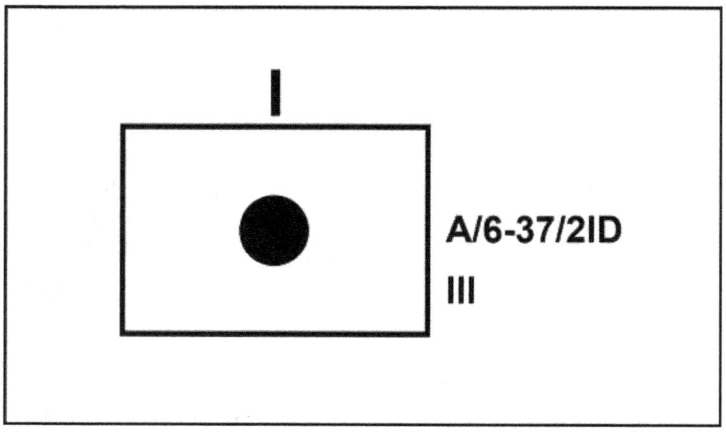

Figure 2-14. Higher echelon amplifier usage example

2-14. **Direction of movement amplifier (Field Q).** The direction of movement amplifier is an arrow or staff identifying the direction of movement or intended movement of an object. For unit and equipment symbols, the amplifier is an angled arrow extending downward from the bottom center of the frame or symbol and pointing in the direction of movement. Figure 2-15 on page 25 shows the template, and figure 2-16 on page 25 is a direction of movement usage construct example of an enemy guerrilla infantry company.

Figure 2-15. Template for direction of movement amplifier

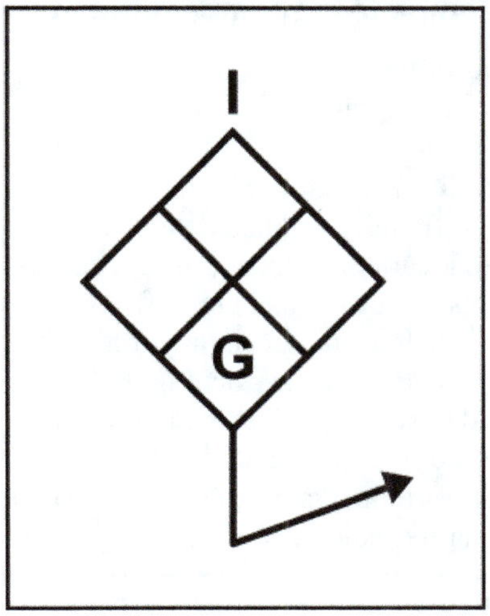

Figure 2-16. Direction of movement amplifier usage example

2-15. **Combat effectiveness amplifiers (Field K).** This amplifier field provides the ability to display the level of combat effectiveness of a unit. Figure 2-17 shows the template for the combat effectiveness amplifier. Combat effectiveness is the ability of a unit to perform its mission, and this amplifier uses the following rating factors for ammunition, personnel, fuel status, and weapons systems:

- Fully operational (FO) – (85 percent or greater)
- Substantially operational (SO) – (70 to 84 percent)
- Marginally operational (MO) – 50 to 69 percent)
- Not operational (NO) – (less than 50 percent)
- Unknown (UNK)

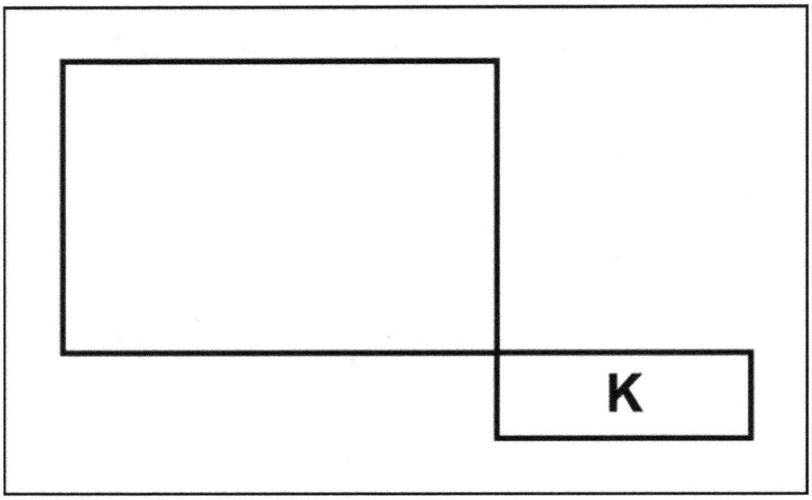

Figure 2-17. Template for combat effectiveness amplifier

2-16. **Offset location indicator amplifier (Field S).** To indicate a precise location or reduce the clutter of headquarters and non-headquarters unit symbols. The offset location line has no restriction in length. There are two types of offset location indicators:

- **Headquarters offset location indicator.** Figure 2-18 shows usage examples of the headquarters offset location indicator for friendly and hostile units. The staff extends vertical from the bottom left-hand corner to the headquarters location point. If several headquarters are at one location, more than one headquarters can be on a single staff. The highest echelon headquarters is placed on top, followed by the next echelons in descending order.
- **Non-headquarters offset location indicator.** Figure 2-19 on page 27 shows an example of how to use the offset friend and hostile units that are not headquarters. For a single unit, the field moves to the bottom center of the frame and the line extends to the exact location point. For grouping of non-headquarters unit symbols, use a bracket with a line extending from the left side of the bracket and ending at the exact location point.

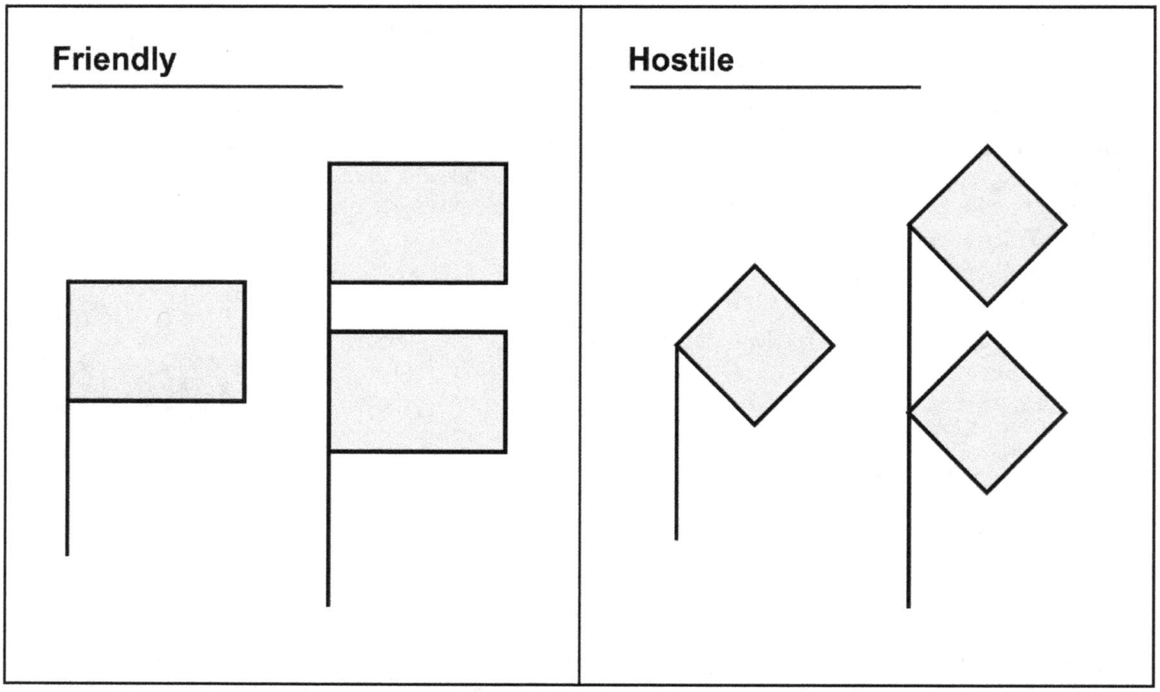

Figure 2-18. Headquarters unit offset location indicators

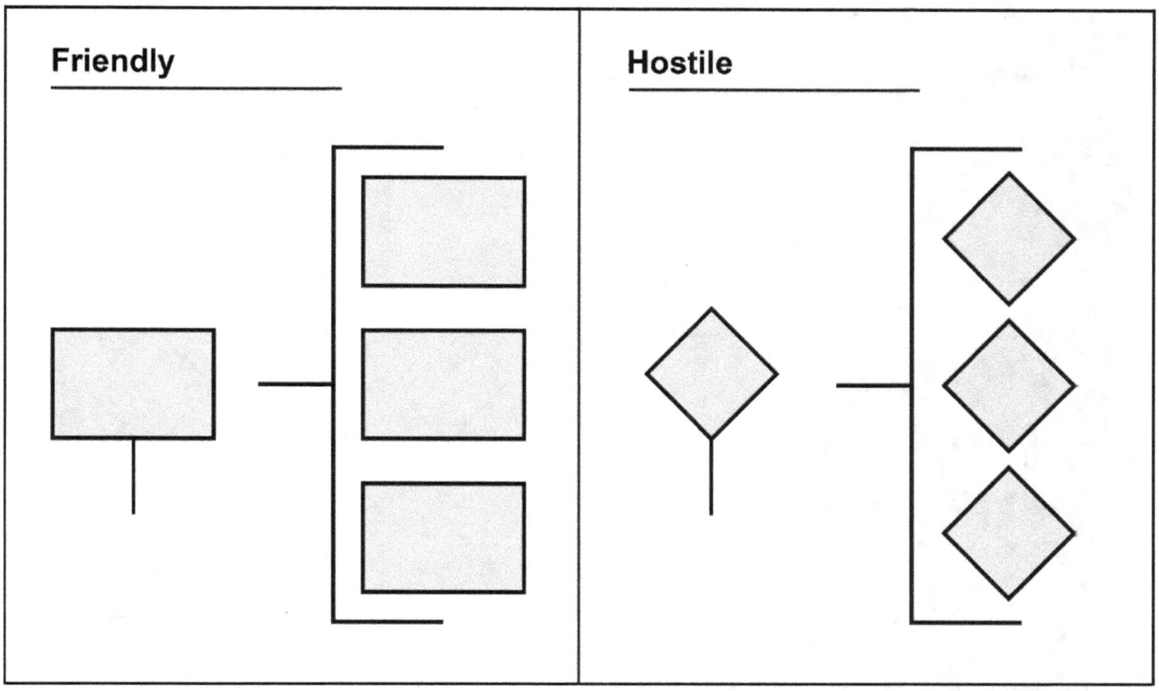

Figure 2-19. Non-headquarters unit location indicators

2-17. **Offset location indicator extension line amplifier (Field S2).** Field S2 is a line that extends from the end of the offset location indicator amplifier "Field S" to the desired location point. This offset location extension line has no length or direction restriction. Figure 2-20 shows examples of how to use the offset location indicator for friendly and hostile headquarters units. Figure 2-21 shows an example of how to use the offset extension line for non-headquarters units.

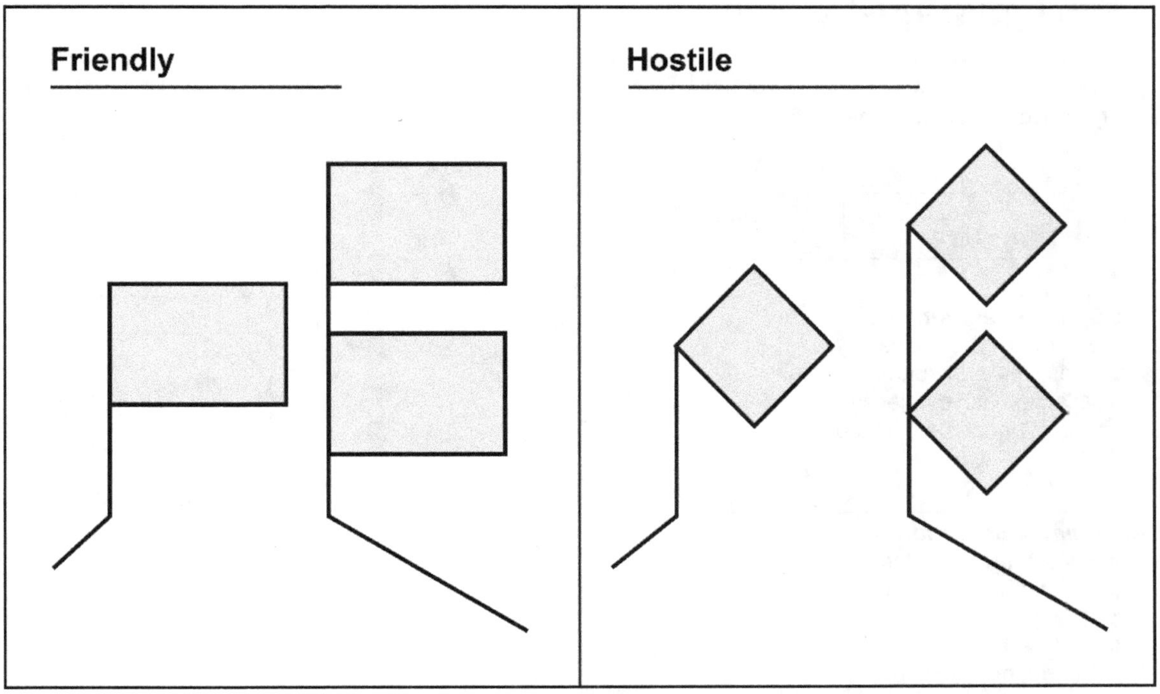

Figure 2-20. Headquarters offset location indicator with extension line

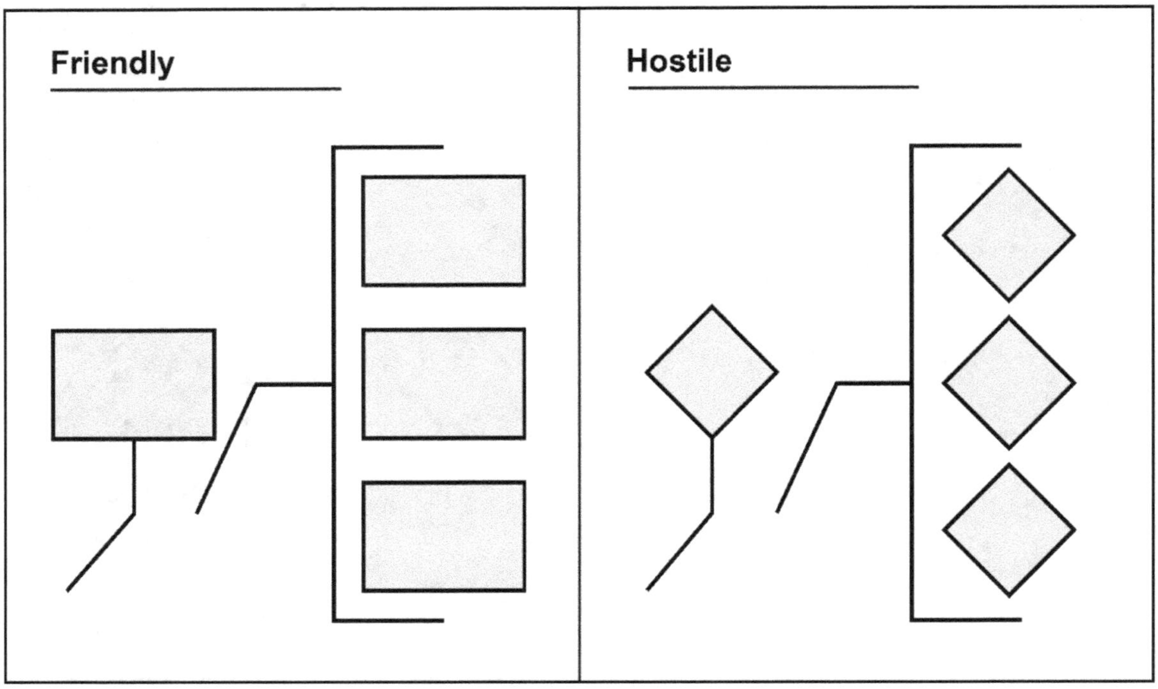

Figure 2-21. Non-headquarters offset location indicator with extension line

MAIN FUNCTION SYMBOLS FOR UNITS

2-18. Most Army Main function symbols for units were determined by table of organization and equipment and modified table of organization and equipment descriptions in the Force Management System of the Army Force Management Support Agency. This section also includes a limited number of North Atlantic Treaty Organization (NATO) and civil authority main function symbols.

2-19. **Main function symbol (Field A).** The Main function symbol is located in the center sector of the octagon, and it reflects the main function of the symbol. (See table 1-5 on page 8 for the construction process for framed symbols.) Table 2-7, on pages 28 through 43, shows the Main function symbols for units.

Table 2-7. Main function symbols for units

Function	*Symbol*	*Symbol Usage Construct Example and Translation*
	Command and Control	
cyberspace operations—The employment of cyberspace capabilities where the primary purpose is to achieve objectives in or through cyberspace. (JP 3-0)	**CYB**	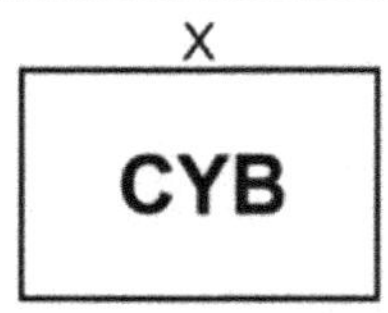 Cyberspace brigade
electromagnetic warfare—Military action involving the use of electromagnetic and directed energy to control the electromagnetic spectrum or to attack the enemy. (JP 3-85) ***Note.*** EW stands for electromagnetic warfare.	**EW**	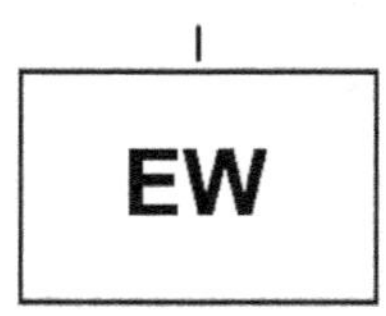 Electromagnetic warfare company

Table 2-7. Main function symbols for units (*continued*)

Function	*Symbol*	*Symbol Usage Construct Example and Translation*
	Command and Control	
information operations force—A force consisting of units, staff elements, and individual military professionals in the Active and Reserve Components, and Department of Defense civilian employees who conduct or directly support the integration of information-related capabilities against adversaries and potential adversaries during military operations, as well as those who train these professionals. (DODD 3600.01)	**IO**	Ø IO Information operations team
interpreter or translator—The capability to translate orally for parties conversing in different languages and turn documents into one's own or other language.		Ø  Interpreter or translator team
isolated personnel—United States military, Department of Defense civilians, and contractor personnel (and others designated by the President or Secretary of Defense) who are separated from their unit (as an individual or a group) while participating in a United States sponsored military activity or mission and are, or may be, in a situation where they must survive, evade, resist, or escape. (JP 3-50)		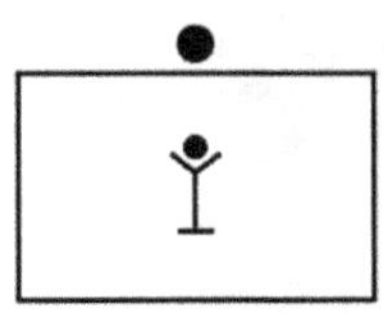 Isolated squad
liaison—That contact or intercommunication maintained between elements of military forces or other agencies to ensure mutual understanding and unity of purpose and action. (FM 6-0)	**LO**	Ø LO Liaison team
multinational operations—A collective term to describe military actions conducted by forces of two or more nations, usually undertaken within the structure of a coalition or alliance. (JP 3-16)	**MN**	XXX **MN** Multinational corps

Table 2-7. Main function symbols for units (*continued*)

Function	*Symbol*	*Symbol Usage Construct Example and Translation*
	Command and Control	
public affairs—Communication activities with external and internal audiences. (JP 3-61)	**PA**	●● PA Public affairs section
signal—Provides and secures the network for commanders to conduct command and control and integrate the other warfighting functions across the range of military operations. (See FM 6-02 for more information on signal.)		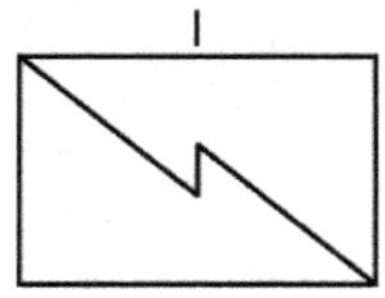 Signal company
space forces—The space and terrestrial systems, equipment, facilities, organizations, and personnel, or combination thereof, necessary to conduct space operations. (JP 3-14)		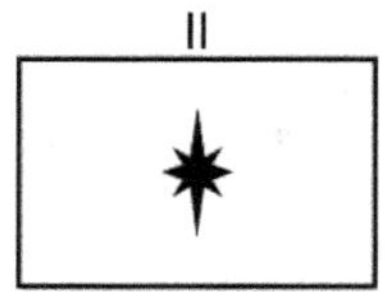 Space battalion
	 Alternate hand-drawn version	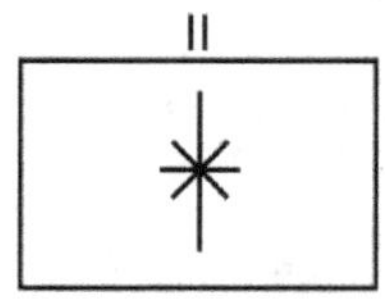
special troops—An organic unit of a modular brigade, division (or equivalent), corps or higher echelon responsible for planning, preparing, executing, and assessing internal support requirements. Typically, it has a headquarters and headquarters and a signal company, but it may include other functional supporting units. (See FM 4-0 and FM 3-96 for more information on special troops.)	**ST**	II ST Special troops battalion
	Fires	
air defense artillery—Weapons and equipment for actively combating air targets from the ground. (JP 3-01)		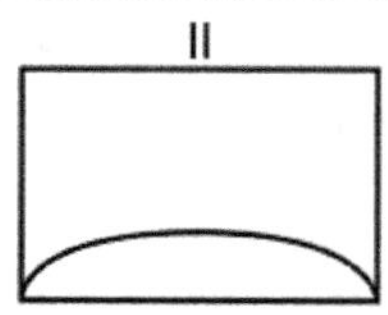 Air defense battery

Table 2-7. Main function symbols for units (*continued*)

Function	***Symbol***	***Symbol Usage Construct Example and Translation***
	Fires	
air and missile defense—Direct [active and passive] defensive actions taken to destroy, nullify, or reduce the effectiveness of hostile air and ballistic missile threats against friendly forces and assets. (JP 3-01)		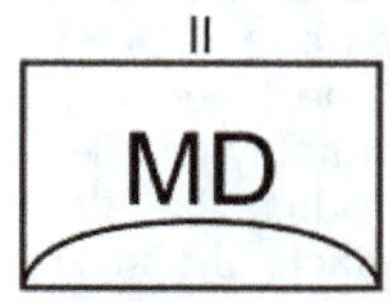 Air and missile defense battalion
air-naval gunfire liaison (ANGLICO)—provides the Marine air-ground task force (MAGTF) commander a liaison capability to plan, coordinate, and conduct the terminal control of fires in support of joint and multinational forces operating within or adjacent to the MAGTF battlespace. Contains Marine and Navy personnel qualified to plan, coordinate, and integrate all fire support assets available to the MAGTF, as well as joint and multinational forces. (See JP 3-09 for more information on ANGLICO)		 ANGLICO team
directed energy—An umbrella term covering technologies that relate to the production of a beam of concentrated electromagnetic energy or atomic or subatomic particles. (JP 3-85)	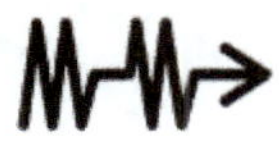	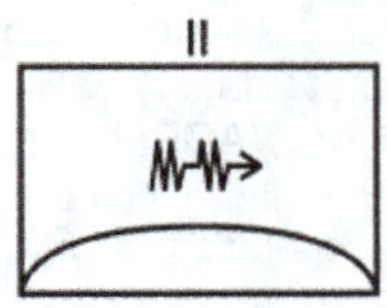 Directed energy air defense battalion
field artillery—The equipment, supplies, ammunition, and personnel involved in the use of indirect fire cannon, rocket, or surface-to-surface missile launchers. (FM 3-09)		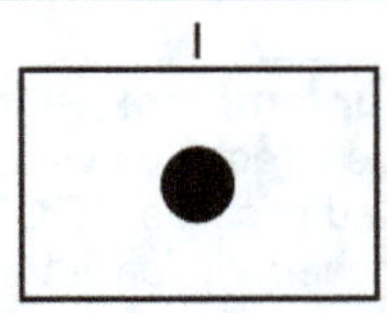 Field artillery battery
missile—A weapon that is self-propelled or directed by remote control, carrying conventional or nuclear explosive.		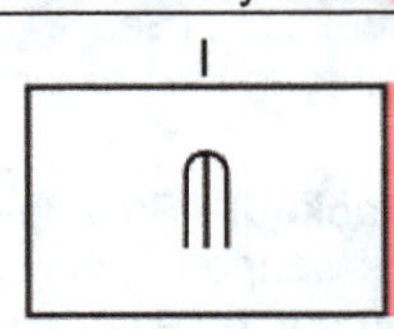 Missile battery

Table 2-7. Main function symbols for units (*continued*)

Function	*Symbol*	*Symbol Usage Construct Example and Translation*
	Fires	
missile defense—Defensive measures designed to destroy attacking enemy missiles or to nullify or reduce the effectiveness of such attack. (JP 3-01)	**MD**	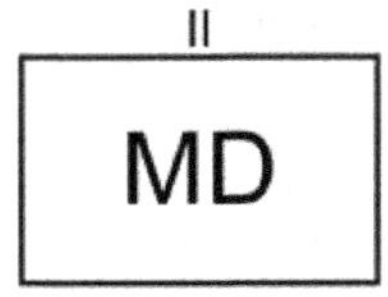 Missile defense battalion
short-range air defense—An air defense artillery (ADA) capability which integrates existing guns, missiles, rockets, and sensors to defend against low-altitude air threats. (See JP 3-01 for more information on short-range air defense.)		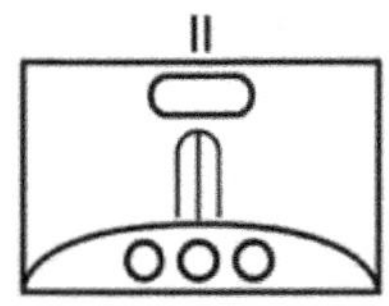 Short-range air defense battalion with ADA capabilities integrated on armored high mobility vehicles (maneuver short-range air defense battalion)
	Intelligence	
intelligence operations—The tasks undertaken by military intelligence units through the intelligence disciplines to obtain information to satisfy validated requirements. (ADP 2-0) ***Note:*** Military intelligence conducts intelligence operations as part of information collection across the Army's strategic roles. (See FM 2-0 for more information on military intelligence.)	**MI**	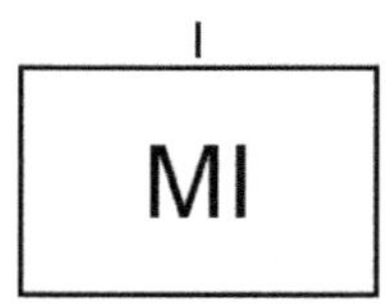 Military intelligence company
	Movement and Maneuver	
anti-armor (anti-tank)—Provides long range direct fires with anti-tank guided missiles (ATGMs), tube-launched, optically tracked, wire-guided (TOW) anti-tank guns, or self-propelled anti-tank guns or ATGM launchers. (See ATP 3-21.91 for information on anti-armor.)		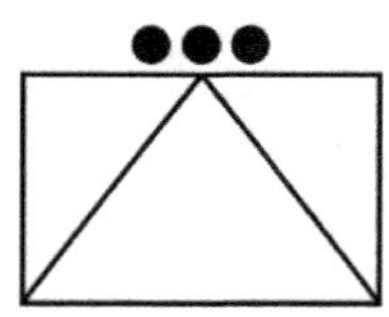 Anti-armor platoon
armor (tracked)—Provides main battle tank weapon system with 120mm smoothbore cannon and increased armor protection. (See ATP 3-90.1 and ATP 3-20.15 for more information on armor.)	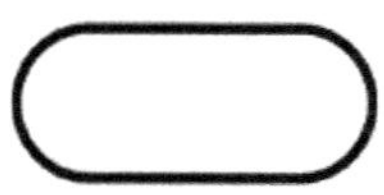	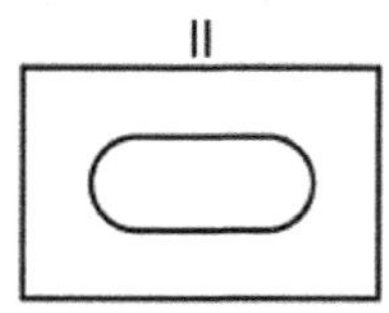 Armor battalion

Table 2-7. Main function symbols for units (*continued*)

Function	*Symbol*	*Symbol Usage Construct Example and Translation*
	Movement and Maneuver	
armored (tracked) cavalry—Conducts reconnaissance and security with armored tracked fighting vehicles to support the brigade's awareness and knowledge in the area of operations. (See ATP 3-20.96 for more information on cavalry.)		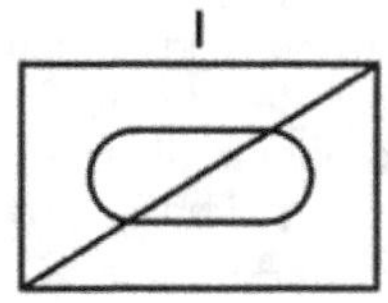 Armored cavalry troop
army aviation or rotary-wing aviation—Conducts attack and air movement functions in support of ground maneuver in the area of operations or area of interest. (See FM 3-04 for more information on aviation.)		 Aviation (rotary-wing) squadron
aviation (rotary-wing) reconnaissance—Conducts aerial reconnaissance and security tasks in close coordination with the brigade's cavalry squadrons. (See FM 3-04 and FM 3-98 for information on aviation reconnaissance.)		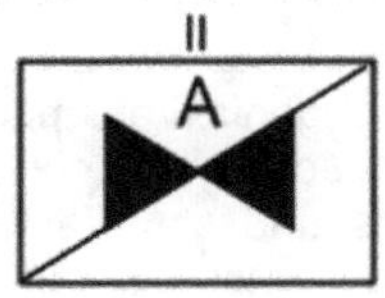 Attack reconnaissance squadron
aviation fixed wing—Conducts air movement of personnel, leaders, critical supplies, equipment, and systems during the conduct of offensive, defensive, stability, and defense support of civilian authorities operations throughout the depth and breadth of the area of operations or are of interest. (See FM 3-04 for more information on aviation fixed wing.)		 Aviation (fixed wing) battalion
cavalry (reconnaissance)—Conducts reconnaissance and security to support friendly forces awareness and knowledge in the area of operations. (See FM 3-98 and ATP 3-20.96 for more information on cavalry reconnaissance.)		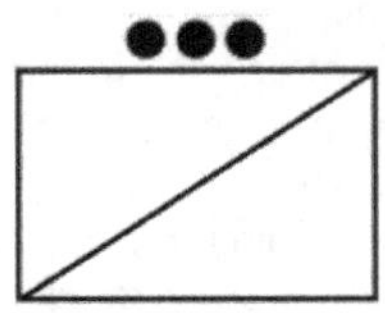 Cavalry platoon
combined arms—Combines the efforts of armor units and mechanized infantry units to execute tactical missions as part of a combined arms operation. (See ATP 3-90.5 and ATP 3-90.1 for more information on combined arms.)		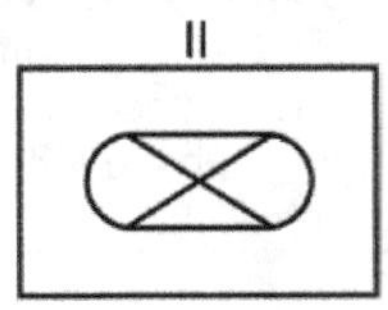 Combined arms battalion

Table 2-7. Main function symbols for units (*continued*)

Function	*Symbol*	*Symbol Usage Construct Example and Translation*
	Movement and Maneuver	
infantry—Provides Soldiers trained, armed, and equipped to fight dismounted by means of fire and movement to destroy, defeat, capture, or repel an enemy assault. (See ATP 3-21.20 for more information on infantry.)		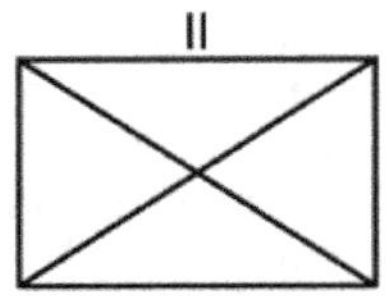 Infantry battalion
mechanized armored (tracked) infantry—Provides armored tracked fighting vehicles to transport and support Soldiers trained, armed, and equipped to fight dismounted by means of fire and movement. (See ATP 3-90.5 and ATP 3-90.1 for more information on mechanized infantry.)	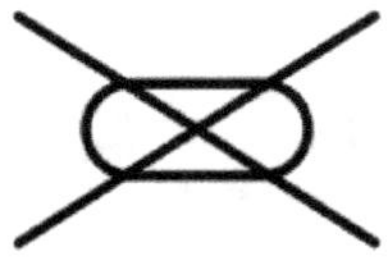	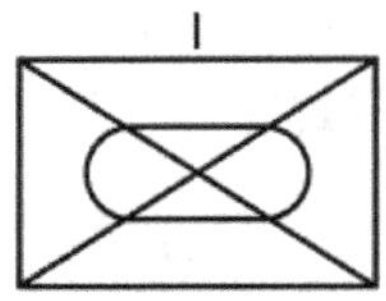 Mechanized armored infantry company
mortar—Organic fire support to battalions, squadrons, companies, and troops, and is available to a commander when other indirect fire support is not available. (See ATP 3-21.90 for more information on mortars.)		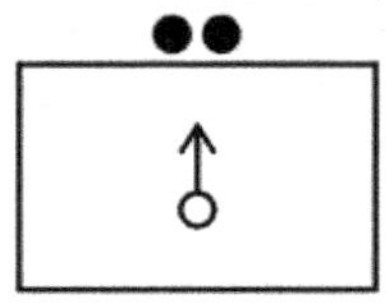 Mortar section
surveillance—The systematic observation of aerospace, cyberspace, surface, or subsurface areas, places, persons, or things by visual, aural, electronic, photographic, or other means. (JP 3-0)		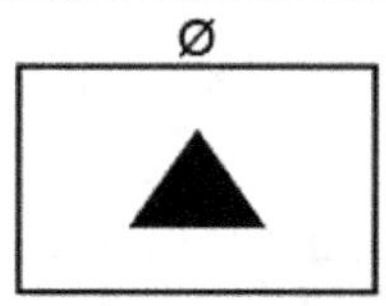 Surveillance team
	Protection	
Bureau of Alcohol, Tobacco, Firearms, and Explosives (ATF)—Civil authority law enforcement agency in the United States Department of Justice that protects communities from violent criminals, criminal organizations, the illegal use and trafficking of firearms, the illegal use and storage of explosives, acts of arson and bombings, acts of terrorism, and the illegal diversion of alcohol and tobacco products. (See ATF.gov website for more information on ATF.)	**ATF**	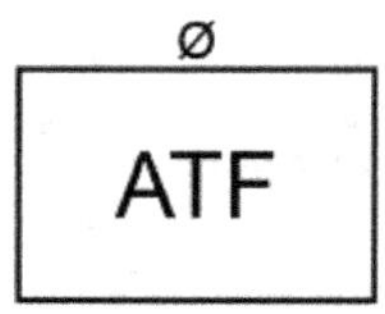 ATF team

Table 2-7. Main function symbols for units (*continued*)

Function	*Symbol*	*Symbol Usage Construct Example and Translation*
	Protection	
chemical, biological, radiological, and nuclear (CBRN)—Recognizes vulnerabilities, identifies, and understands CBRN hazards, and their consequences when they appear, and responds appropriately to protect the force. (See ADP 3-37 for more information on CBRN.)		 CBRN company
chemical, biological, radiological, nuclear, and explosives—Recognizes components that are threats or potential hazards with adverse effects in the operational environment. (See ATP 3-37.11 for more information on CBRN team.)		 CBRN explosive team
chemical, biological, radiological, nuclear reconnaissance—Executes operations to obtain by visual observation or other detection methods, information on the potential or actual CBRN hazards and threats in an area of operations. (See ATP 3-37.11 for more information on nuclear reconnaissance.)		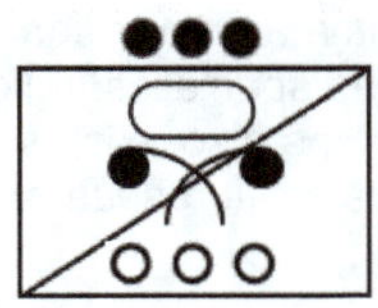 CBRN reconnaissance platoon
Drug Enforcement Administration (DEA)—The civil authority law enforcement agency under the U.S. Department of Justice tasked with combating drug trafficking and distribution within the U.S. It is the lead agency for domestic enforcement of the Controlled Substances Act, sharing concurrent jurisdiction with the Federal Bureau of Investigation, the U.S. Immigration and Customs Enforcement, and U.S. Customs and Border Protection. (See DEA.gov website for more information on the DEA.)	**DEA**	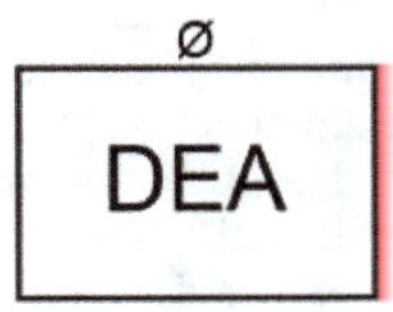 DEA team
engineer—Provides Soldiers with technical skills and equipment to provide freedom of action or land power by mitigating the effects of terrain. (See FM 3-34 for more information on engineers.)		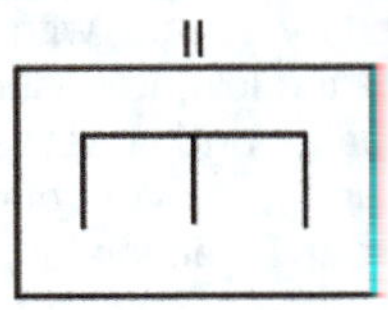 Engineer battalion

BODY-47

Table 2-7. Main function symbols for units (*continued*)

Function	***Symbol***	***Symbol Usage Construct Example and Translation***
	Protection	
Federal Bureau of Investigation (FBI)—The civil authority domestic intelligence and security service of the United States and its principal federal law enforcement agency. Operating under the jurisdiction of the United States Department of Justice, the FBI is also a member of the U.S. Intelligence Community and reports to both the Attorney General and the Director of National Intelligence. (See FBI.gov website for more information.)	**FBI**	ø FBI FBI team
fire department—The civil authority firefighting organization that operates within a municipality, county, state, nation, or special district. In some areas, they may also provide technical rescue, fire protection, fire investigation, emergency medical services, and hazardous material mitigation. Private and specialist firefighting organizations also exist, such as those for aircraft rescue and firefighting.	  Alternate hand-drawn version	 Fire department battalion 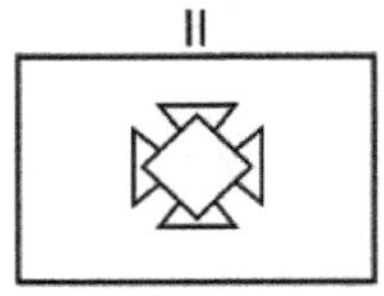
maneuver enhancement—Provides command and control of forces from multiple branches, but especially those that conduct support area and maneuver support operations for the force. (See FM 3-81 for more information on maneuver enhancement.)		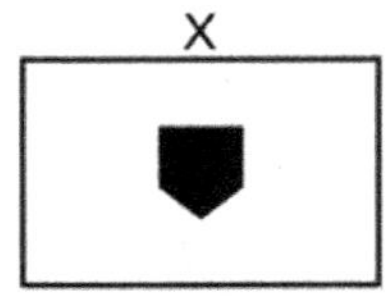 Maneuver enhancement brigade
military police—Provides law enforcement activities to control and protect populations and resources to facilitate the existence of a lawful and orderly environment. (See FM 3-39 for more information on military police.)	**MP**	II MP Military police battalion
police department—A civil authority constituted body of persons empowered and authorized by a state, with the aim to enforce the law, to ensure the safety, health, and possessions of citizens, and to prevent crime and civil disorder. Their lawful powers include arrest and use of force legitimized by the state via its monopoly on violence.		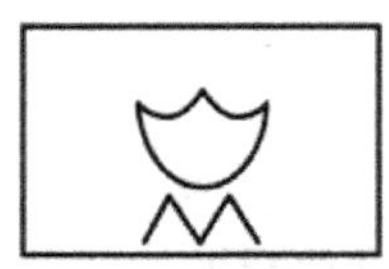 Police unit with pack animal capability

Table 2-7. Main function symbols for units (*continued*)

Function	***Symbol***	***Symbol Usage Construct Example and Translation***
	Protection	
security—Measures taken by a military unit, activity, or installation to protect itself against all acts designed to, or which may, impair its effectiveness. (JP 3-10)	**SEC**	 Security team with working dog capability
United States Marshal Service—A civil authority bureau within the U.S. Department of Justice, operating under the direction of the Attorney General. Responsible for protection of judges and judicial personnel, administration of fugitive operations, management of criminal assets, the operation of the United States Federal Witness Program and the Justice Prisoner and Alien Transportation System, the execution of federal arrest warrants, and the protection of senior government officials through the Office of Protective Operations. (See USMarshals.gov website for more information.)		 Marshal Service unit
United States Secret Service—A federal law enforcement agency under the Department of Homeland Security charged with conducting criminal investigations and protecting U.S. political leaders, their families, and visiting heads of state or government. (See secretservice.gov website for more information.)	**USSS**	 Secret Service team
unmanned aircraft system—That system whose components include the necessary equipment, network, and personnel to control an unmanned aircraft. (JP 3-30)		 Unmanned aircraft system platoon

Table 2-7. Main function symbols for units (*continued*)

Function	***Symbol***	***Symbol Usage Construct Example and Translation***
	Sustainment	
aerial delivery—The air transport of cargo, equipment and/or personnel to a desired location on the ground by aircraft. (ATP 4-48) **rigger**—Support that includes parachute packing, aerial delivery equipment repair, external sling load, rigging equipment and supplies for airdrop, as well as the provision of aerial delivery equipment and systems. (See FM 4-0 and ATP 4-48 for information on aerial delivery and riggers.)		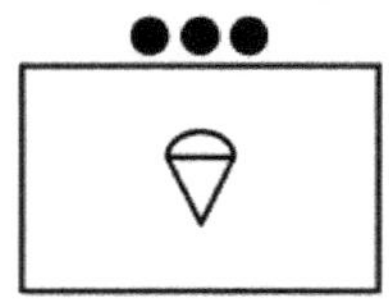 Rigger platoon
Army field support—Integrates and synchronizes delivery of U.S. Army Materiel Command strategic capabilities and enablers to the operational and tactical points of need in support of Army Service component commands and corps during large-scale combat operations. (See FM 4-0 for more information on Army field support.)	**AFS**	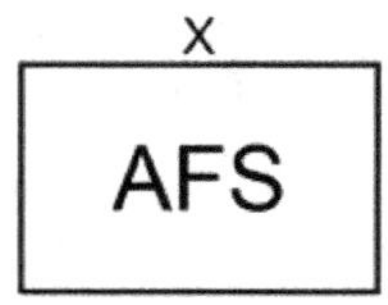 Army field support brigade
ammunition—Provides effective and efficient handling, storing, securing, distributing, and accounting for munitions while ensuring adequate explosives safety guidelines are met. (See FM 4-30 for more information on ammunition.)		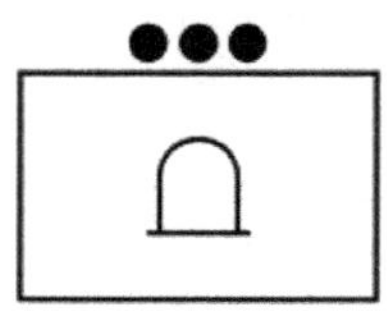 Ammunition platoon
band—Serves as a combat multiplier and plays an integral part in the sustainment of forces engaged in unified land operations. (See ATP 1-19 for more information on the band.)		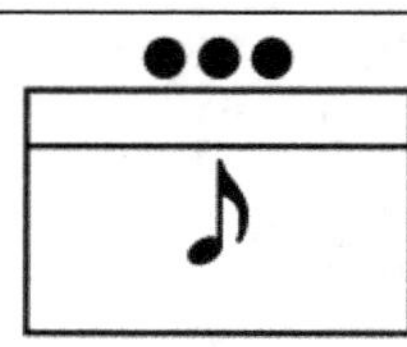 Band performance headquarters detachment
	 Alternate hand-drawn version	
contracting support—The planning, coordination, and execution of contracting authority to legally bind contractors in support of military operations. (JP 4-10)	**KS**	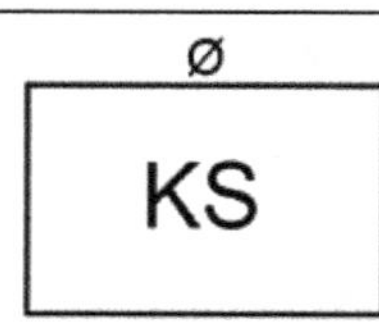 Contracting support team

Table 2-7. Main function symbols for units (*continued*)

Function	*Symbol*	*Symbol Usage Construct Example and Translation*
	Sustainment	
explosive ordnance disposal—The detection, identification, on-site evaluation, rendering safe, exploitation, recovery, and final disposal of explosive ordnance. (FM 4-30)	**EOD**	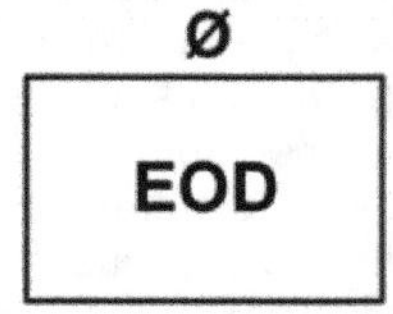 Explosive ordnance disposal (EOD) team
field feeding—The sustained tactical feeding to meet commander's needs as determined by the mission, enemy, terrain and weather, troops and support available, time available and civilian considerations and logistical support available on the battlefield and afloat. (See FM 4-0 and ATP 4-41 for more information on field feeding.)		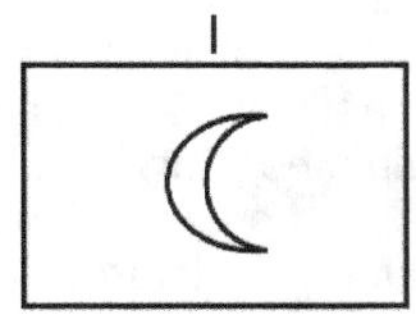 Field feeding company
finance operations—The execution of the joint financial management mission to provide financial advice and guidance, support the procurement process, provide pay support, and provide banking and disbursing support. (See FM 1-06 for more information on finance operations.)		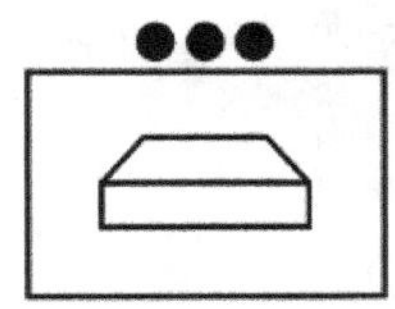 Finance platoon
human resources—Provides operational effectiveness of the Army by anticipating, manning, and sustaining military operations. (See FM 1-0 for more information on human resources.)	**HR**	 Human resources platoon supporting the army theater (human resources sustainment center)
judge advocate general—Provides subject matter experts in the core legal disciplines and are counselors, advocates, and trusted advisors to commanders and Soldiers. (See FM 3-84 for more information on the judge advocate general.)	**JAG**	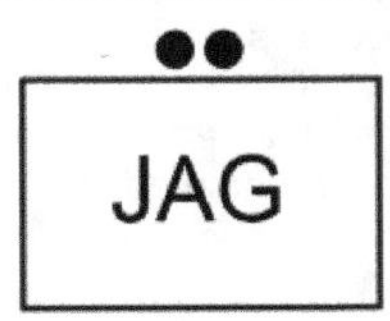 Judge advocate general section

Table 2-7. Main function symbols for units (*continued*)

Function	*Symbol*	*Symbol Usage Construct Example and Translation*
	Sustainment	
maintenance—Ensures unit readiness by maintaining weapons systems and equipment in a fully mission-capable status for immediate and continuous employment in complex and highly lethal environments. (See FM 4-0, FM 4-30, and ATP 4-33 for more information on maintenance.)		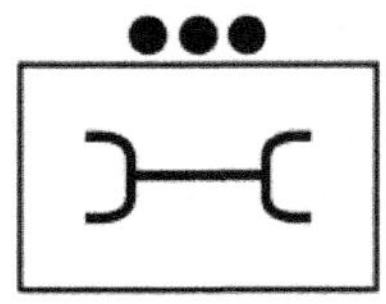 Maintenance platoon
medical—Promotes, improves, conserves, or restores the behavioral and physical well-being of personnel in the Army, and as directed in other Services, agencies, and organizations. (See FM 4-02 for more information on medical.) ***Note.*** To avoid overlapping Main function symbol with modifiers: 1. Most modifiers offset to the right. 2. When modifiers cannot offset to the right, shorten the vertical center line to allow space for the modifier.	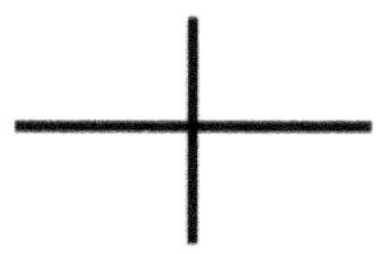	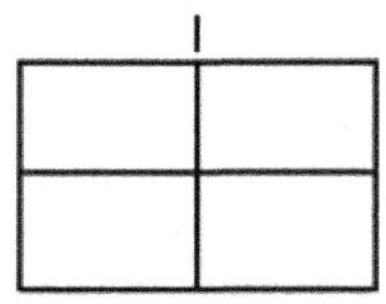 Medical company
		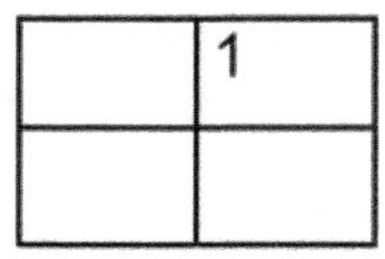 Medical unit with Role 1 capability
		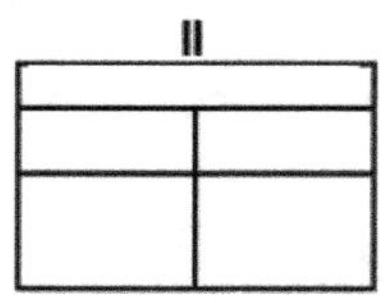 Medical battalion headquarters
medical treatment facility—Any facility established for the purpose of providing medical treatment. This includes battalion aid stations, Role 2 facilities, dispensaries, clinics, and hospitals. (FM 4-02) ***Note.*** To avoid overlapping Main function symbol with modifiers: 1. Most modifiers offset to the right. 2. When modifiers cannot offset to the right, shorten the vertical center line to allow space for the modifier.	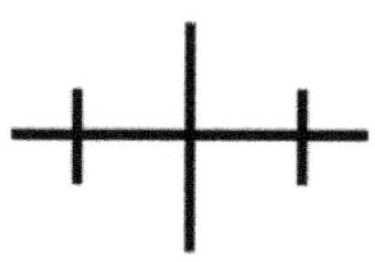	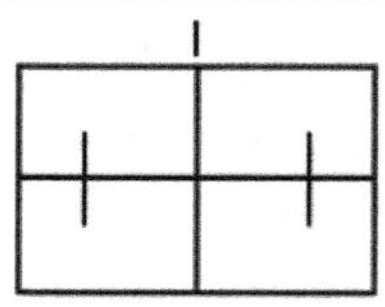 Medical treatment facility company
		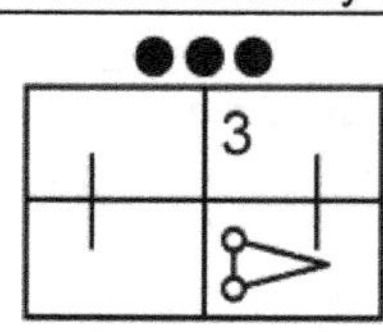 Medical treatment facility platoon with Role 3 surgical capability
mortuary affairs—Provides for the search, recovery, identification, preparation, and disposition of human remains of persons for whom the Services are responsible by status and executive order. (JP 4-0)	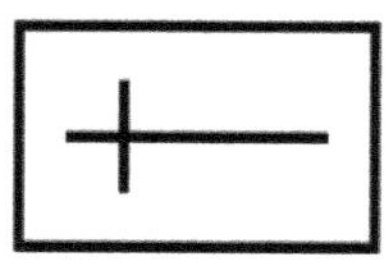	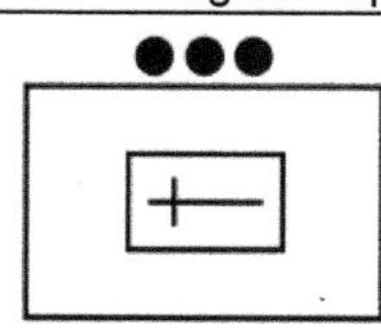 Mortuary affairs platoon

Table 2-7. Main function symbols for units (*continued*)

Function	*Symbol*	*Symbol Usage Construct Example and Translation*
	Sustainment	
ordnance—Provides munitions, maintenance, and explosive ordnance disposal support to generate and maintain combat power and to provide protection to Army, joint, intergovernmental, interagency, and multinational forces. (See FM 4-30 for more information on ordnance.)		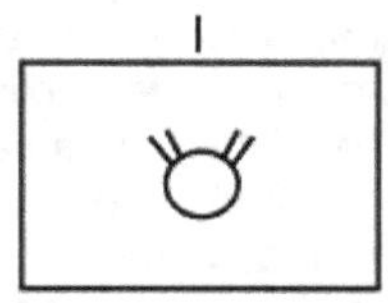 Ordnance company
quartermaster—Provides supply and field services to enable freedom of action, extend operational reach, and prolong endurance. (See FM 4-40 for more information on quartermasters.)	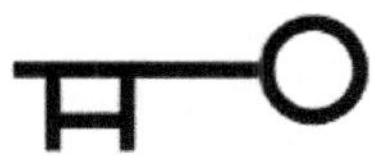	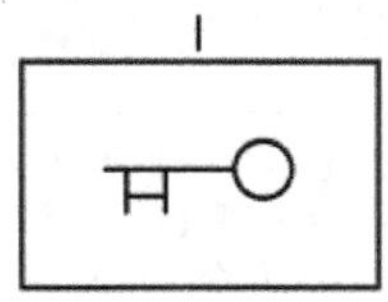 Quartermaster company
religious support—Provides for the free exercise of religion and religious, moral, and ethical advisement and leadership. (See FM 1-05 for more information on religious support.)	**REL**	●● REL Religious support section
shower and laundry operations—Operations that provide shower and laundry services in direct support to the health and morale of all deployed Soldiers. (See ATP 4-42 for more information on shower and laundry operations.)		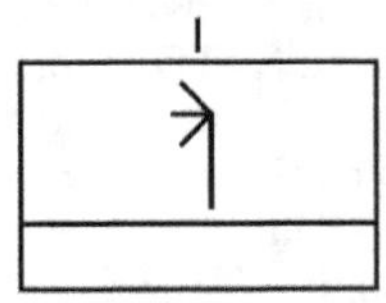 Shower and laundry supply company
support—**1.** The action of a force that aids, protects, complements, or sustains another force in accordance with the directive requiring such action. **2.** A unit that helps another unit in battle. **3.** An element of a command that assists, protects, or supplies other forces in combat. (JP 1 VOL 2)	**SPT**	I FWD SPT Forward support company
sustainment—The provision of logistics, financial management, personnel services, and health service support necessary to main operations until successful mission completion. (ADP 4-0)	**SUST**	X SUST Sustainment brigade

Table 2-7. Main function symbols for units (*continued*)

Function	***Symbol***	***Symbol Usage Construct Example and Translation***
	Sustainment	
transportation—A logistics function that includes movement control and associated activities to incorporate military, commercial, and multinational motor, rail, air, and water mode assets in the movement of units, personnel, equipment, and supplies in support of the concept of operations. (FM 1-02.1)		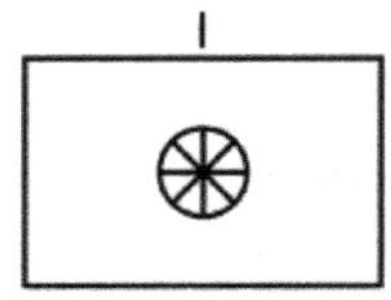 Transportation company
	Special Operations	
civil affairs—Designated Active and Reserve Component forces and units organized, trained, and equipped specifically to conduct civil affairs operations and to support civil-military operations. (JP 3-57)	**CA**	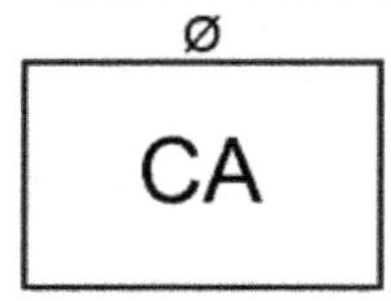 Civil affairs team
civil-military cooperation—(North Atlantic Treaty Organization [NATO]) A joint function comprising a set of capabilities integral to supporting the achievement of mission objectives and enabling NATO commands to participate effectively in a broad spectrum of civil-military interaction with diverse non-military actors. (APP-6)		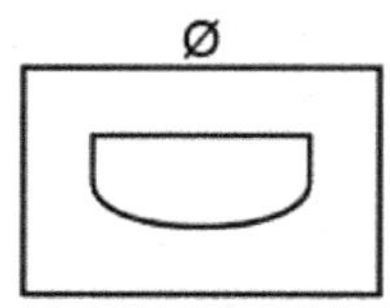 Civil-military cooperation team
Psychological Operations—A robust military influence capability. (See FM 3-53 for more information on psychological operations forces and their operations.)		 Psychological Operations battalion
Rangers—Rapidly deployable airborne light infantry organized and trained to conduct highly complex joint direct action operations in coordination with or in support of other special operations units of all Services. (JP 3-05)	**RGR**	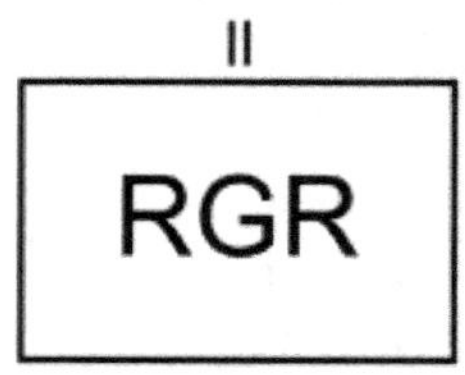 Ranger battalion
search and rescue—The use of aircraft, surface craft, submarines, and specialized rescue teams and equipment to search for and rescue distressed persons on land or at sea in a permissive environment. (JP 3-50)	**SAR**	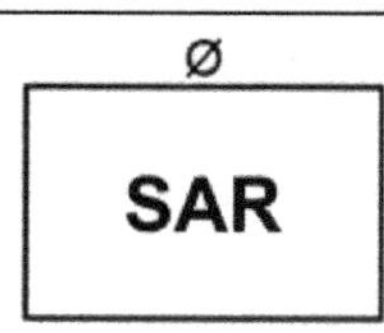 Search and rescue team

Table 2-7. Main function symbols for units (*continued*)

Function	*Symbol*	*Symbol Usage Construct Example and Translation*
	Special Operations	
SEAL team—United States Navy forces organized, trained, and equipped to conduct special operations with an emphasis on maritime, coastal, and riverine environments. (JP 3-05)	**SEAL**	 SEAL team
special forces—United States Army forces organized, trained, and equipped to conduct special operations with an emphasis on unconventional warfare capabilities. (JP 3-05)	**SF**	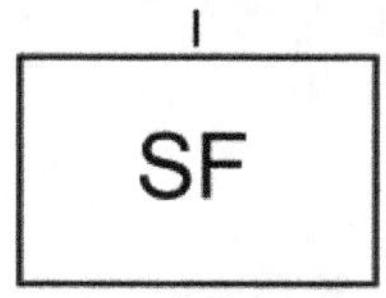 Special forces company
special operations forces—Those Active and Reserve Component forces of the Services designated by the Secretary of Defense and specifically organized, trained, and equipped to conduct and support special operations. (JP 3-05)	**SOF**	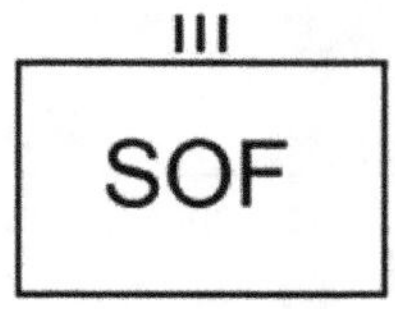 Special operations forces group
	Multidomain Operations	
multidomain operations—The combined arms employment of joint and Army capabilities to create and exploit relative advantages to achieve objectives, defeat enemy forces, and consolidate gains on behalf of joint force commanders. (FM 3-0)	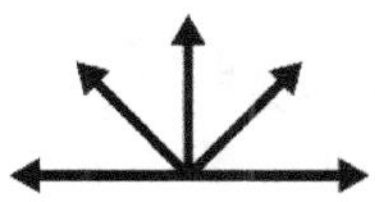	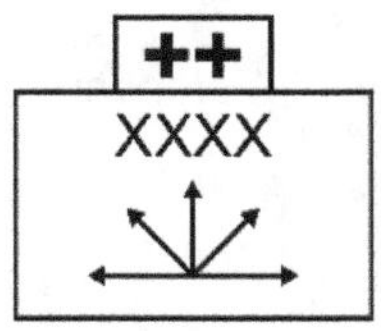 Multidomain operations task force command supporting a theater army

2-20. **Main function symbol for named units (Field AA).** This is a text amplifier field for all special command and control type headquarter-named units, and it allows the placement of a maximum of 9 characters inside the frame. Table 2-8 (on pages 44 through 47) lists some examples of special command and control type headquarter-named unit Main function symbols.

Table 2-8. Main function symbols for named units

Named Unit	*Symbol*	*Symbol Usage Construct Example*
Combatant Command A unified or specified command with a broad continuing mission under a single commander established and so designated by the President, through the Secretary of Defense and with the advice and assistance of the Chairman of the Joint Chiefs of Staff. (JP 1 VOL 2)		
United States Africa Command	**AFRICOM**	AFRICOM
United States Central Command	**CENTCOM**	CENTCOM
United States Cyber Command	**CYBERCOM**	CYBERCOM
United States European Command	**EUCOM**	EUCOM
United States Indo-Pacific Command	**INPACOM**	INPACOM
United States Northern Command	**NORTHCOM**	NORTHCOM
United States Southern Command	**SOUTHCOM**	SOUTHCOM

Table 2-8. Main function symbols for named units (*continued*)

Named Unit	***Symbol***	***Symbol Usage Construct Example***
United States Space Command	**SPACECOM**	SPACECOM (in frame)
Functional Combatant Command A command responsible for a larger functional area requiring single responsibility for effective coordination of the operations therein. (See JP 1 VOL 2 for more information on functional combatant command.)		
United States Special Operations Command	**SOCOM**	SOCOM (in frame)
United States Strategic Command	**STRATCOM**	STRATCOM (in frame)
United States Transportation Command	**TRANSCOM**	TRANSCOM (in frame)
Sub-unified Command Sub-unified commands (subordinate unified commands) may be established by geographical area or functional basis by a combatant commander when authorized to do so by the Secretary of Defense through the Chairman Joint Chief of Staff. (See JP 1 VOL 2 for more information on sub-unified commands.)		
Alaskan Command	**ALCOM**	ALCOM (in frame)
United States Forces Korea	**USFK**	USFK (in frame)

Table 2-8. Main function symbols for named units (*continued*)

Named Unit	*Symbol*	*Symbol Usage Construct Example*
United States Army, Africa Command	USARAF	USARAF
United States Army, Central Command	USARCENT	USARCENT
United States Army, Cyber Command	ARCYBER	ARCYBER
United States Army, Military Surface Deployment and Distribution Command	SDDC	SDDC
United States Army, North	USANORTH	USANORTH
United States Army, Pacific Command	USARPAC	USARPAC
United States Army, Southern Command	USARSO	USARSO
United States Army, Special Operations Command	USASOC	USASOC

Table 2-8. Main function symbols for named units (*continued*)

Named Unit	*Symbol*	*Symbol Usage Construct Example*
	North Atlantic Treaty Organizations	
Allied Command Operations	**ACO**	ACO

SECTOR 1 MODIFIERS FOR UNITS

2-21. This indicator is represented as the upper part of Field A as shown in table 1-4 on page 5. Table 2-9 (on pages 47 through 68) shows sector 1 modifiers for unit capabilities. These modifiers assist in providing additional information specific to the capabilities that the unit organized and equipped to perform.

Table 2-9. Sector 1 modifiers for units

Function	*Modifier*	*Symbol Usage Construct Example and Translation*
air terminals and aerial ports support—The capability to provide airfield functions that may include port clearance, movement control, onward movement, liaison, coordination, operation of holding areas, postal operations, personnel replacement processing, and life and logistic support. (See JP 3-36 for more information on aerial ports and terminals.)	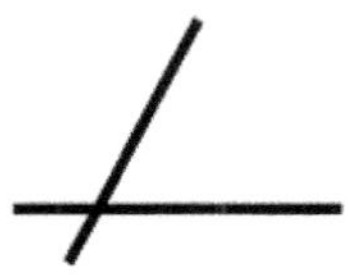	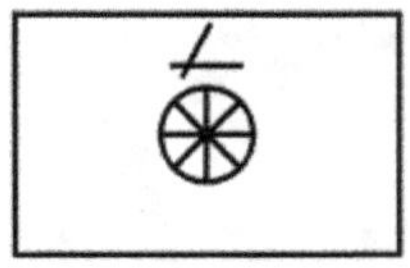 Transportation unit with airfield terminal operations capability
amphibious warfare ship—A combatant ship having organic capability to embark, land, and support landing forces in amphibious operations and which has characteristics enabling long-duration operations on the high seas. (JP 3-02)		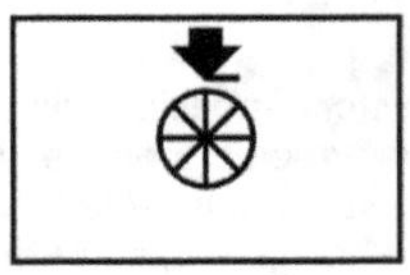 Transportation unit with amphibious warfare ship (generic vessel) capability

Table 2-9. Sector 1 modifiers for units (*continued*)

Function	*Modifier*	*Symbol Usage Construct Example and Translation*
area—A specified geographic surface included within a delineated set of lines (boundaries) used for the purpose of facilitating coordination and de-confliction between adjacent units, formations, or other specific geographical surfaces.	**AREA**	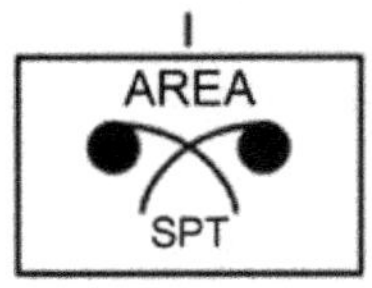 Chemical, biological, radiological, and nuclear (CBRN) company with area support capability
		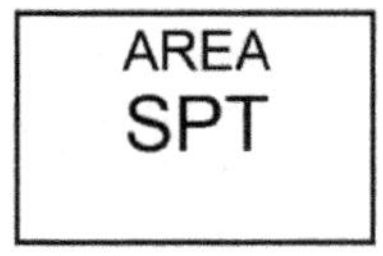 Area support unit
		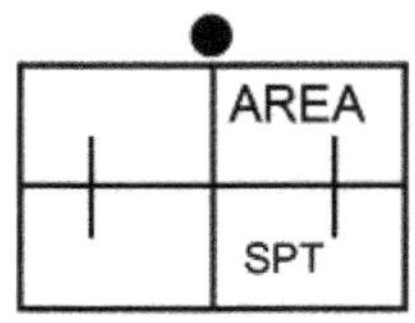 Medical treatment facility squad with area support capability ***Note.*** The modifier offsets to the right to avoid overlapping of symbols
armored (protection)—A vehicle hull equipped or protected with armor. ***Notes.*** 1. As a sector 1 modifier, this symbol represents armored protected and requires a sector 2 wheeled vehicle modifier to complete a non-tracked vehicle type capability. 2. If this sector 1 is used without a sector 2 wheeled vehicle modifier present, it means the unit has armored tracked capability.	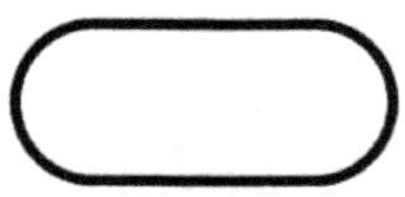	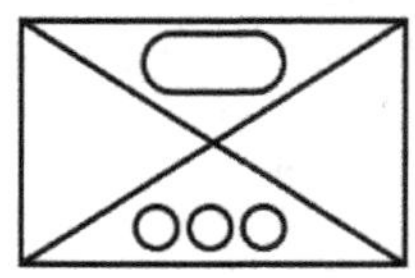 Infantry unit with armor wheeled vehicle capability (Stryker)
assault—A short and violent well-ordered attack against a local objective. (FM 3-90)	**ASLT**	 Aviation unit with assault aviation capability

Table 2-9. Sector 1 modifiers for units (*continued*)

Function	*Modifier*	*Symbol Usage Construct Example and Translation*
attack—A type of offensive operation that defeats enemy forces, seizes terrain, or secures terrain. (FM 3-90)	**A**	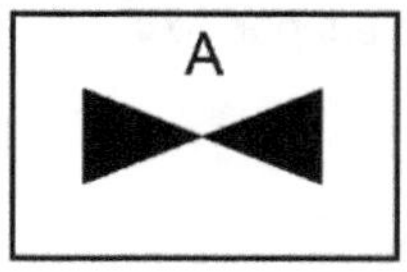 Aviation unit with attack helicopter capability
Army aviation or rotary-wing aviation—Conducts attack and air movement functions in support of ground maneuver in the area of operations or area of interest. (See FM 3-04 for more information on Army aviation.)		 Maintenance platoon with aviation maintenance capability
battalion (echelon of support)—Provides support to a battalion. (See ATP 3-96.1 for more information on battalion echelon of support.)	**II**	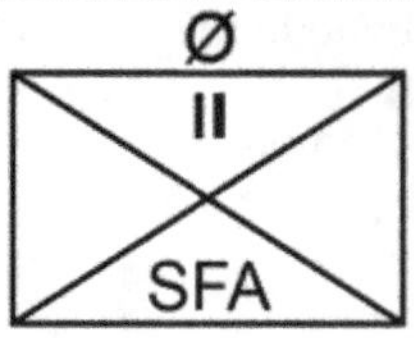 Infantry security force assistance team supporting a battalion
biological—Capability to detect biological warfare agent employment as a measure to provide medical treatment (See FM 3-11 for more information on biological.) A *biological agent* is a microorganism (or a toxin derived from it) that causes disease in personnel, plants, or animals or cause the deterioration of materiel. (JP 3-11)	**B**	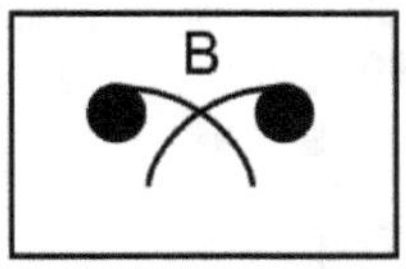 CBRN unit with biological capability
bridging—Assets used to cross a gap; the two types of bridging are standard and nonstandard bridging. (See ATP 3-90.4 for more information on bridging.)	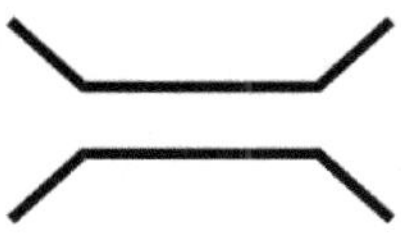	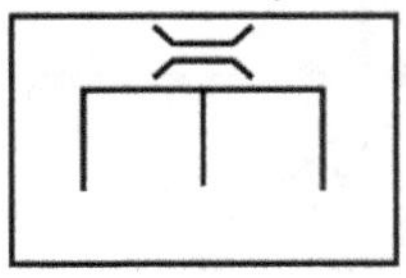 Engineer unit with bridging capability

Table 2-9. Sector 1 modifiers for units (*continued*)

Function	*Modifier*	*Symbol Usage Construct Example and Translation*
brigade (echelon of support)—Provides support to a brigade.	**X**	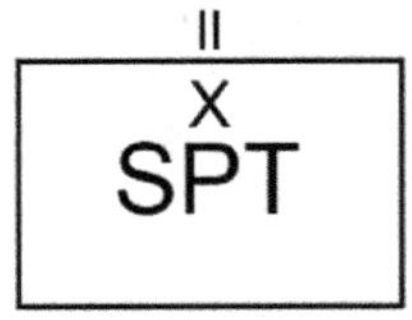 Brigade support battalion 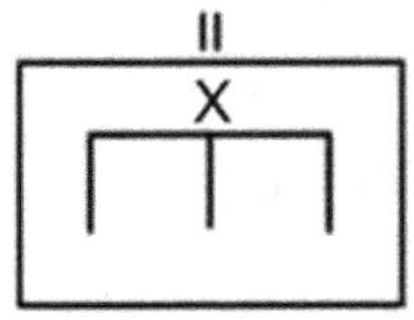 Brigade engineer battalion
chemical detection and decontamination—A capability to detect and to non-intrusively assess chemical munitions. (See JP 3-11 for more information on chemical.)	**C**	 CBRN unit with chemical detection capability
clearing—A mobility (focused on movement) task performed by follow-on engineers and EOD that involves the total elimination or neutralization of an obstacle. (ATP 3-90.4)	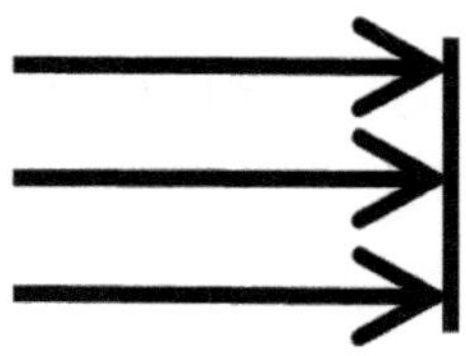	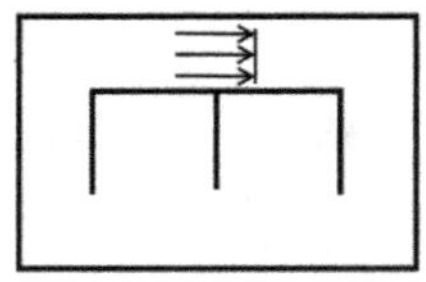 Engineer unit with clearing capability
combat—Can conduct mobility, countermobility, and survivability.	**CBT**	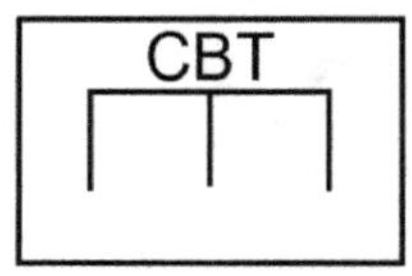 Engineer unit with combat capability

Table 2-9. Sector 1 modifiers for units (*continued*)

Function	***Modifier***	***Symbol Usage Construct Example and Translation***
combat camera—Specially-trained expeditionary forces from Service-designated units capable of providing high-quality directed visual information during military operations. (JP 3-61)		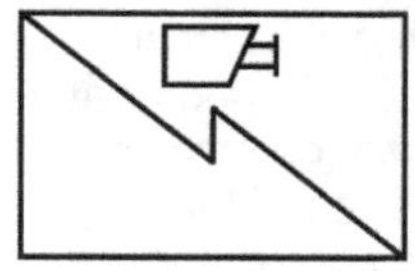 Signal unit with combat camera capability
	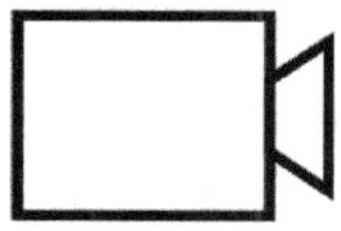 Alternate hand-drawn version	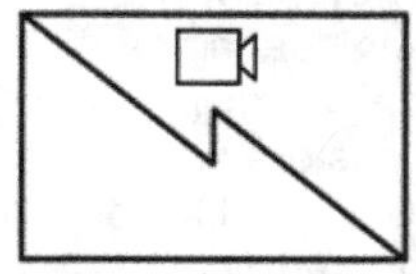
command post node—Provides SECRET Internet Protocol Router Network (also known as SIPRNET), Non-classified Internet Protocol Router Network (also known as NIPRNET), secure and non-secure voice over internet protocol (also known as VoIP), and battlefield video teleconferencing services. (See ATP 6-02.60 for more information on command post node.)	**CPN**	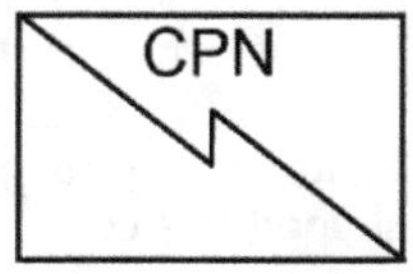 Signal unit with command post node capability
command and control—The exercise of authority and direction by a properly designated commander over assigned and attached forces in the accomplishment of the mission. (JP 1 VOL 2)	**C2**	 Aviation unit with command and control capability
company (echelon of support)—Provides support to a company. (See ATP 3-96.1 for more information on company echelon of support.)	**I**	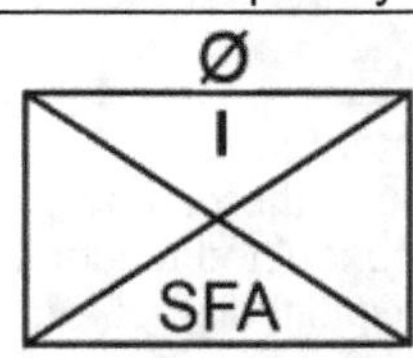 Infantry security force assistance team supporting a company

Table 2-9. Sector 1 modifiers for units (*continued*)

Function	*Modifier*	*Symbol Usage Construct Example and Translation*
composite—A combination of different capabilities and equipment assigned or attached to a unit with common function or purpose. (See FM 4-0 for more information on composite.) ***Note.*** This modifier symbol is interchangeable. Symbol may be used as a sector 1 or sector 2 modifier with the same meaning.	**COMP**	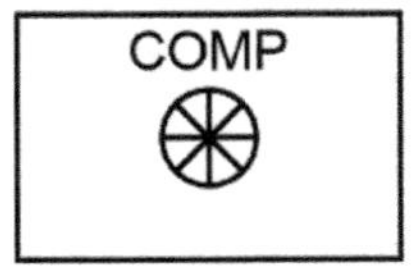 Transportation unit with multiple variations of vehicles
construction support—A general engineering capability focused on improving or repairing austere conditions, infrastructure, and building base camps and new lines of communications. (See FM 3-34 for more information on construction support.) ***Note.*** This modifier symbol is interchangeable as a sector 1 or sector 2 modifier with the same meaning.	**CONST**	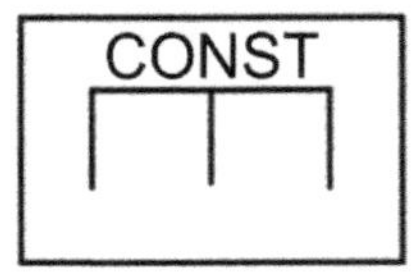 Engineer unit with construction
corps (echelon of support)—Provides support to a corps.	**XXX**	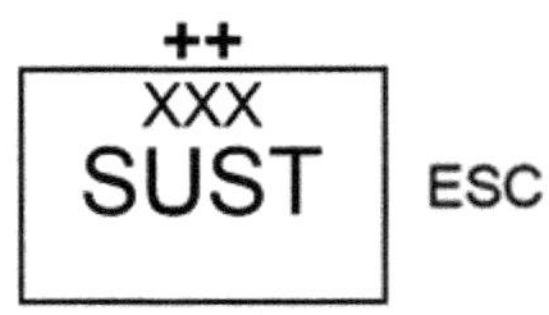 Expeditionary sustainment command supporting a corps 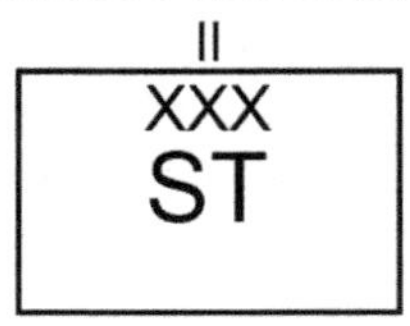 Special troops battalion supporting a corps
counterintelligence—Information gathered and activities conducted to identify, deceive, exploit, disrupt, or protect against espionage, other intelligence activities, sabotage, or assassinations conducted for or on behalf of foreign powers, organizations or persons or their agents, or international terrorist organizations or activities. (JP 2-0)	**CI**	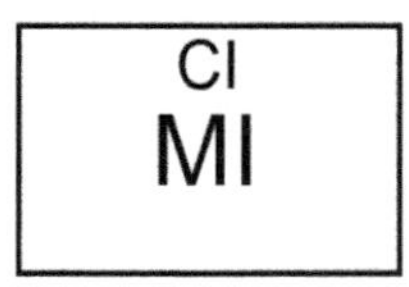 Military intelligence unit with counterintelligence capability

Table 2-9. Sector 1 modifiers for units (*continued*)

Function	*Modifier*	*Symbol Usage Construct Example and Translation*
criminal investigation division—Capabilities are particularly relevant in site exploitation and other evidence collection requirements on the battlefield, training, and assistance to host-nation law enforcement organizations, and collection of police and specific criminal intelligence critical in identifying, understanding, and attacking criminal networks operating against U.S. interests. (See ATP 3-39.12 for more information on criminal investigation division.)	**CID**	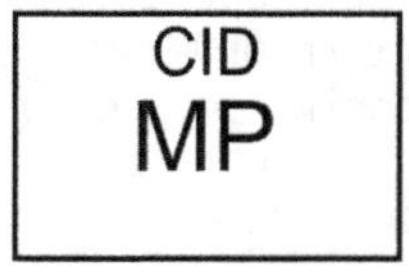 Military police unit with criminal investigation division capability
cross-cultural communication—The capability to communicate with individuals who have differences in culture (including nationality, ethnicity, race, gender). (See FM 3-18 for more information on cross-cultural communication.)	**CCC**	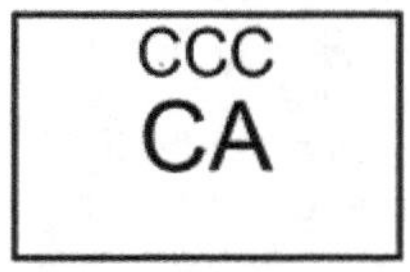 Civil affairs unit with cross-cultural communications capability
crowd and riot control—Capability to use chemical compounds that are developed, in part, for military use (riot control agents and obscurants), but not as weapons. (See FM 3-11 for more information on crowd and riot control.)	**CRC**	 Military police unit with crowd and riot control capability
cyberspace operations—The employment of cyberspace capabilities where the primary purpose is to achieve objectives in or through cyberspace. (JP 3-0)	**CYB**	 Electromagnetic warfare unit with cyberspace operations capability
decontamination—The process of making any person, object, or area safe by destroying, neutralizing, making harmless, or absorbing and removing chemical or biological agents or by removing radioactive material clinging to or around it. (JP 3-11) ***Note.*** This modifier symbol is interchangeable as a sector 1 or sector 2 modifier with the same meaning.	**D**	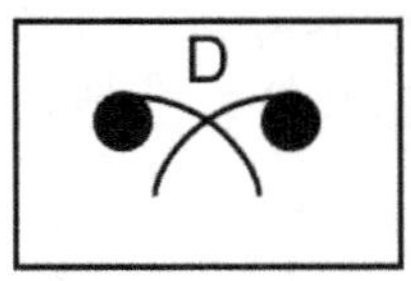 CBRN unit with decontamination capability

Table 2-9. Sector 1 modifiers for units (*continued*)

Function	*Modifier*	*Symbol Usage Construct Example and Translation*
detention—Capability to shelter, sustain, guard, protect, and account for populations or groups (detainees or U.S. military prisoners) because of military or civil conflict or to facilitate criminal prosecution. (See FM 3-63 for more information on detention.)	**DET**	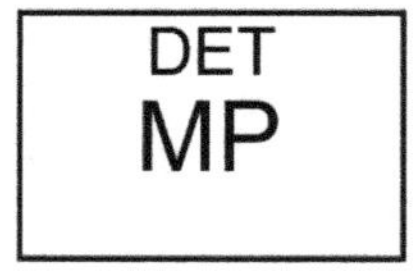 Military police unit with detention capability
digital—An electronic and especially computerized technology capability.	**DIG**	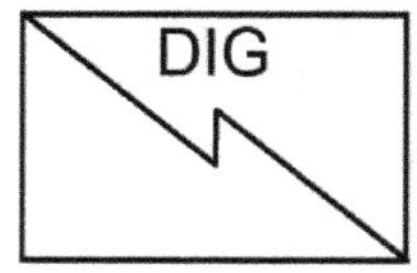 Signal unit with digital capability
diving—Capability to conduct scuba and surface diving operations to a depth of 190 feet in a maritime environment in support of combat, general, and geospatial engineering. (See FM 3-34 for more information on diving.)		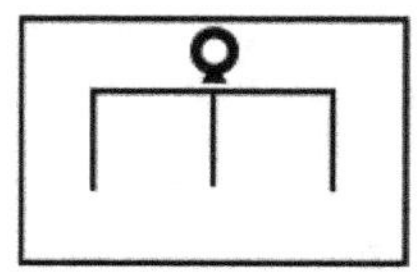 Engineer unit with diving capability
division (echelon of support)—Provides support to a division.	**XX**	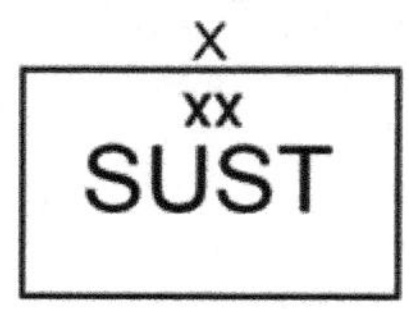 Sustainment brigade supporting a division
dog (working dog)—A canine capability that enhances security, police operations, and force protection missions (including counter improvised explosive device operations and assured mobility). (See ATP 3-39.34 for more information on working dogs.)	**DOG**	 Military police unit with working dog capability
drilling—Capability to detect and assess water sources and drilling water wells. (See FM 3-34 for more information on drilling.)		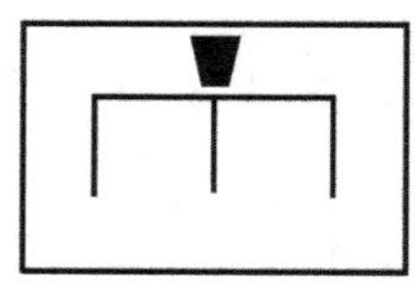 Engineer unit with drilling capability

Table 2-9. Sector 1 modifiers for units (*continued*)

Function	*Modifier*	*Symbol Usage Construct Example and Translation*
electromagnetic warfare—Military action involving the use of electromagnetic and directed energy to control the electromagnetic spectrum or to attack the enemy. (JP 3-85)	**EW**	 Cyberspace unit with electromagnetic warfare capability
electric power production—Capability to produce electrical power by converting fuels or other energy sources to electricity. (See ATP 3-34.45 for more information on electric power production.)		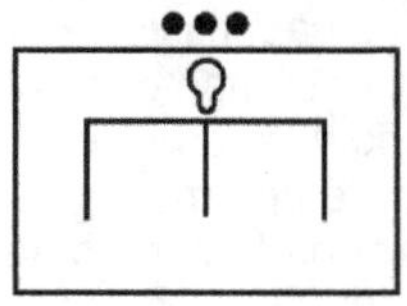 Engineer platoon with electric generation capability
Enhanced—Provides network installation, troubleshooting, quality assurance testing, and handoff coordination to enable the transition from tactical to semi-permanent automation support. (See FM 6-02 for more information on enhanced.)	**ENH**	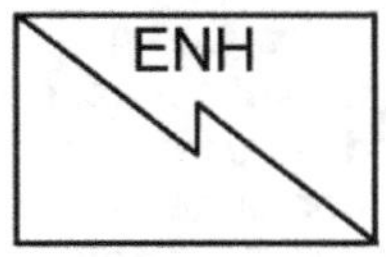 Signal unit with enhanced capability
explosive ordnance disposal—The detection, identification, on-site evaluation, rendering safe, exploitation, recovery, and final disposal of explosive ordnance. (FM 4-30)	**EOD**	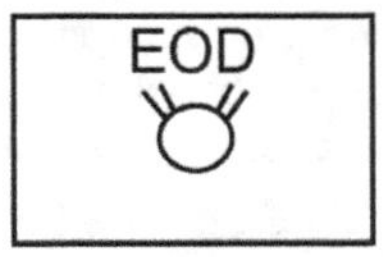 Ordnance unit with explosive ordnance disposal (EOD) capability 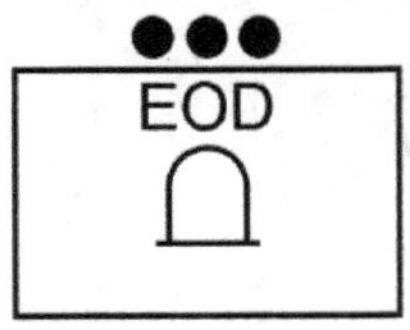 Ammunition platoon with explosive ordnance disposal capability
forward—An inherent designed capability to function effectively in positions located in geographic proximity of an objective.	**FWD**	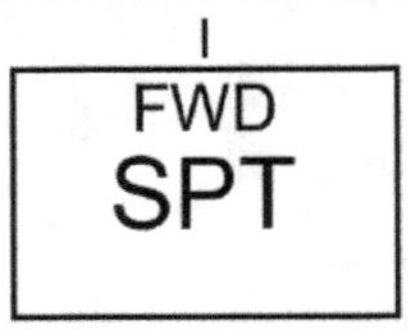 Forward support company

Table 2-9. Sector 1 modifiers for units (*continued*)

Function	*Modifier*	*Symbol Usage Construct Example and Translation*
fire direction center—That element of a command post, consisting of gunnery and communications personnel and equipment, by means of which the commander exercises fire direction and/or fire control. (JP 3-09.3)	**FDC**	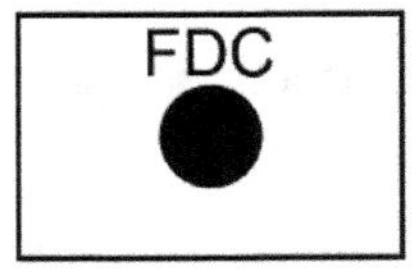 Artillery fire direction center unit
general engineering—Those engineering capabilities and activities, other than combat engineering, that provide infrastructure and modify, maintain, or protect the physical environment. (JP 3-34)	**GEN**	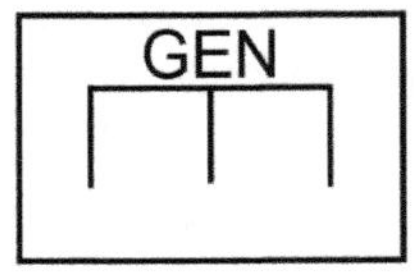 Engineer unit with general engineering capability
geospatial information—Information that identifies the geographic location and characteristics of natural or constructed features and boundaries on or about the Earth, including: data and Information derived from, among other things, remote sensing, mapping, and surveying technologies; and mapping, charting, geomatics data, and related products and services. (JP 2-0)		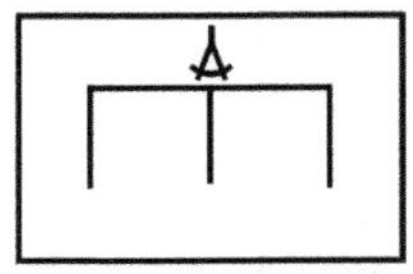 Engineer unit with geospatial capability

Table 2-9. Sector 1 modifiers for units (*continued*)

Function	*Modifier*	*Symbol Usage Construct Example and Translation*
headquarters or headquarters element—A place from which a commander performs the functions of command.	———	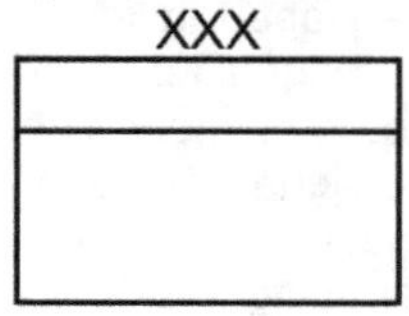 Corps headquarters
		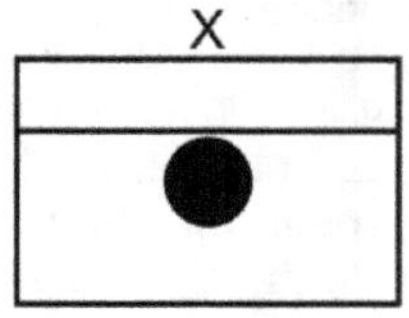 Field artillery brigade headquarters
		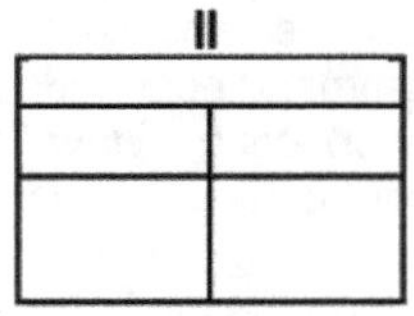 Medical battalion headquarters ***Note.*** To avoid overlapping of symbols, shorten center vertical Main function symbol line.
independent support command (echelons of support)—Provides support to an independent support command.	**++**	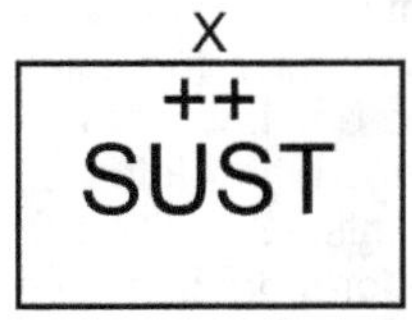 Sustainment brigade supporting an independent support command
intermodal operations—The process of using multiple modes (air, sea, highway, rail) and conveyances (truck, barge, containers, pallets) to move troops, supplies and equipment through expeditionary entry points and the network of specialized transportation nodes to sustain land forces. (ADP 4-0) ***Note.*** This modifier symbol is interchangeable. Symbol may be used as a sector 1 or sector 2 modifier with the same meaning.	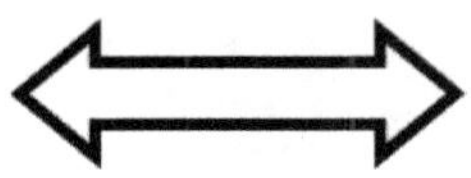	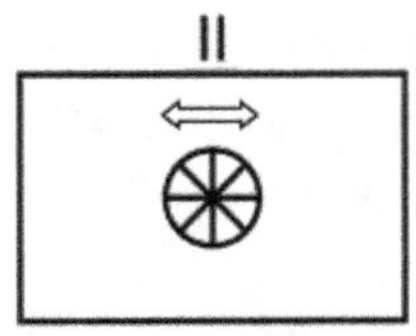 Transportation battalion with intermodal capability

Table 2-9. Sector 1 modifiers for units (*continued*)

Function	*Modifier*	*Symbol Usage Construct Example and Translation*
jamming—Capability to broadcast a signal tuned to frequencies with enough power to override signals at the receivers (spot jamming or barrage jamming). (See ATP 6-02.70 for more information on jamming.) ***Note.*** This modifier symbol is interchangeable as a sector 1 or sector 2 modifier with the same meaning.		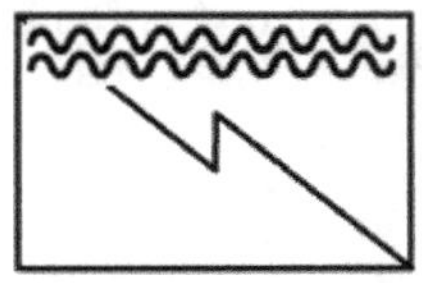 Signal unit with jamming capability ***Note.*** To avoid overlapping of symbols, shorten the line of Main function symbol to allow space for the modifiers.
joint node network—Provides connection to the regional hub node, uses a dedicated frequency division multiple access satellite communications link, and shares bandwidth among command post nodes using network centric waveform satellite communications. (See ATP 6-02.60 for more information on the joint node network.)	**JNN**	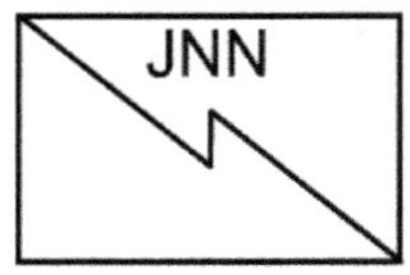 Signal unit with joint node network capability
maintenance—Ensures unit readiness by maintaining weapons systems and equipment in a fully mission-capable status for immediate and continuous employment in complex and highly lethal environments. (See FM 4-0 and ATP 4-33 for more information on maintenance.) ***Note.*** This modifier is interchangeable as a sector 1 or sector 2 modifier with the same meaning.		 Aviation unit with organic maintenance capability

Table 2-9. Sector 1 modifiers for units (*continued*)

Function	***Modifier***	***Symbol Usage Construct Example and Translation***
medical evacuation—The timely and effective movement of the wounded, injured, or ill to and between medical treatment facilities on dedicated and properly marked medical platforms with en route care provided by medical personnel. (ATP 4-02.2)		 Aviation unit with medical evacuation capability 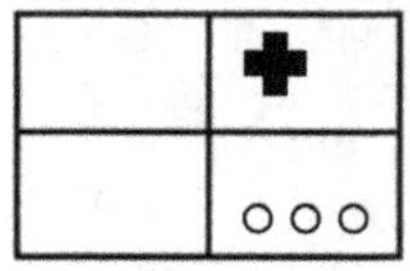 Medical unit with wheeled high mobility vehicle medical evacuation capability ***Note.*** The modifier offsets to the right to avoid overlapping of symbols.
medical role 1—Unit-level medical care capability provided by the combat medic or medical treatment provided by the battalion aid station. (See FM 4-02 for more information on medical role 1.) ***Note.*** Modifier offsets to the right to avoid overlapping with the Main function symbol.	**1**	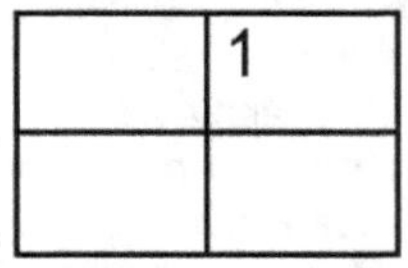 Medical unit with role 1 capability
medical role 2—Capability to provide care by area support squads or medical treatment platoons of medical companies with greater medical capabilities available than Role 1. (See FM 4-02 for more information on medical role 2.) ***Note.*** Modifier offsets to the right to avoid overlapping with the Main function symbol.	**2**	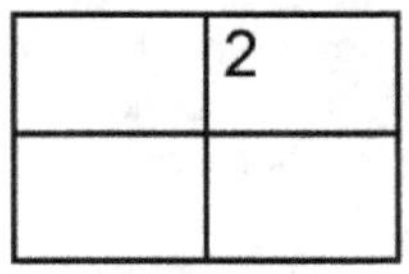 Medical unit with role 2 capability
medical role 3—Capability to provide care to all categories of patients, to include resuscitation, initial wound surgery, damage control surgery, and postoperative treatment. (See FM 4-02 for more information on medical role 3.) ***Note.*** Modifier offsets to the right to avoid overlapping of symbols.	**3**	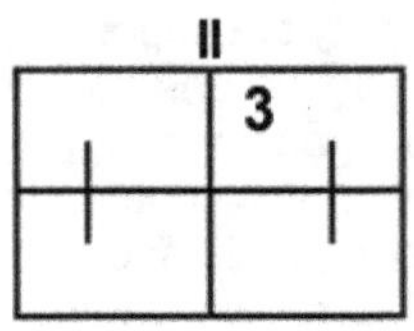 Medical treatment facility battalion with role 3 capability

Table 2-9. Sector 1 modifiers for units (*continued*)

Function	*Modifier*	*Symbol Usage Construct Example and Translation*
medical role 4—Medical care capability found in continental United States (DTG)-based medical treatment facilities, robust outside continental United States medical treatment facilities, and other safe havens. (See FM 4-02 for more information on medical role 4.) ***Note.*** Modifier offsets to the right to avoid overlapping of symbols.	**4**	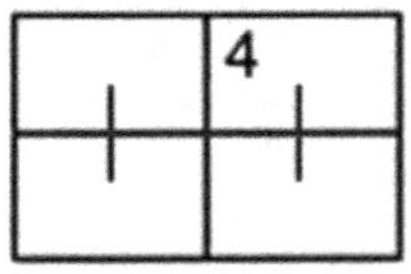 Medical treatment facility with role 4 capability
meteorological—Capability to provide weather and weather forecasting data. (See ATP 2-22.7 and ATP 3-34.80 for more information on meteorological.)	**MET**	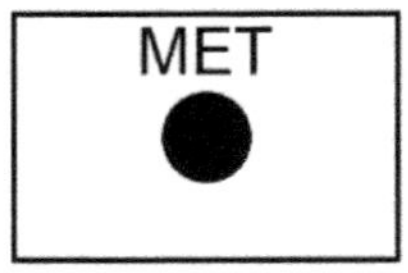 Field artillery unit with meteorological capability
multidomain operations—The combined arms employment of joint and Army capabilities to create and exploit relative advantages to achieve objectives, defeat enemy forces, and consolidate gains on behalf of joint force commanders. (FM 3-0)	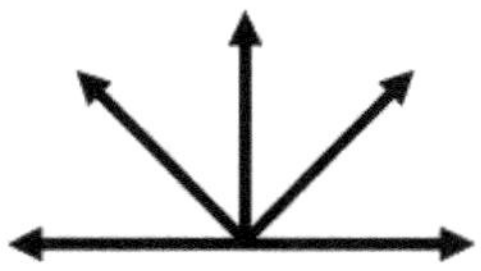	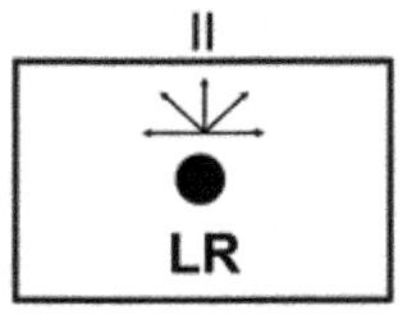 Long-range field artillery battalion with multidomain capabilities
watercraft (generic vessel)—The capability to conduct heavy lifting associated with water transport operational maneuver and intra-theater lift of units, equipment, and supplies. (See ATP 4-15 for more information on watercraft.) ***Note.*** This modifier symbol is interchangeable. Symbol may be used as a sector 1 or sector 2 modifier with the same meaning.		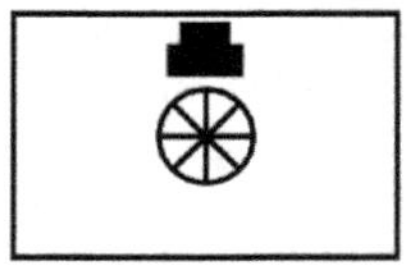 Transportation unit with watercraft capability
mobile advisor and support—The capability to support, breaching operations, clearing operations, gap-crossing operations, traffic control plan development, main and alternate supply route regulation and enforcement, passage of lines, straggler movement control. (See ATP 3-39.30 for more information on mobility support.)		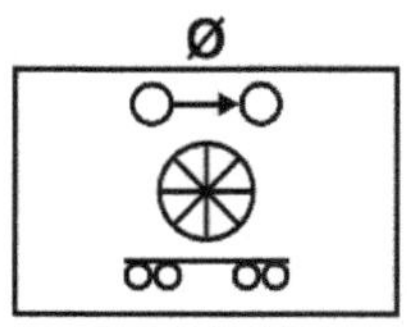 Transportation rail team with mobile advisor and support capability

Table 2-9. Sector 1 modifiers for units (*continued*)

Function	*Modifier*	*Symbol Usage Construct Example and Translation*
mobility support—The capability to support breaching operations, clearing operations, gap-crossing operations, traffic control plan development, main and alternate supply route regulation and enforcement, passage of lines, and straggler movement control. (See ATP 3-39.30 for more information on mobility support.)	**MS**	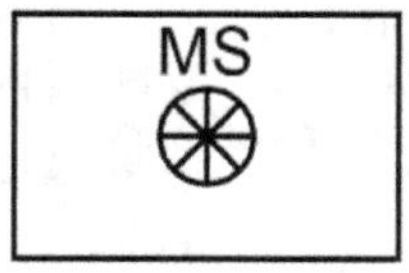 Transportation unit with mobility support capability
mortar—Organic fire support to battalions, squadrons, companies, and troops that is available to a commander when other indirect fire support is not available. (See ATP 3-21.90 for more information on mortars.)		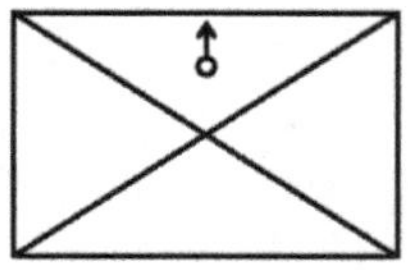 Infantry unit with mortar capability 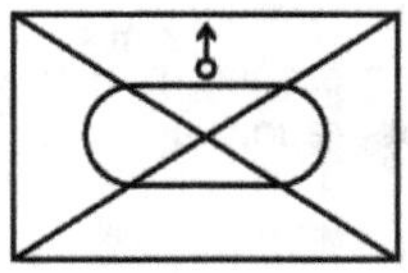 Mechanized armored infantry unit with mortar capability
multinational operations—A collective term used to describe military actions conducted by forces of two or more nations, usually undertaken within the structure of a coalition or alliance. (JP 3-16)	**MN**	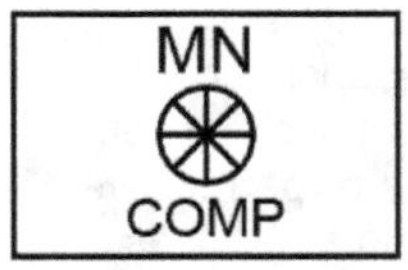 Multinational transportation unit with multiple variations of vehicles
multiple rocket launcher—Multiple launch rocket system or high mobility artillery rocket system capability. (See ATP 3-09.60 for more information on the multiple launch rocket system or the high mobility artillery rocket system.)		 Artillery unit with multiple rocket launcher capability

Table 2-9. Sector 1 modifiers for units (*continued*)

Function	*Modifier*	*Symbol Usage Construct Example and Translation*
Maritime terminal support—The capability to provide support functions to fixed, unimproved, bare beach, and degraded port facilities, and at off-shore anchorages. (See JP 4-18 for more information on maritime terminal support.)		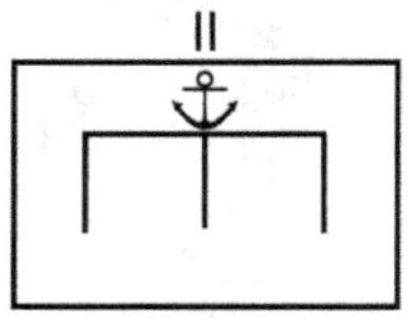 Engineer battalion with maritime terminal support capability
	 Alternate hand-drawn version	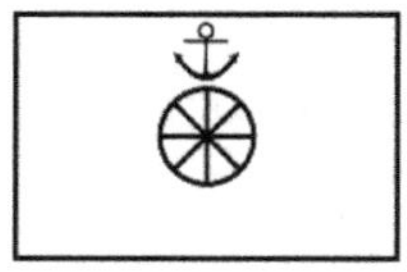 Transportation unit with maritime terminal support capability
network operations—Activities conducted to operate and defend the global information grid. (See JP 6-0 for more information on network operations.)	**NET**	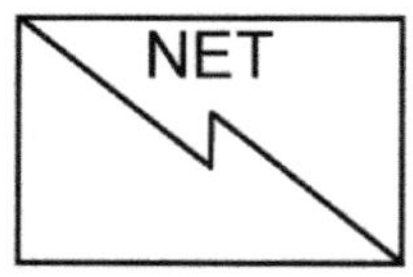 Signal unit with network operations capability
nuclear—The capability of assessing, exploiting, characterizing, and disabling facilities associated with the nuclear fuel cycle in semi-permissive or permissive environments. Advises commanders on the risks associated with these facilities, provide detailed information related to potential material proliferation, and make recommendations on how to dispose of nuclear material. (See FM 3-11 for more information on nuclear.)	**N**	 CBRN unit with nuclear capability

Table 2-9. Sector 1 modifiers for units (*continued*)

Function	*Modifier*	*Symbol Usage Construct Example and Translation*
operations—The capability to execute the principal planning and operating functions of a unit.	**OPS**	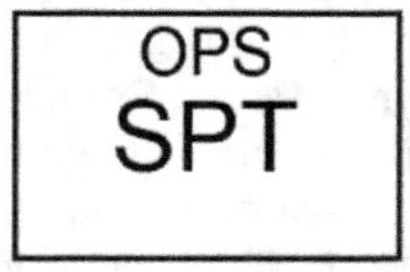 Support operations unit
		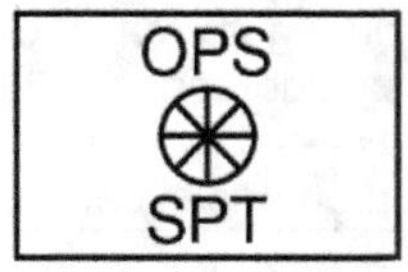 Transportation support operations unit
palletized load system—A fitted integral self-loading and unloading cargo system capability. (See FM 4-01 for more information on the palletized load system.)	**PLS**	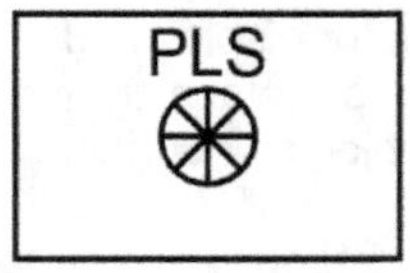 Transportation unit with palletized load system capability
petroleum, oil, and lubricants (POL)—Capability to receive, stock, or distribute petroleum products. (See ATP 4-43 for more information on POL.)		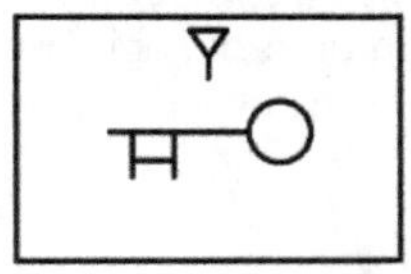 Quartermaster unit with POL storage capability
		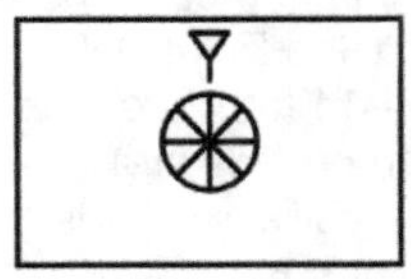 Transportation unit with POL transport capability
pipeline—A capability that consists of pipeline sets, pipeline pump stations, and pipeline support equipment. Its primary function is to transport fuel from one area to another. (See ATP 4-43 for more information on pipeline.)	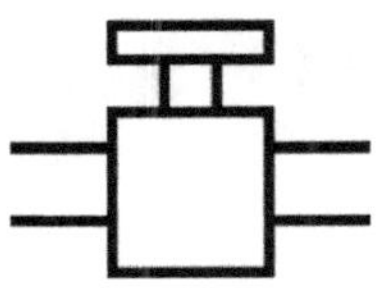	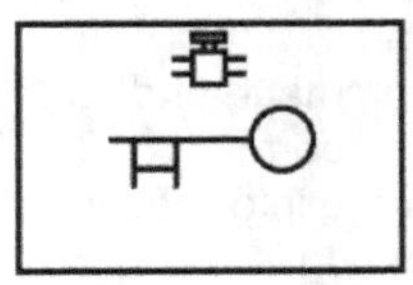 Quartermaster unit with pipeline capability

Table 2-9. Sector 1 modifiers for units (*continued*)

Function	*Modifier*	*Symbol Usage Construct Example and Translation*
postal service—The capability to operate as an extension of the United States Postal Service consistent with public law and federal regulations beyond the boundaries of U.S. sovereignty and provides postal services for all DOD personnel where there is no United States Postal Service available during normal and contingency operations. (See FM 1-0 and FM 4-0 for more information on postal service.)		 Human resources unit with postal service capability
radar—A device or system usually consisting of a synchronized radio transmitter and receiver that emits radio waves and processes their reflections for display and is used especially for detaching and locating objects or surface features. (See ATP 3-09.12 and ATP 3-27.5 for more information on radar.)		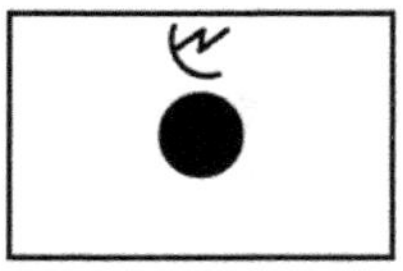 Artillery unit with radar capability
radiological—Capability to coordinate radiological survey missions. (See FM 3-11 for more information on radiological.)	**R**	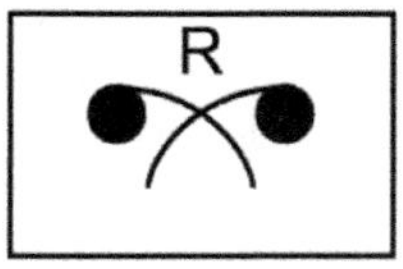 CBRN unit with radiological capability
railway—Provides rail network capability and infrastructure assessments and coordinates contracts. (See FM 4-0, FM 4-01, and ATP 4-14 for more information on railway.) ***Note.*** This modifier symbol is interchangeable as a sector 2 or sector 2 modifier with the same meaning.		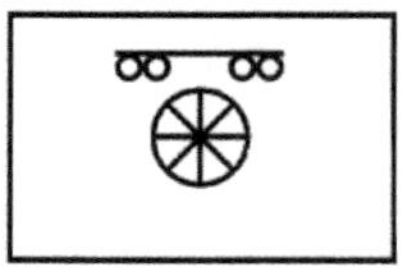 Transportation unit with railway capability
retransmission—The capability to extend the range of single-channel radio networks to support command and control in retrograde operations. (See FM 6-02 for more information on retransmission.)	**RTNS**	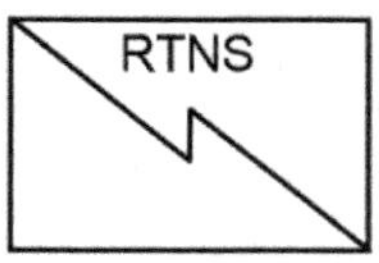 Signal unit with retransmission capability

Table 2-9. Sector 1 modifiers for units (*continued*)

Function	*Modifier*	*Symbol Usage Construct Example and Translation*
riverine—Provides water transport to move troops and equipment. (See ATP 4-15 for more information on riverine.)		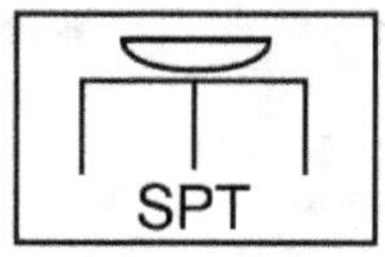 Engineer unit with riverine support capability
robotic (guided and autonomous)—The capability to employ operational equipment that has robot guided or autonomous control mechanism characteristics.		 Infantry unit with robotic capability
search and rescue—The use of aircraft, surface craft, submarines, and specialized rescue teams and equipment to search for and rescue distressed persons on land or at sea in a permissive environment. (JP 3-50)	**SAR**	 Aviation (rotary) unit with search and rescue capability
sensor—The capability to observe or receive a signal or observable from a person or object. (See FM 2-0 for more information on sensors.)		 Electromagnetic warfare unit with sensor capability
signals intelligence—Intelligence derived from communications, electronic, and foreign instrumentation signals. (JP 2-0)		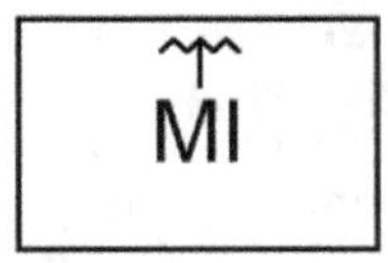 Military intelligence unit with signals intelligence capability
single rocket launcher (shoulder-launched munitions)—Unguided free-flight rocket and a launcher that contains all features and controls necessary to aim, fire, and engage targets. (See TM 3-23.25 for more information on single rocket launchers.)		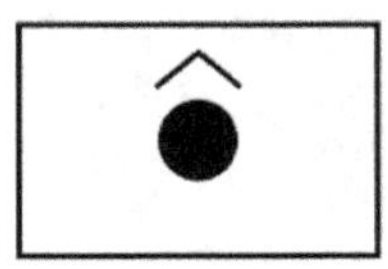 Artillery unit with single rocket launcher capability

Table 2-9. Sector 1 modifiers for units (*continued*)

Function	*Modifier*	*Symbol Usage Construct Example and Translation*
sniper—A specialized trained marksman with tactical skills and techniques who is highly capable at conducting detailed surveillance and shooting at exposed enemy's forces from a concealed vantage point. (See TC 3-22.10 for more information on snipers.)		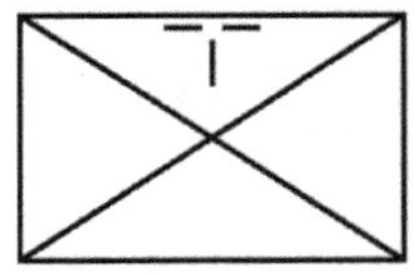 Infantry unit with sniper capability
survey—Provides position and azimuth determining Global Positioning System capability. (See ATP 3-09.12 for more information on surveys.)		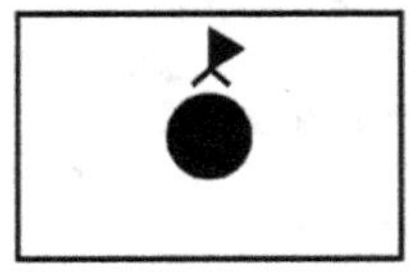 Artillery unit with survey capability
tactical satellite communications—Capabilities enabling communications on the move for better situational awareness and understanding. (See ATP 3-05.60 and ATP 6-02.54 for more information on tactical satellite communications.)		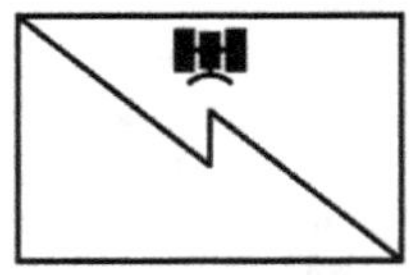 Signal unit with tactical satellite communications capability
	 alternate hand-drawn version	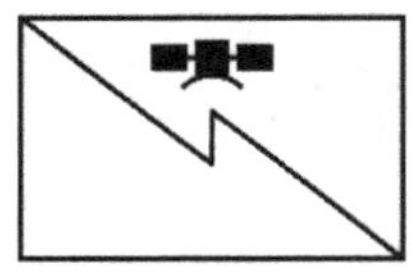
target acquisition—The detection, identification, and location of a target in sufficient detail to permit the effective employment of capabilities that create the required effects. (JP 3-60)	**TA**	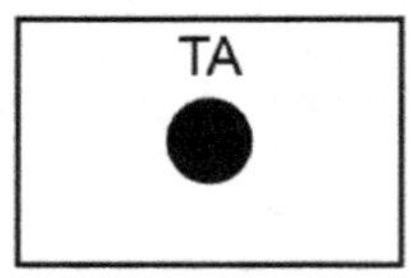 Artillery unit with target acquisition capability
theater army (echelon of support)—Provides support to a theater army.	**XXXX**	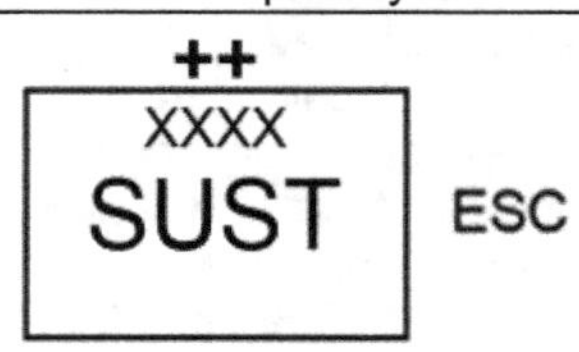 Expeditionary sustainment command supporting a theater army

Table 2-9. Sector 1 modifiers for units (*continued*)

Function	*Modifier*	*Symbol Usage Construct Example and Translation*
theater of operations (echelon of support)—Provides support to a theater.	**XXXXXX**	 Theater sustainment command supporting a theater of operations
unmanned aircraft system—That system whose component include the necessary equipment, network, and personnel to control an unmanned aircraft. (JP 3-30)		 Aviation (fixed-wing) unit with unmanned aircraft system (UAS) capability
		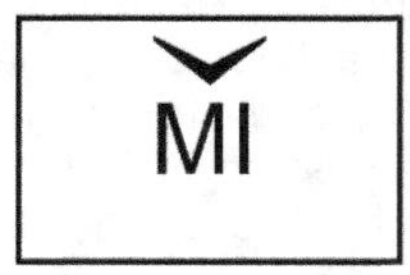 Military intelligence unit with UAS capability
		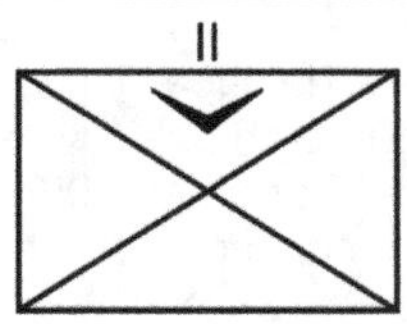 Infantry battalion with UAS capability
utility—Equipment designed or adapted for general purpose use.	**U**	 Aviation (rotary-wing) unit with utility type helicopter capability
water—The capability to receive, stock, produce, or distribute water. (See FM 4-0 and ATP 4-44 for more information on water.)	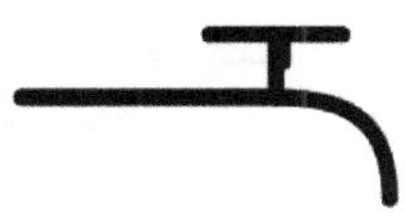	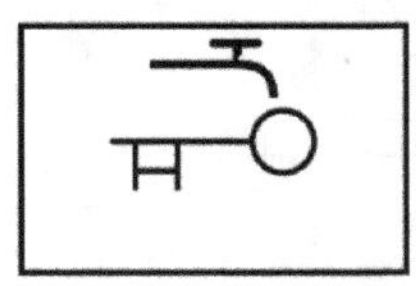 Quartermaster unit with water production capability

Table 2-9. Sector 1 modifiers for units (*continued*)

Function	***Modifier***	***Symbol Usage Construct Example and Translation***
weapons—Heavy weapons system capability (machine guns and anti-armor weapons) to provide additional combat power. (See ATP 3-21.10 and ATP 3-21.20 for more information on weapons.)	**WPN**	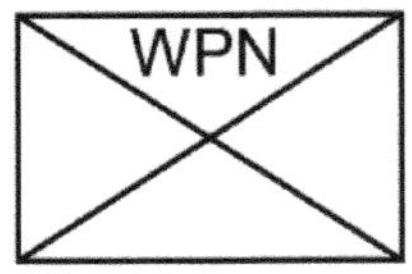 Infantry unit with machine gun and anti-armor capability

SECTOR 2 MODIFIERS FOR UNITS

2-22. This indicator is represented as the lower part of Field A as shown in table 1-4 on page 5. Table 2-10 (on pages 68 through 80) shows sector 2 symbols. Sector 2 modifiers provide additional information specific to the capabilities of a unit.

Table 2-10. Sector 2 modifiers for units

Function	***Modifier***	***Symbol Usage Construct Example and Translation***
air assault—The movement of friendly assault forces by rotary-wing or tiltrotor aircraft to engage and destroy enemy forces or to seize and hold key terrain. (JP 3-18)		 Aviation unit with air assault capability
airborne—The capability to parachute into an objective area. (See JP 3-18 and FM 3-99 for more information on airborne.)		 Infantry unit with airborne capability 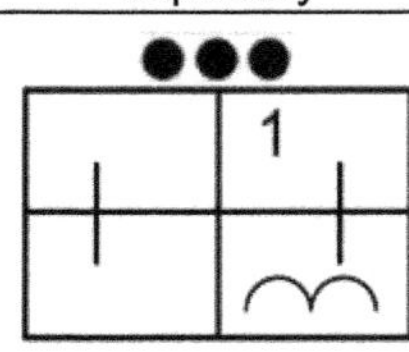 Medical treatment facility platoon with Role 1 and airborne capability ***Note.*** The modifier offsets to the right to avoid overlapping of symbols.

Table 2-10. Sector 2 modifiers for units (*continued*)

Function	*Modifier*	*Symbol Usage Construct Example and Translation*
air defense—Defensive measures designed to destroy attacking enemy aircraft or aerodynamic missiles, or to nullify or reduce the effectiveness of such attack. (JP 3-01)		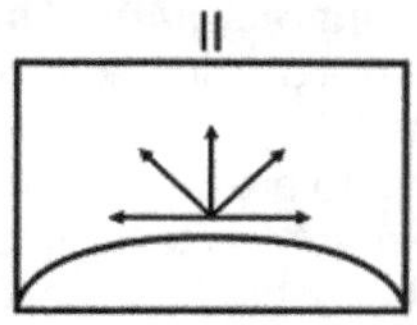 Multidomain battalion with air defense capability
amphibious—The capability to conduct amphibious operations within the littorals. (See JP 3-02 for more information on amphibious.)		 Infantry unit with amphibious capability
analysis—The capability to conduct a detailed examination of anything complex in order to understand its nature or to determine its essential features.		 Electromagnetic warfare unit with analysis capability
armored tracked—A vehicle equipped or protected with armor, and with a continuous band of treads or track plates for self-propelled mobility. ***Note.*** As a sector 2 modifier, this symbol represents an armored self-propelled tracked vehicle capability.		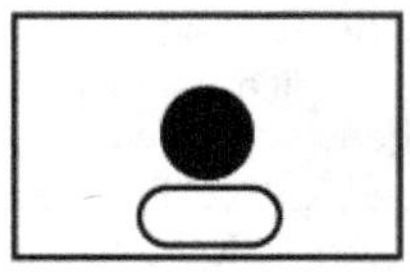 Field artillery unit with armored self-propelled (tracked) capability
barge, not self-propelled—Class C vessel that are usually subject to wind, tide, and sea state. When afloat, they have a constant requirement for tending, even when not being actively employed for their designed purpose. (See ATP 4-15 for more information on barges.)	**YB**	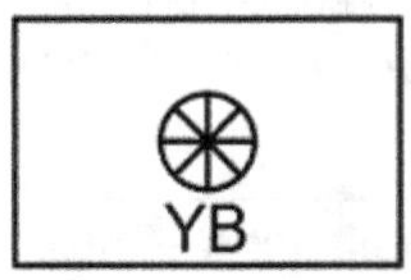 Transportation unit with barge (not self-propelled) capability
blood support—A capability to receive, account, store, and distribute blood and blood products. (See ATP 4-02.1 for information on blood support.)		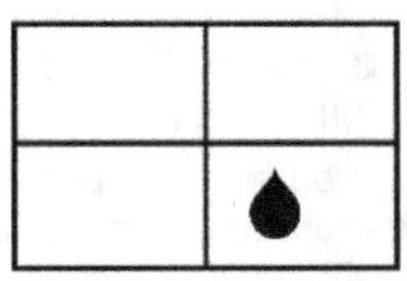 Medical unit with blood support capability ***Note.*** The modifier offsets to the right to avoid overlapping of symbols.

Table 2-10. Sector 2 modifiers for units (*continued*)

Function	*Modifier*	*Symbol Usage Construct Example and Translation*
combat and operational stress control—A coordinated program of actions taken by military leadership to prevent, identify, and manage reactions to traumatic events that may affect exposed organizations and individuals during unified land operations. (FM 4-02)	Ψ	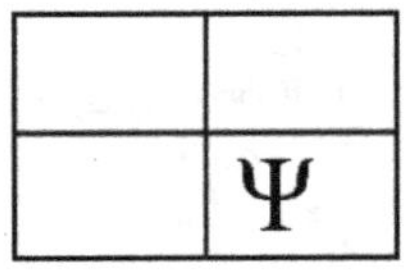 Medical unit with combat and operational stress control capability ***Note.*** The modifier offsets to the right to avoid overlapping of symbols.
composite—A combination of different capabilities and equipment assigned or attached to a unit with a common function or purpose. (See FM 4-0 for more information on composite.) ***Note.*** This modifier symbol is interchangeable. Symbol may be used as a sector 1 or sector 2 modifier with the same meaning.	**COMP**	 Transportation unit with multiple variations of vehicles
construction support—A general engineering capability focused on improving or repairing austere conditions, infrastructure, and building base camps and new lines of communications. (See FM 3-34 for more information on construction support.) ***Note.*** This modifier symbol is interchangeable as a sector 1 or sector 2 modifier with the same meaning.	**CONST**	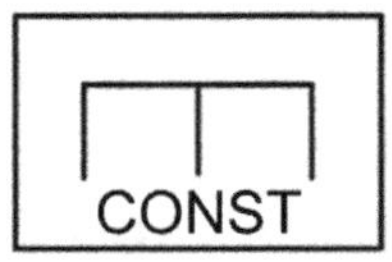 Engineer unit with construction
control—An action taken to eliminate a hazard or reduce its risk. (ATP 5-19) ***Note.*** This symbol demonstrates the capability and authority to exercise restraining or directing influence (regulating over a specific function).	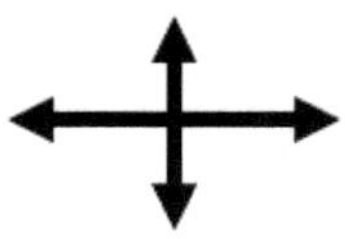	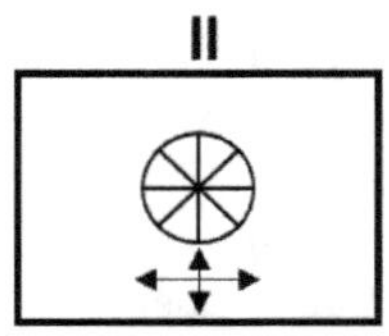 Transportation unit with movement control capability
decontamination—The process of making any person, object, or area safe by absorbing, destroying, neutralizing, making harmless, or removing chemical or biological agents, or by removing radioactive material clinging to or around it. (JP 3-11) ***Note.*** This modifier symbol is interchangeable as a sector 1 or sector 2 modifier with the same meaning.	**D**	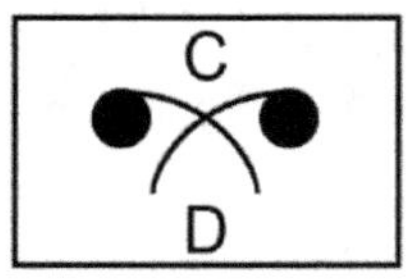 Chemical, biological, radiological, and nuclear (CBRN) unit with chemical decontamination capability

Table 2-10. Sector 2 modifiers for units (*continued*)

Function	***Modifier***	***Symbol Usage Construct Example and Translation***
dental services—To provide consultation, early treatment of severe oral and maxillofacial injuries; and augment medical personnel (as necessary) during mass casualty operations. (See FM 4-02 for more information on dental services.) ***Note.*** This modifier offsets to the right to avoid overlapping with the Main function symbol.		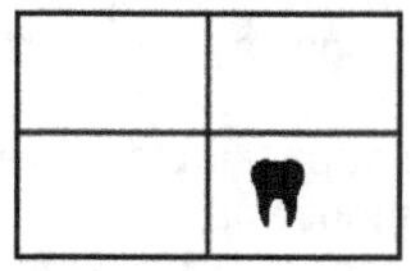 Medical unit with dental service capability
direction finding—A procedure for obtaining bearings of radio frequency emitters by using a highly directional antenna and a display unit on an intercept receiver or ancillary equipment. (JP 3-85)		 Electromagnetic warfare unit with direction finding capability
guerrilla—An irregular, predominantly indigenous member of a guerilla force organized similar to military concepts and structure in order to conduct military and paramilitary operations in enemy-held, hostile, or denied territory. (ATP 3-05.1)	**G**	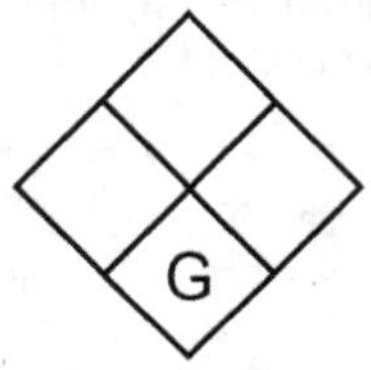 Enemy guerrilla infantry unit 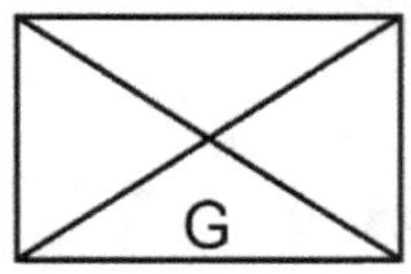 Friendly guerrilla infantry unit
heavy —Capable of large capacity or output.	**H**	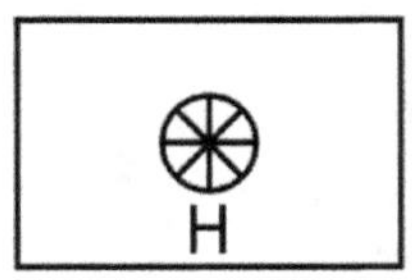 Transportation unit with heavy vehicle capability
high altitude—Above 8,000 feet.	**HA**	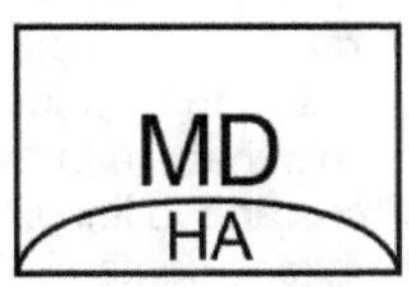 Air and missile defense unit with high altitude capability

Table 2-10. Sector 2 modifiers for units (*continued*)

Function	***Modifier***	***Symbol Usage Construct Example and Translation***
intermodal operations—The process of using multiple modes (air, sea, highway, rail) and conveyances (truck, barge, containers, pallets) to move troops, supplies and equipment through expeditionary entry points and the network of specialized transportation nodes to sustain land forces. (ADP 4-0) ***Note.*** This modifier symbol is interchangeable. Symbol may be used as a sector 1 or sector 2 modifier with the same meaning.		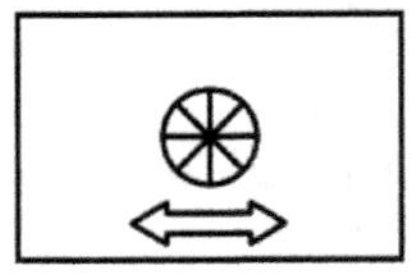 Transportation unit with intermodal capability
jamming—The capability to broadcast a signal tuned to frequencies with enough power to override signals at the receivers (spot jamming or barrage jamming). (See ATP 6-02.70 for more information on jamming.) ***Note.*** This modifier symbol is interchangeable. Symbol may be used as a sector 1 or sector 2 modifier with the same meaning.		 Electromagnetic warfare unit with jamming capability
laboratory—A place equipped for experimental study in a science or for testing and analysis. (See FM 4-02 for more information on laboratories.)	**LAB**	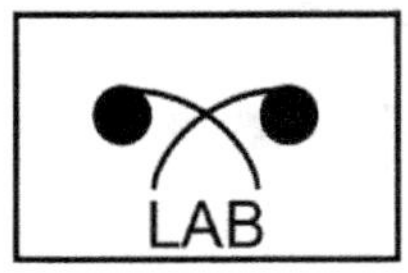 CBRN unit with laboratory capability 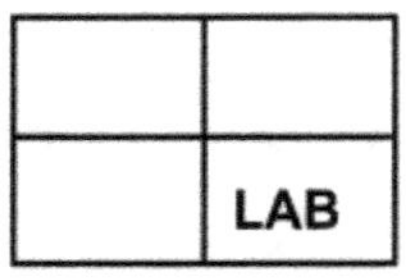 Medical unit with laboratory capability ***Note.*** The modifier offsets to the right to avoid overlapping of symbols.
landing craft—Provides vessels (Landing Craft Utility 2000 or Landing Craft Mechanized-9) to increase access to austere points on the littorals that are unavailable to land forces and to link ship and shore operations centers. (See ATP 4-15 for more information on landing craft.)	**LC**	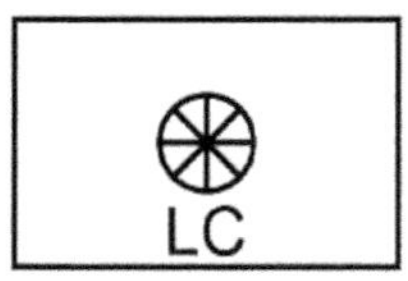 Transportation unit with landing craft capability

Table 2-10. Sector 2 modifiers for units (*continued*)

Function	*Modifier*	*Symbol Usage Construct Example and Translation*
light—The capability designed to carry a comparatively small load.	**L**	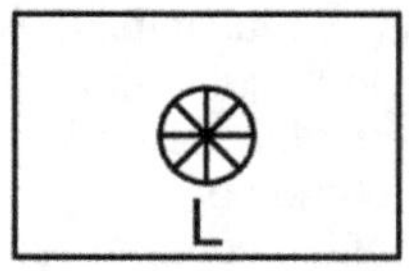 Transportation unit with light lift capability
light and medium—The capability designed to carry a comparatively small and intermediate size loads.	**L/M**	 Transportation unit with light and medium vehicle capability
long range—The capability relating to or fit for long distances.	**LR**	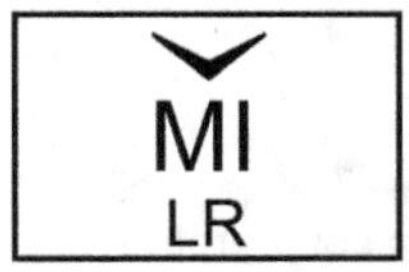 Military intelligence unit with long-range unmanned aircraft system capability
low altitude	**LA**	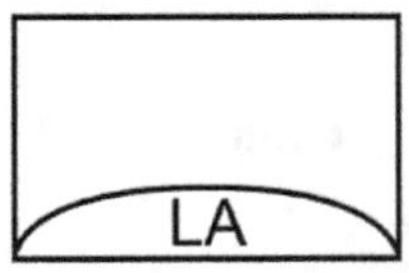 Air defense unit with low altitude capability
low to medium altitude	**LMA**	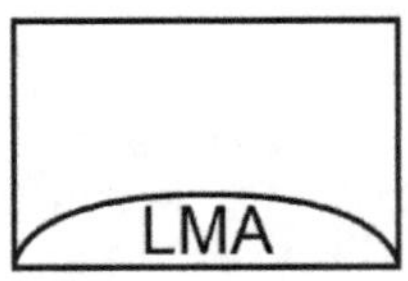 Air defense unit with low to medium altitude capability

Table 2-10. Sector 2 modifiers for units (*continued*)

Function	*Modifier*	*Symbol Usage Construct Example and Translation*
maintenance—Ensures unit readiness by maintaining weapons systems and equipment in a fully mission-capable status for immediate and continuous employment in complex and highly lethal environments. (See FM 4-0, FM 4-30, and ATP 4-33 for more information on maintenance.) ***Note.*** This modifier symbol is interchangeable as a sector 1 or sector 2 modifier with the same meaning.		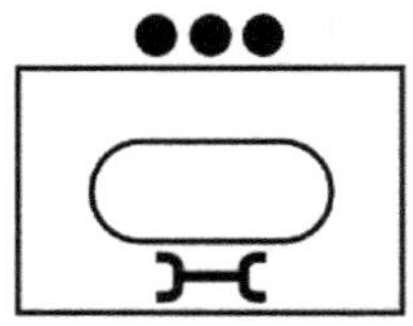 Armor (tracked) platoon with organic maintenance capability
medical bed—A capability to identify, provide, and locate a type of medical device used to provide patients with comfortable and secure sleeping accommodations while they are receiving medical treatment.	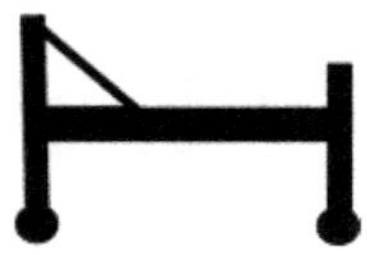	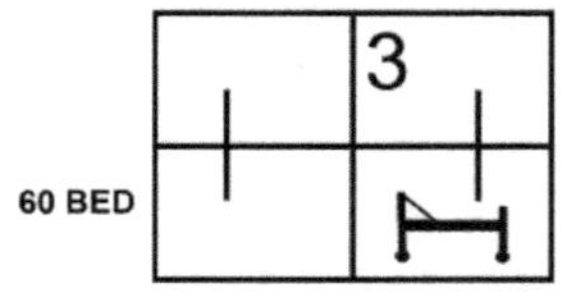 Medical unit with Role 3 and 60 medical bed capability ***Note.*** The modifiers offset to the right to avoid overlapping of symbols.
medium—The capability designed to carry a comparatively intermediate size load.	**M**	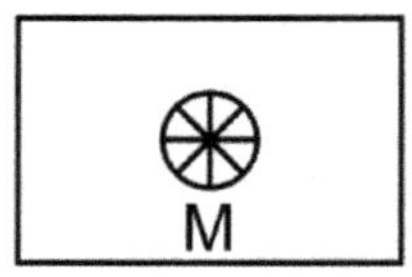 Transportation unit with medium vehicle capability
medium altitude	**MA**	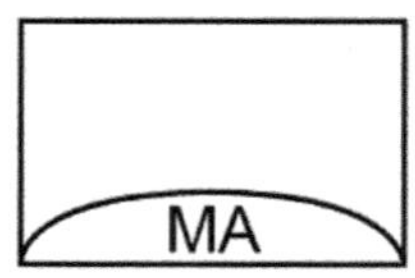 Air defense unit with medium altitude capability
medium range—The capability relating to or fit for intermediate distances.	**MR**	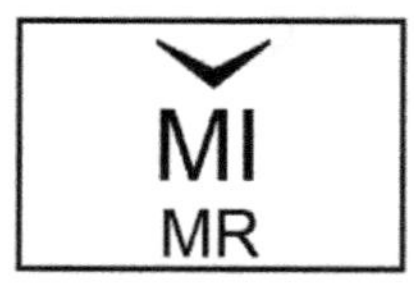 Military intelligence unit with medium range unmanned aircraft system capability

Table 2-10. Sector 2 modifiers for units (*continued*)

Function	*Modifier*	*Symbol Usage Construct Example and Translation*
medium to high altitude	**MHA**	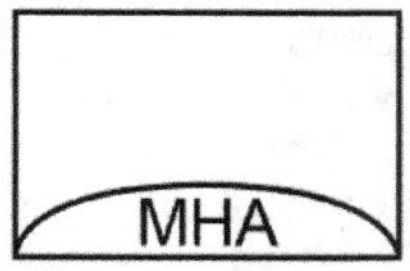 Air defense unit with medium to high altitude capability
mountain—The capability to conduct mountain warfare. (See ATP 3-90.97 for more information on mountains.)		 Infantry unit with mountain capability
multifunctional—Performing or capable of performing more than one function.	**MF**	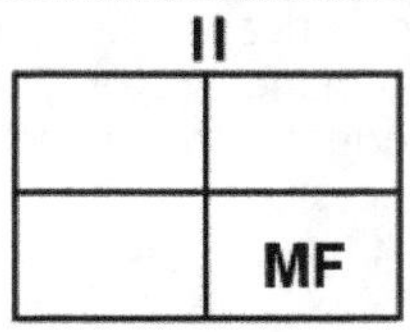 Medical battalion, multifunctional ***Note.*** The modifier offsets to the right to avoid overlapping of symbols.
ocean-going tugboat—Class A 128-foot large tug capable of coastal and ocean towing and docking and undocking operations with large ocean vessels. (See ATP 4-15 for more information on tugboats.)	**AT**	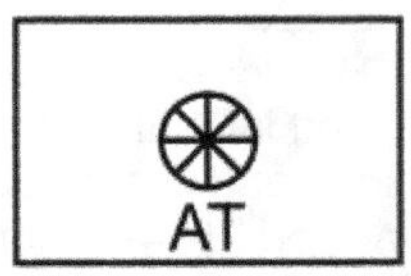 Transportation unit with tug (ocean-going) capability
optometry—A capability to provide optometry care, optical fabrication, and repair support. (See FM 4-02 for more information on optometry.)	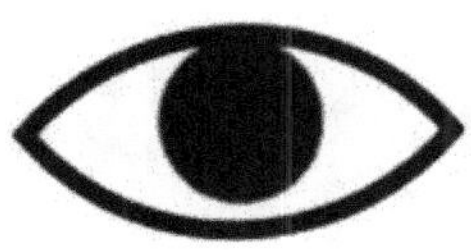	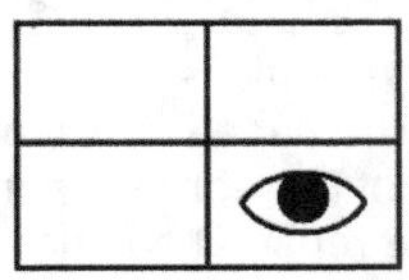 Medical unit with optometry capability ***Note.*** The modifier offsets to the right to avoid overlapping of symbols.
over-snow (prime mover)—Capability to travel in cold weather conditions using powerful all-terrain vehicles designed for snow conditions.		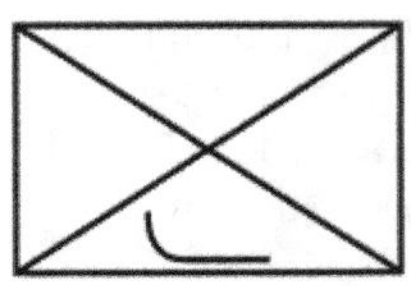 Infantry unit with over-snow capability

Table 2-10. Sector 2 modifiers for units (*continued*)

Function	***Modifier***	***Symbol Usage Construct Example and Translation***
pack animal—An animal transport system capability that enhances mobility when the area of operations restricts normal methods of transport or resupply. (See ATP 3-18.13 for more information on pack animals.)		 Infantry unit with pack animal capability
medical surveillance—The ongoing, systematic collection, analysis, and interpretation of data derived from instances of medical care or medical evaluation, and the reporting of population-based information for characterizing and countering threats to a population's health, well-being, and performance. (See JP 4-02 for more information on medical surveillance.)		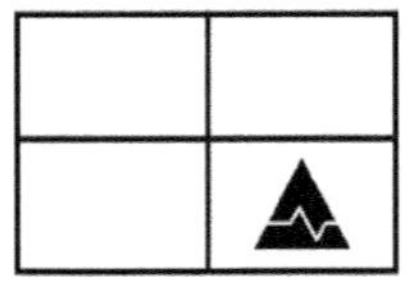 Medical unit with preventative medicine capability ***Note.*** The modifier offsets to the right to avoid overlapping of symbols.
	 Alternate hand-drawn version	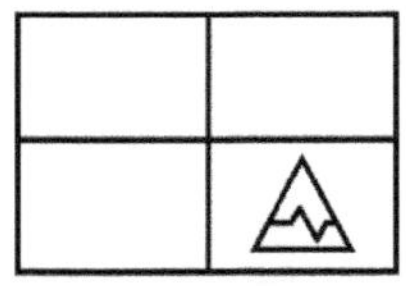
railway—Provides rail network capability and infrastructure assessments and coordinates contracts. (See FM 4-0, FM 4-01, and ATP 4-14 for more information on railways.) ***Note.*** This modifier symbol is interchangeable as a sector 1 or sector 2 modifier with the same meaning.		 Transportation unit with railway capability
recovery—Actions taken to extricate damaged or disabled equipment for return to friendly control or repair at another location. (JP 3-34)		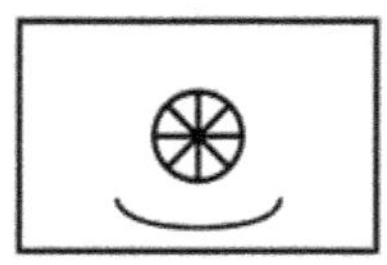 Transportation unit with recovery capability
riverine—Capability to provide inland water and water surface transport to move troops and equipment. (See ATP 4-15 for more information on riverine.)		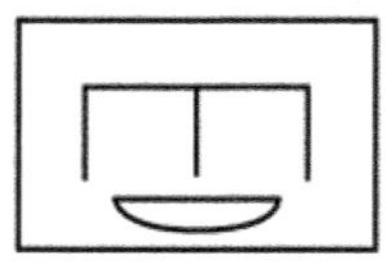 Engineer unit with riverine capability

Table 2-10. Sector 2 modifiers for units (*continued*)

Function	*Modifier*	*Symbol Usage Construct Example and Translation*
search—To look into or over carefully or thoroughly in an effort to find or discover someone or something.		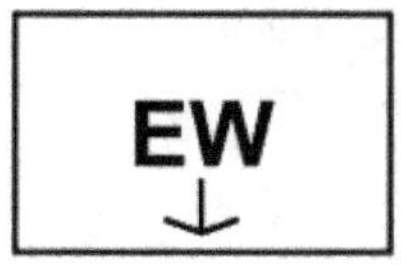 Electromagnetic warfare unit with search capability
security force assistance—(DOD) The Department of Defense activities that support the development of the capacity and capability of foreign security forces and their supporting institutions. (JP 3-22) (Army) The unified action to generate, employ, and sustain local, host nation, or regional security forces in support of legitimate authority. (FM 3-07)	**SFA**	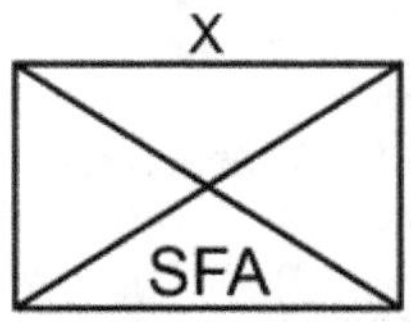 Infantry security force assistance brigade
surgical—Capability to provide life or limb saving operative treatment using specialized instruments to repair or stabilize a patient. (See FM 4-02 for more information on surgical.) ***Note.*** This modifier offsets to the right to avoid overlapping with the Main function symbol.	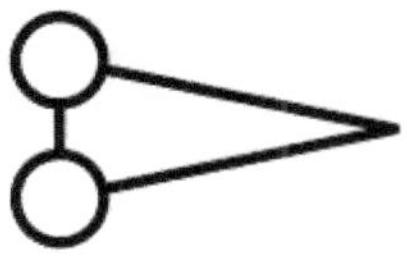	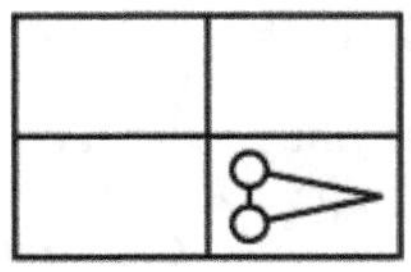 Medical unit with surgery capability
service craft yard—A maritime capability that may include various types of sustainment floating platforms and or small powerful watercraft capable of towing and pushing other vessels.	**YY**	 Transportation unit with seaport service craft yard capability
short range—Capability relating to or fit for short distances	**SR**	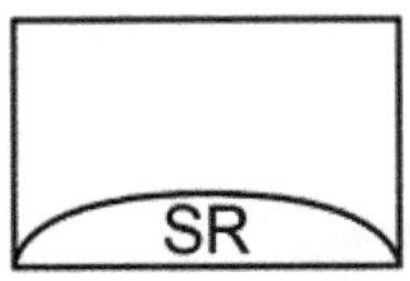 Air defense unit with short range capability
ski—Provides specialized equipment used by trained individuals with skills to negotiate arduous snow-covered terrain. (See ATP 3-90.97 for more information on ski.)		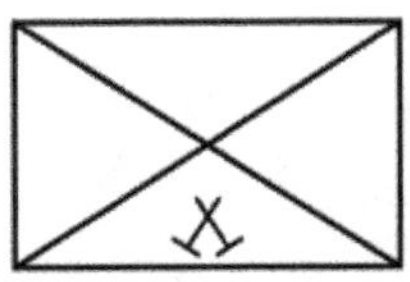 Infantry unit with ski capability

Table 2-10. Sector 2 modifiers for units (*continued*)

Function	*Modifier*	*Symbol Usage Construct Example and Translation*
sled—A snow or ice transportation vehicle capability.		 Infantry unit with sled capability
Surface Deployment and Distribution Command—Global ocean and inland waterway port manager and surface transportation service provider as part of U.S. Transportation Command's Joint Deployment and Distribution Enterprise. (See FM 4-0 for more on the Surface Deployment and Distribution Command.)	**SDDC**	 Transportation seaport unit with surface deployment and distribution support capability
supply—(DOD) The procurement, distribution, maintenance while in storage, and salvage of supplies, including the determination of kind and quantity of supplies. **a. producer phase**—That phase of military supply that extends from determination of procurement schedules to acceptance of finished supplies by the Services. **b. consumer phase**—That phase of military supply that extends from receipt of finished supplies by the Services through issue for use or consumption. (JP 4-0) (Army) The process of providing all items necessary to equip, maintain, and operate a military command. (FM 1-02.1)		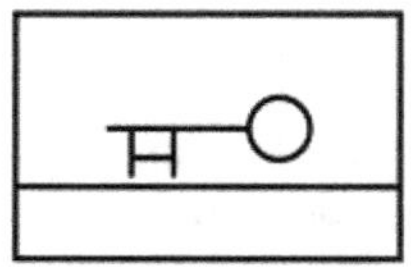 Quartermaster unit with supply capability
		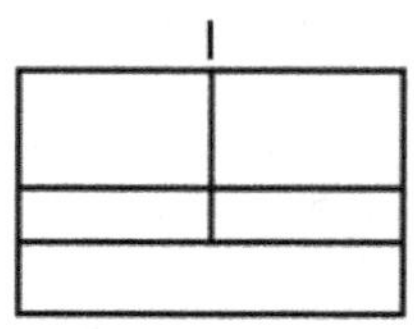 Medical company with supply capability ***Note.*** To avoid overlapping of symbols, shorten the vertical center line of Main function symbol to allow space for the modifier.

Table 2-10. Sector 2 modifiers for units (*continued*)

Function	***Modifier***	***Symbol Usage Construct Example and Translation***
support—1. The action of a force that aids, protects, complements, or sustains another force in accordance with the directive requiring such action. **2.** A unit that helps another unit in battle. **3.** An element of a command that assists, protects, or supplies other forces in combat. (JP 1 VOL 2) ***Note.*** This modifier symbol is interchangeable as a sector 1 or sector 2 modifier with the same meaning.	**SPT**	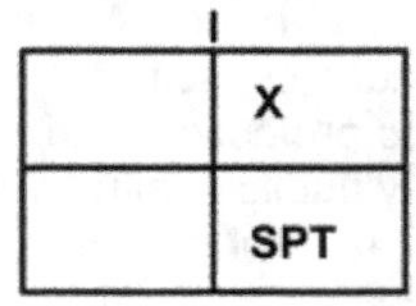 Medical company with brigade support capability ***Note.*** The modifiers offset to the right to avoid overlapping of symbols. 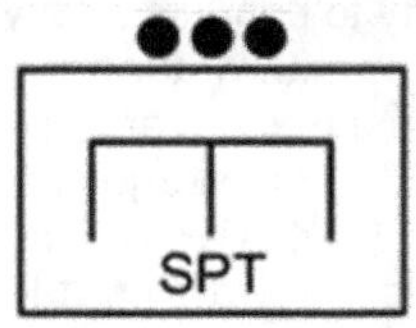 Engineer platoon with support capability
towed—Prime mover capability to pull a piece of equipment unable to move on its own.		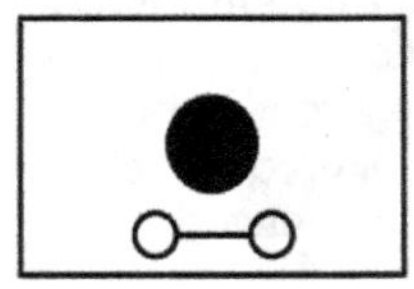 Field artillery unit with towed capability
tug, harbor—Tug 900 Class B vessel that supports movement of barges and lighterage of various types in harbors, port areas, and during logistics over-the-shore anchorage. (See ATP 4-15 for more information on harbor tugs.)	**YT**	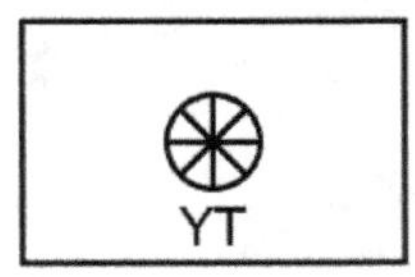 Transportation unit with tug (harbor) capability
vertical take-off and landing—Capability to take off and land vertically.	**VTOL**	 Fixed-wing unit with vertical take-off and landing capability
very heavy—Cannon artillery larger than 210mm. (See FM 3-09 for more information.)	**VH**	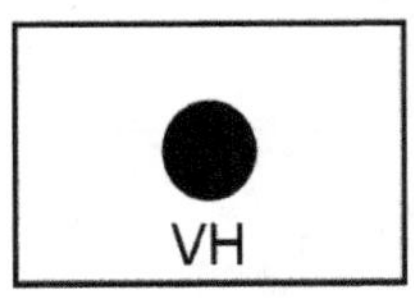 Artillery unit with very heavy artillery capability

Table 2-10. Sector 2 modifiers for units (*continued*)

Function	***Modifier***	***Symbol Usage Construct Example and Translation***
veterinary service—A capability that provides consultation, animal care, food protection, and veterinary public health services. (See FM 4-02 for more information on veterinary service.) ***Note.*** This modifier offsets to the right to avoid overlapping with the Main function symbol.	**V**	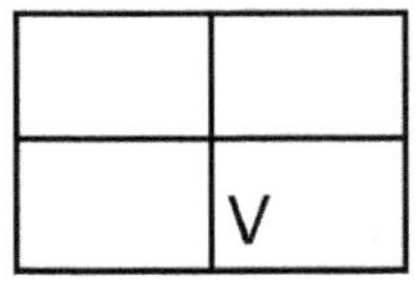 Medical unit with veterinary service capability
watercraft (generic vessel)—Capability to conduct heavy lifting associated with water transport operational maneuver and intra-theater lift of units, equipment, and supplies. (See ATP 4-15 for more information on watercraft.) ***Note.*** This modifier symbol is interchangeable. Symbol may be used as a sector 1 or sector 2 modifier with the same meaning.		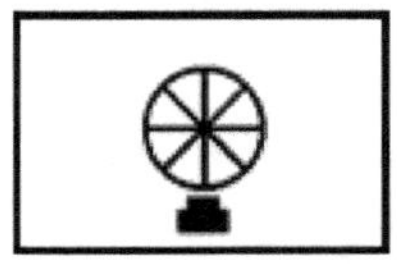 Transportation unit with watercraft capability
wheeled (high mobility)—A wheeled capability that can traverse various types of off-road terrain, unimproved routes, and paved road networks.		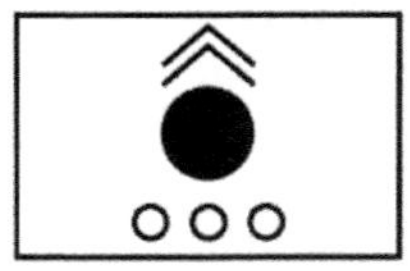 Artillery unit with wheeled (high mobility) and multiple launch rocket system capability 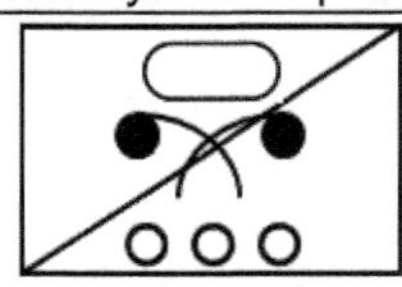 CBRN reconnaissance unit with armored wheeled (high mobility) vehicle capability

2-23. Table 2-11 on pages 80 through 82 provides unit symbol construct examples and their translations.

Table 2-11. Unit symbol construct examples

Unique Unit Designation	***Symbol Construct***
1st Aviation Battalion (attack), 25th Regiment, 25th Combat Aviation Brigade, 25th Infantry Division	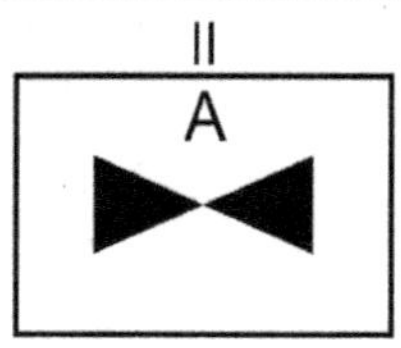 1-25/25CAB 25ID

Table 2-11. Unit symbol construct examples (*continued*)

Unique Unit Designation	*Symbol Construct*
2nd Infantry Battalion (mountain capability),1st Regiment, 2nd Brigade Combat Team, 10th Mountain Division, 3rd Corps	 2-1/2BCT/10 III
2nd Military Intelligence Battalion (unmanned aircraft system capability), 66th Military Intelligence Brigade	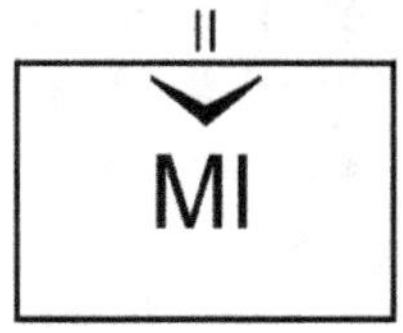 2/66
2nd Ranger Battalion, 75th Ranger Regiment	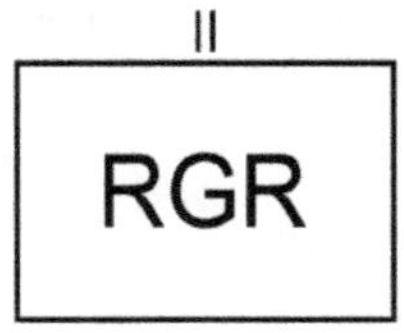 2/75
4th Squadron, 10th Cavalry Regiment, 3rd Armored Brigade Combat Team, 4th Infantry Division	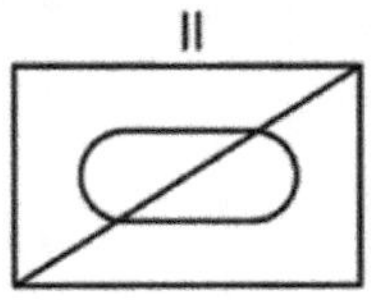 4-10/3ABCT 4ID
14th Brigade Engineer Battalion, 2nd Stryker Brigade Combat Team, 2nd Infantry Division	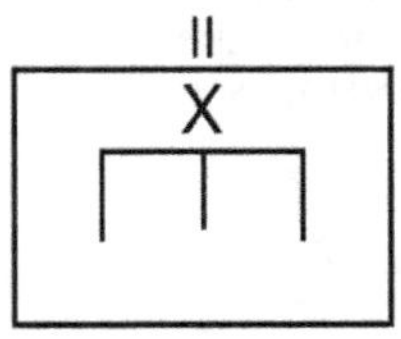 14/2SBCT 2ID
299th Brigade Support Battalion, 2nd Brigade Combat Team, 1st Infantry Division	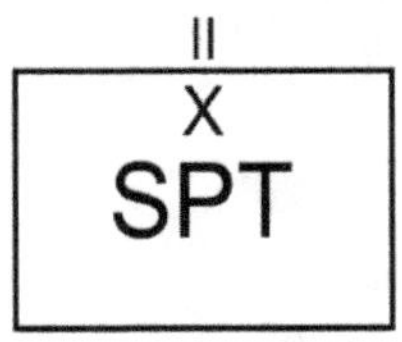 299/2BCT 1ID
4th Infantry Battalion (armored high mobility vehicle capability), 23rd Infantry Regiment, 2nd Stryker Brigade Combat Team, 2nd Infantry Division	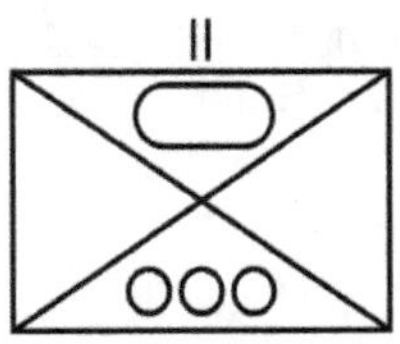 4-23/2BCT 2ID
B Troop, 6th Squadron, 4th Cavalry Regiment, 3rd Infantry Brigade Combat Team, 1st Infantry Division	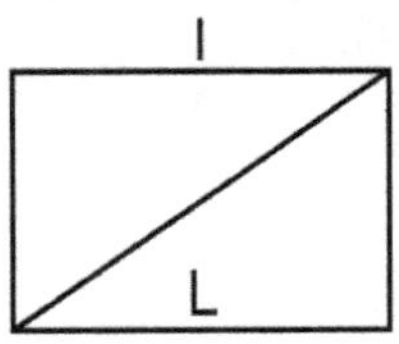 B/6-4/3IBCT 1ID

BODY-93

Table 2-11. Unit symbol construct examples and translations (*continued*)

Unique Unit Designation	***Symbol Construct***
6th Field Artillery Battalion (self-propelled multiple rocket launcher capability), 37th Field Artillery Regiment, 2nd Infantry Division	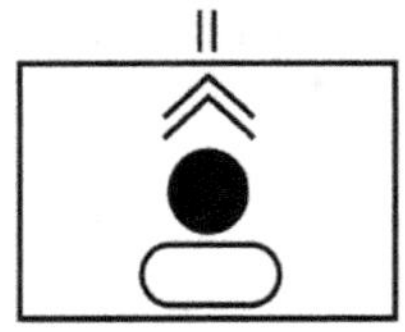 6-37 2ID
181 Transportation Battalion, 3rd Expeditionary Sustainment Command, 3rd Corps	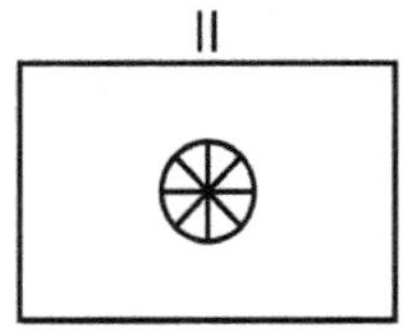 181/3ESC III
62nd Engineer Battalion (bridging and construction capability), 36th Engineer Brigade	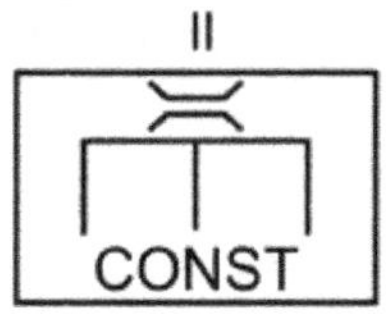 62/36 ENGBDE
67th Forward Support Company, 201st Brigade Support Battalion, 3rd Infantry Brigade Combat Team, 1st Infantry Division	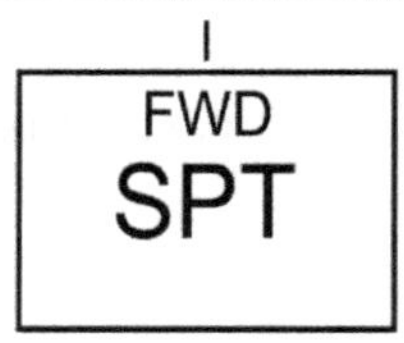 67/201/3IBCT 1ID
C Field Artillery Battery (air assault capability), 3rd Battalion, 319 Field Artillery Regiment	 C/3-319
F Company, 1st Battalion, 61st Infantry Regiment, 65th Infantry Brigade	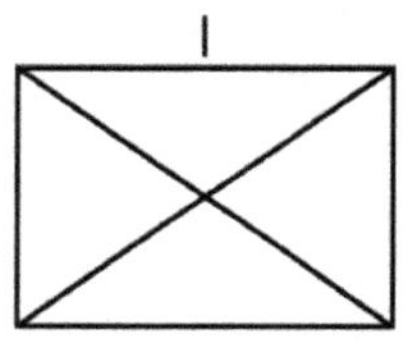 F/1-61/65
Operational Detachment Alpha 595 (pack animal capability), C Company, 3rd Battalion, 5th Special Forces Group	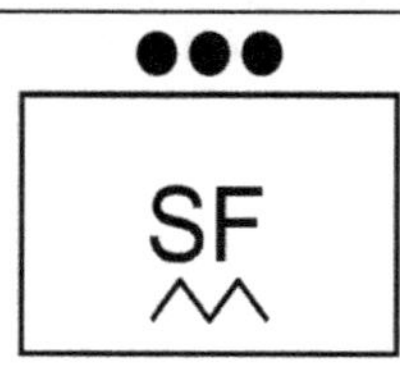 ODA 595/C/3 5 SFG
III Marine Expeditionary Force	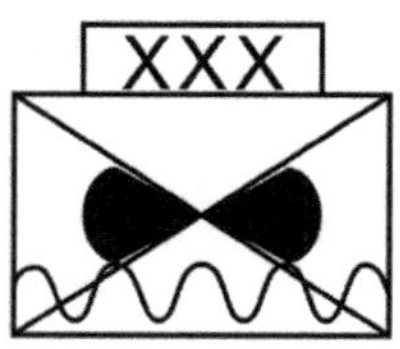 III

Chapter 3
Activity and Installation Symbols

This chapter discusses Main function symbols and sector 1 and 2 modifiers used with activities and installation frames.

ACTIVITY FRAME SHAPES

3-1. This frame is used to identify friendly, enemy, or criminal actions that can reveal civic, ethnic, religious, social, or other grouping activities in an area of interest or operation that may affect unified land operations. The frame shape follows the construct guidelines in chapter 1, main and modifier symbols, the activities frame shapes for standard identities in table 3-1, and amplifier placement locations in figure 3-1.

Table 3-1. Activity standard identity frame shapes

	Friendly	*Hostile*	*Neutral*	*Unknown*
Activity	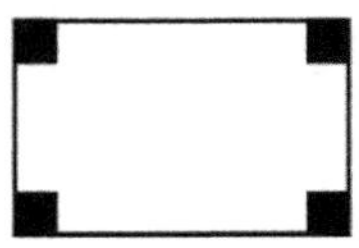			
	Assumed Friend	*Suspect*		*Pending*
				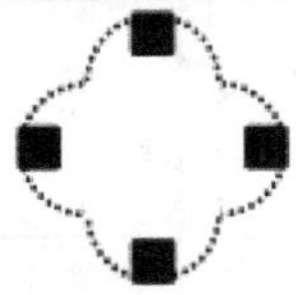

3-2. **Amplifier field.** Figure 3-1 shows the placement of main and modifier symbols within the frame and amplifiers around the friendly symbol frame. Table 3-2 provides descriptions and formats of each amplifier.

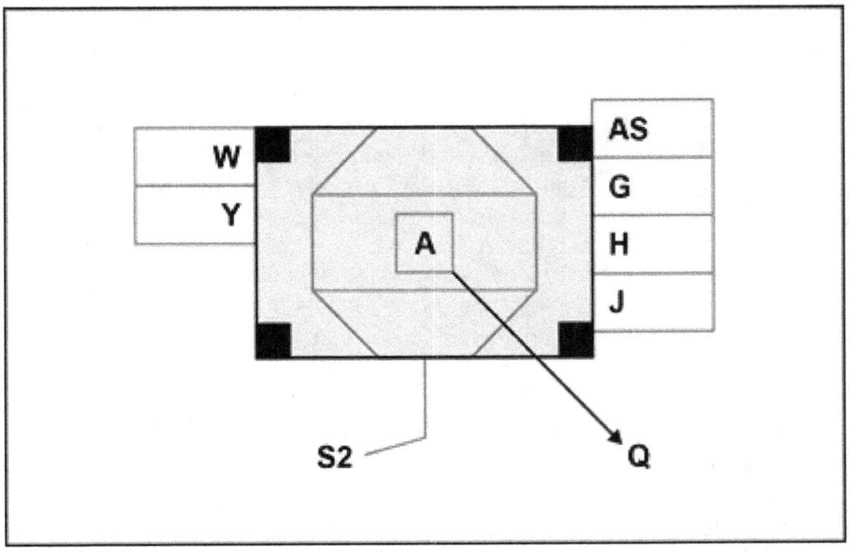

Figure 3-1. Placement of activity main, modifiers, symbol, and amplifiers

Table 3-2. Descriptions of main, modifier, and amplifier fields for activity frames

Field	*Field Title*	*Description*
A	Main and modifier symbols	The innermost part of a symbol that represents the main function and its capabilities (modifiers 1 and 2)
G	Staff comments	A text amplifier content is implementation specific. ***Note.*** This field allows a maximum of 20 characters.
H	Additional information	A text amplifier content is implementation specific. ***Note.*** This field allows a maximum of 20 characters.
J	Evaluation rating	A text amplifier that consists of a single-letter reliability rating and a single digit credibility rating. **Reliability Ratings:** A-completely reliable. B-usually reliable. C-fairly reliable. D-not usually reliable. E-unreliable. F-reliability cannot be judged. **Credibility Ratings:** 1-confirmed by other sources. 2-probably true. 3-possibly true. 4-doubtfully true. 5-improbable. 6-truth cannot be judged. ***Note.*** This field allows a maximum of 2 characters.
Q	Direction of movement	A graphic amplifier that identifies the direction of movement or intended movement of an object.
S^2	Offset location indicator	A graphic amplifier used to indicate the offset or precise location of a single symbol.
W	Date-time group	An alphanumeric designator for displaying a date-time group (DDHHMMSSZMONYYYY) or "O/O" for an order. The date-time group is composed of a group of six numeric digits with at time zone suffix and the standardized three-letter abbreviation for the month followed by four digits representing the year. The first pair of digits represents the day; the second pair, the hour; the third pair, the minutes. For automated systems, two digits may be added before the time zone suffix and after the minutes to designate seconds. ***Note.*** This field allows a maximum of 16 characters.
Y	Location	A text amplifier that displays a symbol's location in degrees, minutes, and decimal minutes (or in military grid reference system, global area reference system, or other applicable display formats). ***Note.*** This field allows a maximum of 22 characters.
AS	Country	A three-letter code that indicates the country of origin of the organization. In stability activities, this field can be used for factions or groups. ***Note.*** This field allows a maximum of 3 characters.

3-3. **Evaluation rating amplifier (Field J).** The evaluation rating amplifier is a 2-digit alphanumeric code that allows for adding reliability and credibility rating. The reliability levels are A through F, and the credibility levels are 1 through 6. Table 3-2 provides the meaning of each letter and number code for this amplifier. Figure 3-2 on page 85 provides an example of an evaluation rating amplifier usage construct of an enemy leader activity that is completely reliable and confirmed by other sources.

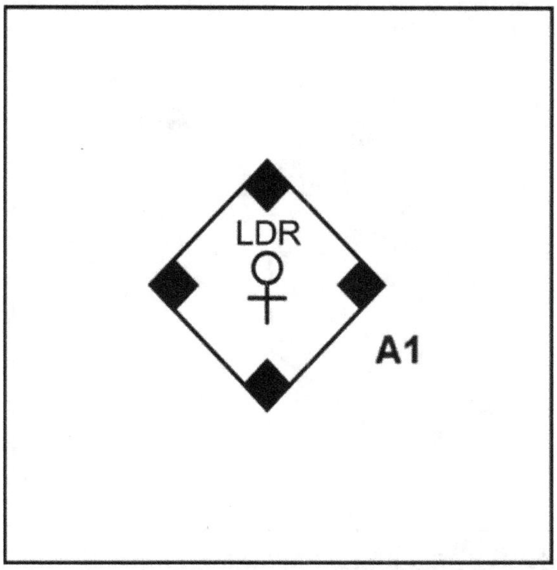

Figure 3-2. Evaluation rating amplifier usage construct

3-4. **Activity direction of movement amplifier (Field Q).** The activity direction of movement amplifier is an arrow extending from the center of the frame or Main function symbol. The arrow extends in the direction of movement or intended movement of activity symbol. Figure 3-3 is an example of the usage construct.

Figure 3-3. Activity direction of movement amplifier usage construct

3-5. **Activity offset location indicator amplifier (Field S2).** The offset location amplifier for activities is placed differently from units and installations. Field S2 is located at the center of the bottom of the activity frame and the offset location indicator line protrudes from this point (without an arrow) to the desired location. The line may be extended or bent as needed. Figure 3-4 is an example of the usage construct of a friendly organized group meeting activity.

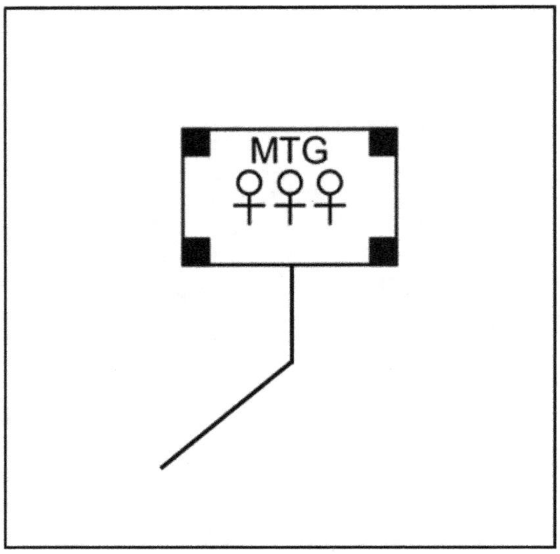

Figure 3-4. Activities offset location indicator amplifier usage construct

INSTALLATION FRAME SHAPES

3-6. This frame shape is used to identify friendly military, civilian, or enemy installations, facilities, campuses, terminals, depots, caches, and specific buildings in an area of interest or operation that supports or may support a common interest during unified land operations. The frame shape follows the construct guidelines in chapter 1 and the activities frame shapes for standard identities in table 3-3.

Table 3-3. Installation standard identity frame shapes

Standard Identity	*Friendly*	*Hostile*	*Neutral*	*Unknown*
Installation	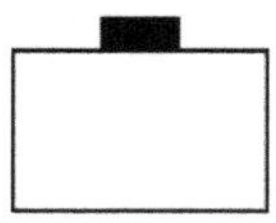	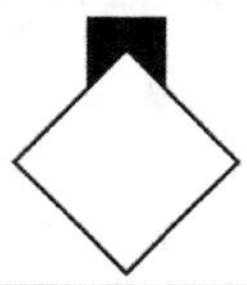	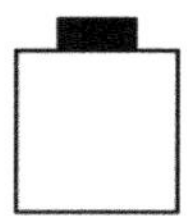	
	Assumed Friend	*Suspect*		*Pending*
		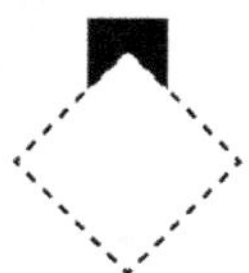		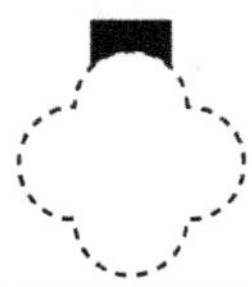

3-7. **Amplifier fields.** Figure 3-5 shows the placement of the main function, modifier symbols in the frame and amplifiers around the land installation or facility frame. Table 3-4 on page 87 provides descriptions and formats of each amplifier.

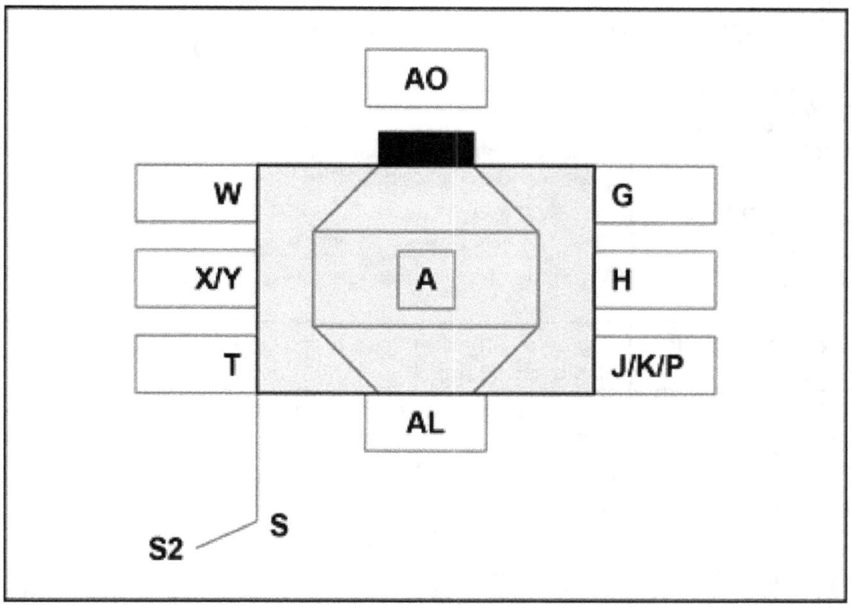

Figure 3-5. Placement of installation main and modifier symbol and amplifiers

Table 3-4. Descriptions of main, modifier, and amplifier fields for installation frames

Field	*Field Title*	*Description*
A	Main and modifier symbols	The innermost part of a symbol that represents the main function and its capabilities (modifiers 1 and 2).
G	Staff comments	A text amplifier content is implementation specific. ***Note***. This field allows a maximum of 20 characters.
H	Additional information	A text amplifier content is implantation specific. ***Note***. This field allows a maximum of 20 characters.
J	Evaluation rating	A text amplifier that consists of a single-letter reliability rating and a single digit credibility rating. **Reliability Ratings:** A-completely reliable. B-usually reliable. C-fairly reliable. D-not usually reliable. E-unreliable. F-reliability cannot be judged. **Credibility Ratings:** 1-confirmed by other sources. 2-probably true. 3-possibly true. 4-doubtfully true. 5-improbable. 6-truth cannot be judged. ***Note***. This field allows a maximum of 2 characters.
K	Combat effectiveness	A text amplifier that indicates effectiveness. The entries are— Fully operational (FO). Substantially operational (SO). Marginally operational (MO). Not operational (NO). Unknown (UNK). ***Note***. This field allows a maximum of 5 characters.

Table 3-4. Descriptions of main, modifier, and amplifier fields for installation frames (*continued*)

Field	*Field Title*	*Description*
P	Identification, friend or foe Selective identification feature	A text amplifier displaying one or more identification, friend or foe, or selective identification feature identification modes and codes. Display priority is mode 5, mode, mode 4, mode 3, and mode 2. ***Note.*** This field allows a maximum of 15 characters.
S	Headquarters staff indicator	A graphic amplifier that identifies a headquarters.
S2	Offset location indicator	A graphic amplifier used to indicate the offset or precise location of a single point symbol.
T	Unique identifier	An amplifier field reserved for command and control systems that uniquely identifies a particular symbol with a track number. Prefix = TN: #####. Example: TN: 13579. ***Note.*** This field allows a maximum of 30 characters.
W	Date-time group	An alphanumeric designator for displaying a date-time group (DDHHMMSSZMONYYYY) or "O/O for an order. The date-time group is composed of a group of six numeric digits with a time zone suffix and the standardized three-letter abbreviation for the month followed by four digits representing the year. The first pair of digits represents the day; the second pair, the hour; the third pair, the minutes. For automated systems, two digits may be added before the time zone suffix and after the minutes to designate seconds. ***Note.*** This field allows a maximum of 16 characters.
X	Altitude or depth	A text amplifier that displays either altitude, flight level, depth for submerged objects, or height of equipment or structures on the ground. Measurement units shall be displayed in the string. Examples: 1500MSL FL150 ***Note.*** This field allows a maximum of 14 characters.
Y	Location	A text amplifier that displays a symbol's location in degrees, minutes, and decimal minutes (or in military grid reference system, global area reference system, or other applicable display formats). ***Note.*** This field allows a maximum of 22 characters.
AL	Operational condition	A graphic amplifier that indicates operational condition or capacity. Operational condition amplifier, if used, shall be comprised of only one color. Example: Aircraft: Red—damaged, Green—fully capable example: Missile: Red—imminent threat, Green—no threat
AO	Engagement bar	A graphic amplifier placed immediately atop the symbol. May denote 1) local/remote status, 2) engagement status, and 3) weapon type. Format: A:BBB-CC, where A = remote/local BBB = engagement status CC = weapon status

3-8. **Operational condition amplifier (AL).** An AL is used to display the level of operational condition of an installation symbol. Figure 3-6 on page 89 shows the template for the combat effectiveness amplifier. Table 3-5 on page 89 shows operational condition amplifiers and construct examples.

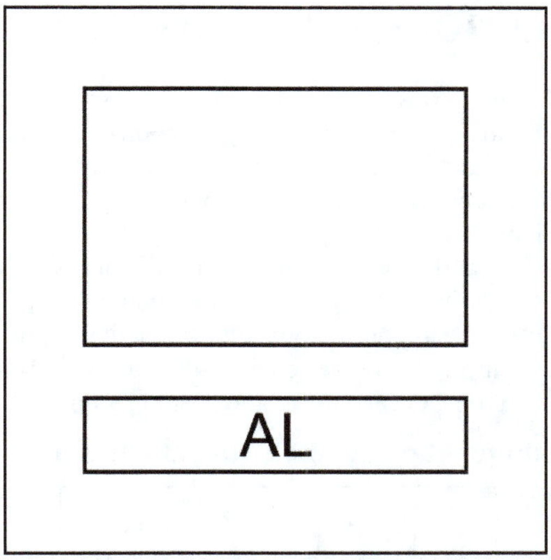

Figure 3-6. Template for operational condition amplifier

Table 3-5. Operational condition amplifiers and construct examples

Function	***Amplifier***	***Example of amplifier construct usage***
Fully operational		
Damaged but substantially operational		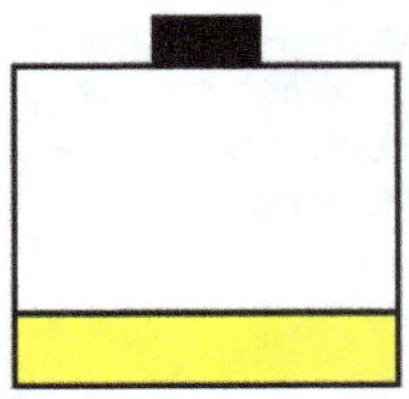
Destroyed		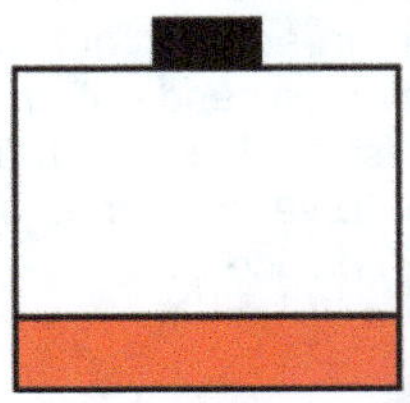
Full to capacity		

MAIN FUNCTION SYMBOLS FOR ACTIVITIES AND INSTALLATIONS

3-9. Main function symbols (Field A) reflect the primary function of the symbol. The Main function symbols for activities and installations include some military symbols used in chapter 2 and unique civilian symbols used in defense support of civil authorities and stability operations. Activity symbols provide the means to construct military and civilian symbols to identify individual and group activities (including isolated personnel, civic, religious, social, and other groups), and installation symbols identify military and civilian infrastructure. The use of unique civilian symbols is a recognition of the larger role of military forces beyond war fighting and reflect stability and support to civil authority activities around the world. (See FM 3-57 for more information on civil authorities.) Table 3-6 (on pages 90 through 99) shows the Main function symbols for civilian individuals, organizations, events, installations, and facilities.

Table 3-6. Main function symbols for activities and installations

Function	***Symbol***	***Symbol Usage Construct Example and Translation***
airport—A place from which aircraft operate that usually has paved runways and maintenance facilities and often serves as a terminal	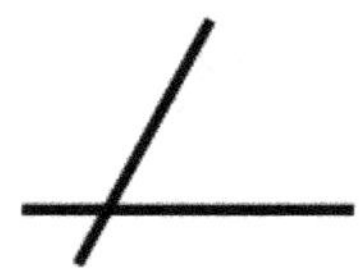	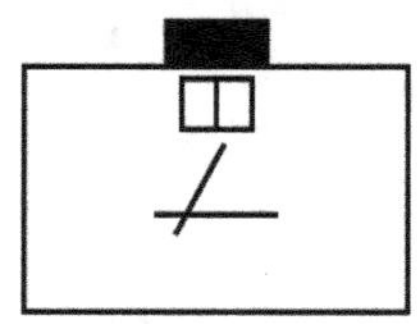 Airport facility with cargo capability
ammunition—The projectiles with their fuses, propelling charges, or primers fired from guns.		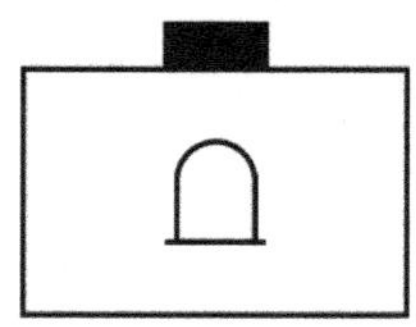 Ammunition facility
booby trap—A concealed explosive device contrived to go off when some harmless-looking object is touched.		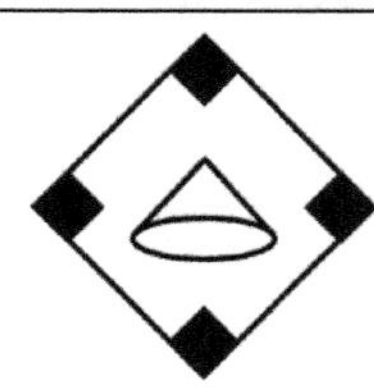 Enemy booby trap activity
broadcast transmitter antenna—A device for radiating or receiving radio waves to send out or transmit (something, such as a program) to various types of electronic devices.		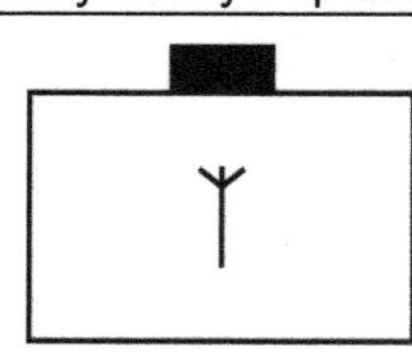 Broadcast transmitter antenna installation

Table 3-6. Main function symbols for activities and installations (*continued*)

Function	*Symbol*	*Symbol Usage Construct Example and Translation*
Bureau of Alcohol, Tobacco, Firearms, and Explosives (ATF)—Civil authority law enforcement agency in the United States' Department of Justice that protects communities from violent criminals, criminal organizations, the illegal use and trafficking of firearms, the illegal use and storage of explosives, acts of arson and bombings, acts of terrorism, and the illegal diversion of alcohol and tobacco products. (See ATF.gov website for more information on ATF.)	**ATF**	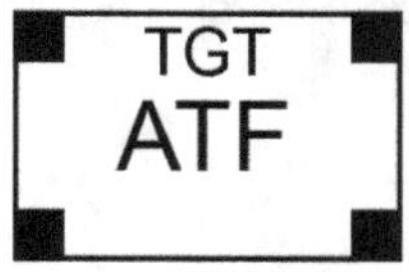 ATF targeted activity
chemical, biological, radiological, or nuclear (CBRN)—Recognizes vulnerabilities, identifies, and understands CBRN hazards, and their consequences when they appear, and responds appropriately to protect the force. (See ADP 3-37 for more information on CBRN.)		 Enemy CBRN facility
Coast Guard—A maritime security, search and rescue, and law enforcement service branch responsible for patrolling coasts and international waters using cutters, aircraft, and intelligence to detect, intercept, and disrupt dangerous and illegal activities such as drug smuggling and human trafficking. (See https://www.uscg.mil/ for more information)		 Coast Guard air station installation
cyber-server—A computer in a network that is used to provide services (such as access to files or shared peripherals or the routing of email) to other computers in the network.		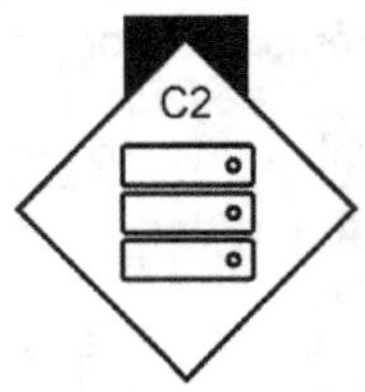 Enemy cyber-server command and control facility

Table 3-6. Main function symbols for activities and installations (*continued*)

Function	*Symbol*	*Symbol Usage Construct Example and Translation*
cyberspace operations—The employment of cyberspace capabilities where the primary purpose is to achieve objectives in or through cyberspace. (JP 3-0)	**CYB**	 Cyberspace operations center facility
data—Capability to transmit or process information.in digital form.	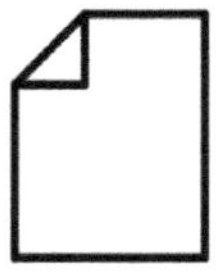	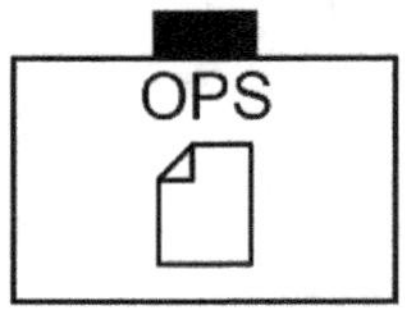 Data operations center facility
drugs—A substance other than food intended to affect the structure or function of the body.	**DRUG**	 Illegal drug activity
Drug Enforcement Administration (DEA)—The civil authority law enforcement agency under the U.S. Department of Justice tasked with combating drug trafficking and distribution within the U.S. It is the lead agency for domestic enforcement of the Controlled Substances Act, sharing concurrent jurisdiction with the Federal Bureau of Investigation, the U.S. Immigration and Customs Enforcement, and U.S. Customs and Border Protection. (See DEA.gov website for more information on the DEA.)	**DEA**	 DEA search activity
electric generation—The process of generating electricity from sources of primary energy.		 Electric generation facility

Table 3-6. Main function symbols for activities and installations (*continued*)

Function	*Symbol*	*Symbol Usage Construct Example and Translation*
emergency management—The managerial function charged with creating the framework within which communities reduce vulnerability to hazards and cope with disasters.		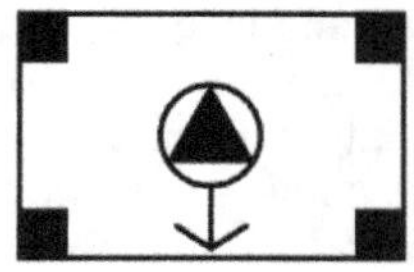 Emergency management search activity
		 Emergency management operations facility
emergency medical services (EMS)—A first responder unit with specially trained medical technicians certified to provide basic emergency services (such as cardiopulmonary resuscitation) before and during transportation to a hospital.		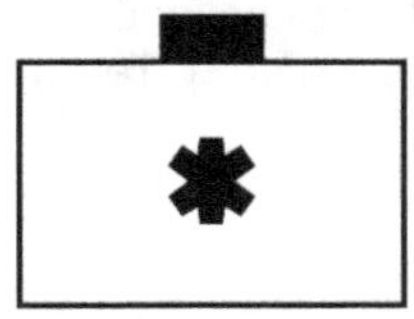 EMS facility
explosion—A large-scale, rapid, or spectacular expansion or bursting out or forth.		 Enemy explosion activity
extortion—The act or practice of extorting especially money or other property.	**$**	 Enemy extortion activity
Federal Bureau of Investigation (FBI)—The civil authority domestic intelligence and security service of the United States and its principal federal law enforcement agency. Operating under the jurisdiction of the United States Department of Justice, the FBI is also a member of the U.S. Intelligence Community and reports to both the Attorney General and the Director of National Intelligence. (See FBI.gov website for more information.)	**FBI**	 FBI raid activity

Table 3-6. Main function symbols for activities and installations (*continued*)

Function	*Symbol*	*Symbol Usage Construct Example and Translation*
fire department—The civil authority firefighting organization that operates within a municipality, county, state, nation, or special district. In some areas, they may also provide technical rescue, fire protection, fire investigation, emergency medical services, and hazardous material mitigation. Private and specialist firefighting organizations also exist, such as those for aircraft rescue and firefighting.		 Fire department activity
	 Alternate hand-drawn version	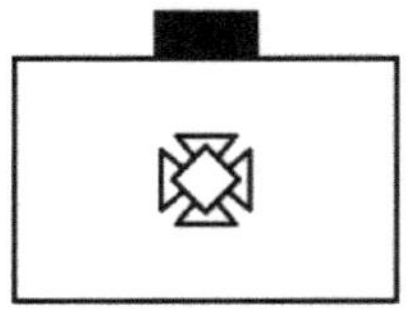 Fire department facility
governmental organization—An organization controlled and financed by its national government. (NATOTerm)	**GO**	 Governmental organization facility
grenade—a small projectile that contains an explosive or a chemical agent (such as tear gas, a flame producer, or a smoke producer) and that is thrown by hand or projected (as by a rifle or special launcher).	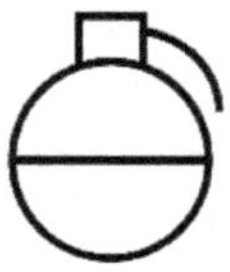	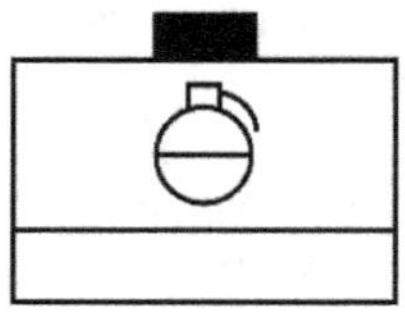 Grenade cache
group (organized)—A number of individuals assembled together or having some unifying relationship.		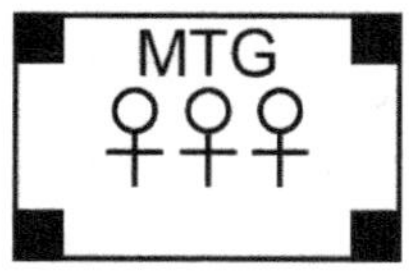 Group (organized) meeting activity
group of attempted crime victims—Attempt to conduct an illegal act on several individuals that have some unifying relationship.		 Attempted criminal activity on a group of individuals
group of assault victims—Several individuals that have some unifying relationship injured and killed by a violent physical attack.		 Enemy assault on group of victims

Table 3-6. Main function symbols for activities and installations (*continued*)

Function	***Symbol***	***Symbol Usage Construct Example and Translation***
improvised explosive device (IED)—A weapon fabricated or emplaced in an unconventional manner incorporating destructive, lethal, noxious, pyrotechnic, or incendiary chemicals. (JP 3-42)	**IED**	 IED activity
individual—A single human being as distinct from a group, class, or family.		 Individual suspicious activity
attempted crime victim—Attempt to conduct an illegal act on an individual.		 Attempted criminal activity on individual
Assault victim—An individual injured or killed by a violent physical attack.		 Enemy assault on individual victim
industrial building—Commercial properties used for business purposes in manufacturing, warehousing, research, and refrigeration.	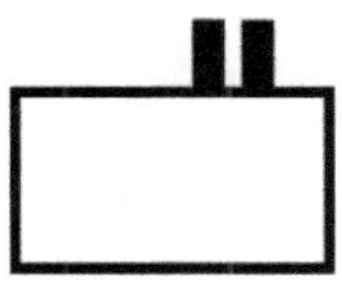	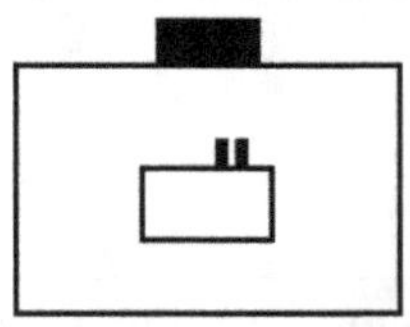 Industrial building facility
internet service provider—Provides services for accessing, using, managing, or participating in the Internet.	**ISP**	 Internet service provider facility

Table 3-6. Main function symbols for activities and installations (*continued*)

Function	*Symbol*	*Symbol Usage Construct Example and Translation*
isolated personnel—United States military, Department of Defense civilians, and Department of Defense contractor personnel authorized to accompany the Armed Forces of the United States (and others designated by the President or Secretary of Defense) who are separated from their unit (as an individual or a group) while participating in a United States-sponsored military activity or mission and are, or may be, in a situation where they must survive, evade, resist, or escape. (JP 3-50)		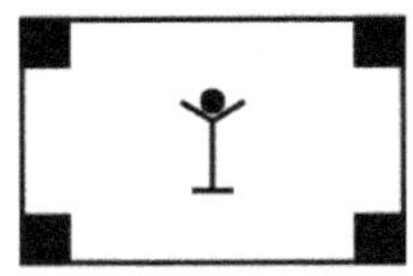 14 14 isolated individuals activity
law enforcement—A civil authority that encompasses police, courts, and correction who act in an organized manner to enforce the law by discovering, deterring, rehabilitating, or punishing people who violate the rules and norms of a government.		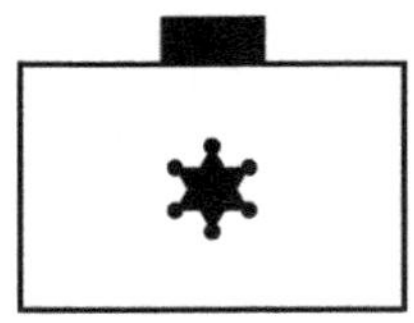 Law enforcement facility
maritime terminal—A facility for berthing ships simultaneously at piers, quays, and/or working anchorages. (See JP 4-18 for more information on maritime terminals.)		 Maritime terminal facility
	 Alternate hand-drawn version	
mass demonstration—A very large public gathering of people to support or to protest something.	**MASS**	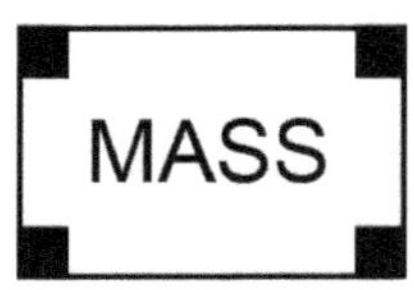 Mass demonstration activity
mass grave—A burial site containing multiple human corpses, which may or may not be identified prior to burial.		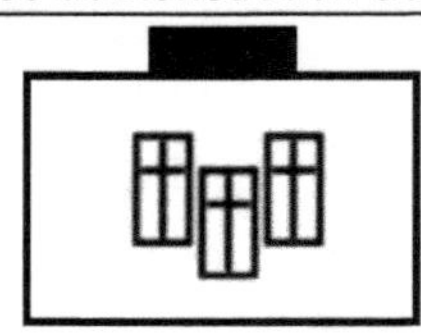 Mass grave facility

BODY-108

Table 3-6. Main function symbols for activities and installations (*continued*)

Function	*Symbol*	*Symbol Usage Construct Example and Translation*
medical treatment—The attempted remediation of a health problem, usually following a medical diagnosis.	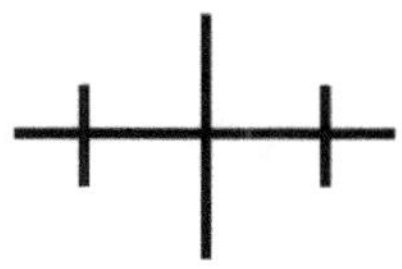	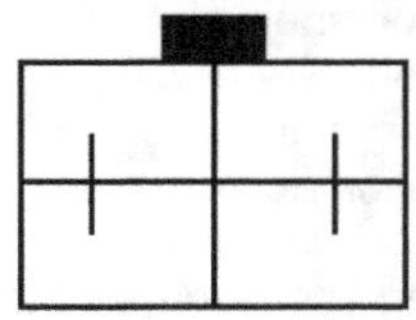 Medical treatment facility (hospital)
psychological actions—Lethal and nonlethal actions planned, coordinated, and conducted to produce a psychological effect in a foreign individual, group, or population. (FM 3-53)		 Psychological action (PSYACT) 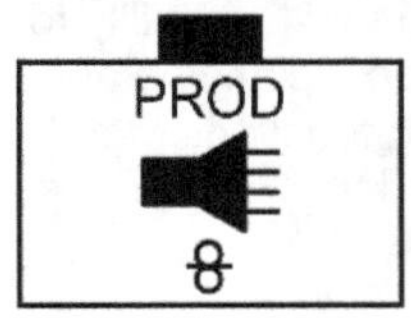 Psychological Operations print production facility
mine—A pit or excavation in the earth from which mineral substances are taken.		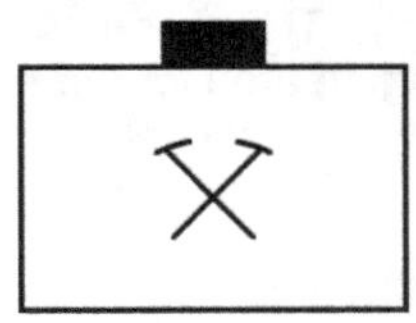 Mining facility  Illegal mining activity
network—A system of computers and peripherals that are able to communicate with each other.	**NET**	 Continuity of operations network facility

Table 3-6. Main function symbols for activities and installations (*continued*)

Function	*Symbol*	*Symbol Usage Construct Example and Translation*
nongovernmental organization—(DOD) A private, self-governing, not-for-profit organization dedicated to alleviating human suffering; and/or promoting education, health care, economic development, environmental protection, human rights, and conflict resolution; and/or encouraging the establishment of democratic institutions and civil society. Also called NGO. (JP 3-08)	**NGO**	 Nongovernmental facility  Nongovernmental activity
operations—Capability to execute the principal planning and operating functions of an organization.	**OPS**	 Operations facility (operations center)
patrolling—The action of traversing an area for the purpose of observation, security, or combat.		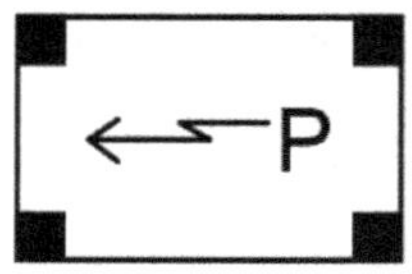 Patrolling activity
petroleum—A complex mixture of hydrocarbons with small amounts of other substances, and is prepared for use as gasoline, naphtha, or other products by various refining processes.		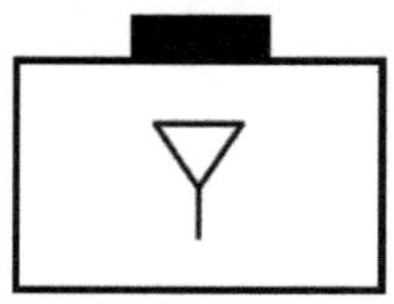 Petroleum facility
police department—A civil authority constituted body of persons empowered and authorized by a state with the aim to enforce the law, to ensure the safety, health, and possessions of citizens, and to prevent crime and civil disorder. Their lawful powers include arrest and use of force legitimized by the state via its monopoly on violence.		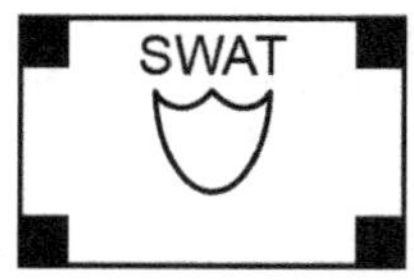 Police Department special weapons and tactics (SWAT) activity

Table 3-6. Main function symbols for activities and installations (*continued*)

Function	*Symbol*	*Symbol Usage Construct Example and Translation*
print media—Means of mass communication in the form of printed publications, such as newspapers and magazines.	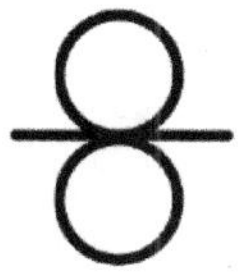	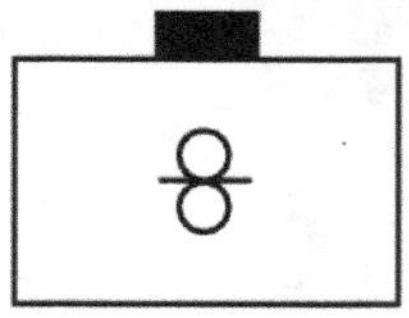 Print media center facility
religious—Relating to or manifesting faithful devotion to an acknowledged ultimate reality or deity.	**REL**	 Religious facility
security—Measures taken by a military unit, activity, or installation to protect itself against all acts design to, or which may, impair its effectiveness. (JP 3-10)	**SEC**	 Security operations center facility
transportation—A logistics function that includes movement control and associated activities to incorporate military, commercial, and multinational motor, rail, air, and water mode assets in the movement of units, personnel, equipment, and supplies in support of the concept of operations. (FM 1-02.1)		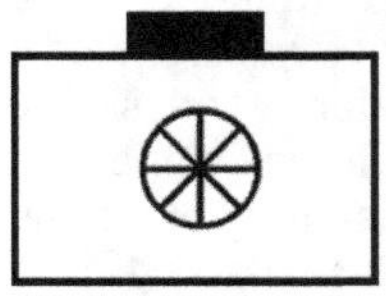 Transportation facility
telecommunications—The transmission of information by various types of technologies over wire, radio, optical, or other electromagnetic systems.		 Telecommunications facility
water (potable)—Suitable for human consumption.	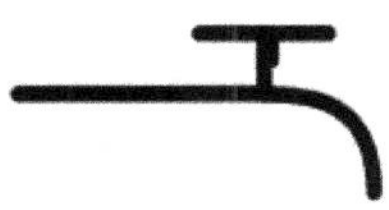	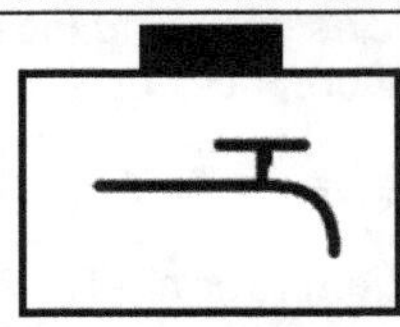 Potable water facility

SECTOR 1 MODIFIERS FOR ACTIVITIES AND INSTALLATIONS

3-10. Table 3-7 (on pages 100 through 105) shows sector 1 modifiers (Field A). Sector 1 modifiers further identify affiliation, capability, special characteristic, or specialty.

Table 3-7. Sector 1 modifiers for activities and installations

Function	*Symbol*	*Symbol Usage Construct Example and Translation*
biological—Used in or produced by applied biology.	**B**	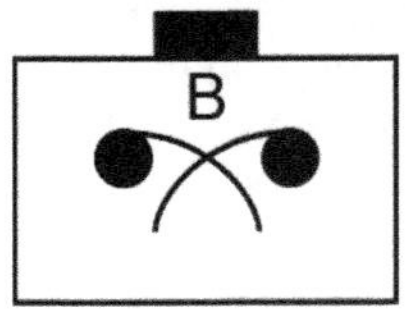 Biological facility
chemical—A substance obtained by a chemical process or producing a chemical effect.	**C**	 Chemical facility
coerced recruitment—The process of adding new individuals by force or threat. ***Note.*** This modifier symbol is interchangeable. Symbol may be used as a sector 1 or sector 2 modifier with the same meaning.	**CR**	 Group coerced recruitment activity
command and control—The exercise of authority and direction by a properly designated commander over assigned and attached forces in the accomplishment of the mission. Also called C2. (JP 1, Volume 2)	**C2**	 Enemy cyber-server command and control facility
continuity of operations—An effort within individual executive departments and agencies to ensure that primary mission essential functions continue to be performed during a wide range of emergencies, including localized acts of nature, accidents, and technological or attack-related emergencies.	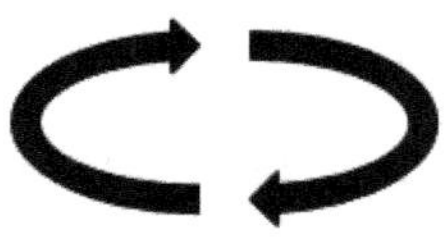	 Continuity of operations network facility
electric generation—The capability of producing electrical energy.		 Electric generation network facility

Table 3-7. Sector 1 modifiers for activities and installations (*continued*)

Function	*Symbol*	*Symbol Usage Construct Example and Translation*
foreign fighters—Recruited or self-motivated individuals who travel to a conflict zone to train and/or fight in support of a particular group.	**FF**	 Foreign fighter group activity
gang—Organized associates, friends, or members of a family with a defined leadership that identifies with or claims control over area in a community and participates in illegal, and possibly violent, behavior.	**GANG**	 Group gang activity
hydroelectric—Electricity generated from hydropower.	**HY**	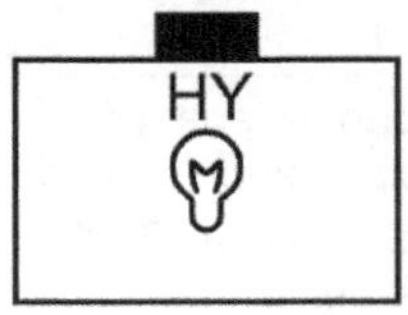 Electric generation hydroelectric powered facility
leader—A person who has commanding authority or influence. *Note.* This modifier symbol is interchangeable. Symbol may be used as a sector 1 or sector 2 modifier with the same meaning.	**LDR**	 Enemy individual leader activity
meeting—An assembly for a common purpose.	**MTG**	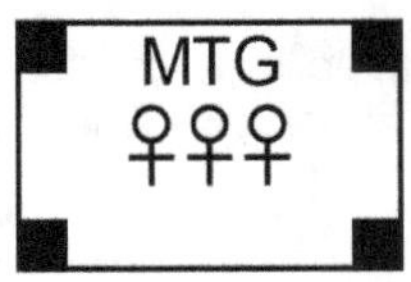 Friendly group meeting activity
natural gas—A naturally occurring mixture of gaseous hydrocarbons consisting primarily of methane in addition to various smaller amounts of other higher alkanes.	**NG**	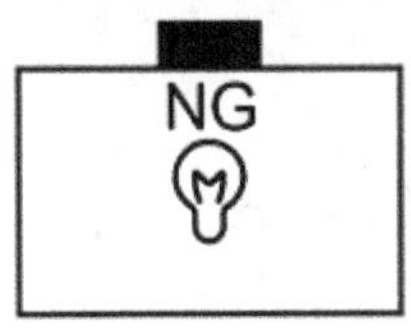 Electric generation natural gas-powered facility

Table 3-7. Sector 1 modifiers for activities and installations (*continued*)

Function	*Symbol*	*Symbol Usage Construct Example and Translation*
nuclear—Relating to or powered by nuclear energy.	**N**	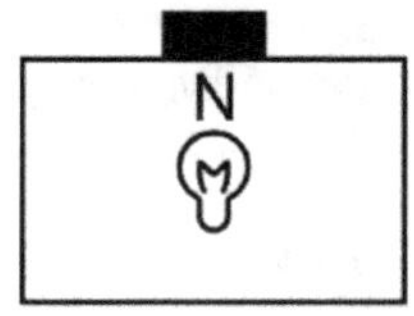 Electric generation nuclear-powered facility  Nuclear facility
operations—Capability to execute the principal planning and operating functions of an organization.	**OPS**	 Network operations center facility
petroleum—Relating to or powered by petroleum-based products.		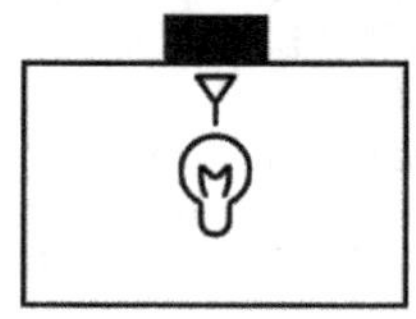 Electric generation petroleum powered plant facility
raid—Capability to conduct a surprise attack by a small force.	**RAID**	 Bureau of Alcohol, Tobacco, Firearms, and Explosives raid activity
radiological—Of or relating to nuclear radiation.	**R**	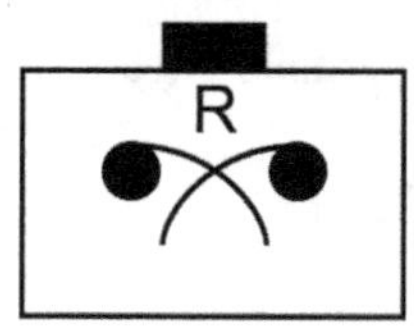 Radiological facility

Table 3-7. Sector 1 modifiers for activities and installations (*continued*)

Function	*Symbol*	*Symbol Usage Construct Example and Translation*
religious—Relating to or manifesting faithful devotion to an acknowledged ultimate reality or deity.	**REL**	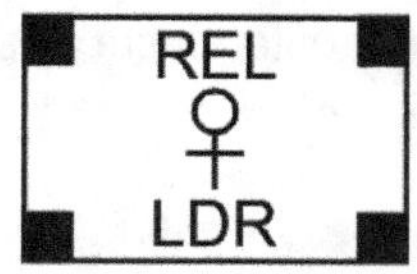 Individual religious leader activity
security—Measures taken by a military unit, activity, or installation to protect itself against all acts designed to, or which may, impair its effectiveness. (JP 3-10)	**SEC**	 Cyber security operations center facility
speaker—The presiding officer of a deliberative assembly.	**SPK**	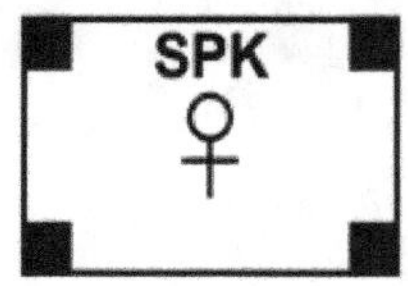 Individual speaker activity
targeted—Something or someone fired at or marked for attack.	**TGT**	 Individual targeted activity
telephone—A device by which sound (such as speech) is converted into electrical impulses and transmitted (as by wire or radio waves) to one or more specific receivers.	**T**	 Telecommunications telephone facility
television—An electronic system of transmitting transient images of fixed or moving objects together with sound over a wire or through space by apparatus that converts light and sound into electrical waves and reconverts them into visible light rays and audible sound.	**TV**	 Telecommunications television facility

Table 3-7. Sector 1 modifiers for activities and installations (*continued*)

Function	*Symbol*	*Symbol Usage Construct Example and Translation*
terrorist—An advocate or practitioner of terrorism as a means of coercion.	**TER**	 Organized group terrorist activity  Individual terrorist activity
training—The skill, knowledge, or experience acquired by one that trains.	**TNG**	 Police training facility
trafficking—Deal or trade in something illegal.	**TFK**	 Drug trafficking activity
water—Capability to receive, stock, produce, or distribute water	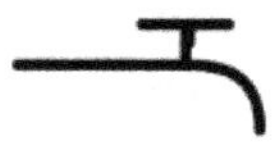	 Water network facility
willing recruitment—the process of adding new individuals without coercion. ***Note.*** This modifier symbol is interchangeable. Symbol may be used as a sector 1 or sector 2 modifier with the same meaning.	**WR**	 Group willing recruitment activity
wired communication—The transmission of data over a wire-based communication technology.	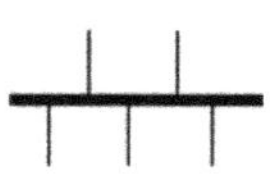	 Wired network facility

Table 3-7. Sector 1 modifiers for activities and installations (*continued*)

Function	*Symbol*	*Symbol Usage Construct Example and Translation*
wireless communication—The transfer of information (telecommunication) between two or more points without the use of an electrical conductor, optical fiber, or other continuous guided medium for the transfer.		 Wireless network facility
yard—Capability to provide organized limited stagging, storing, distribution, or maintenance.	YRD	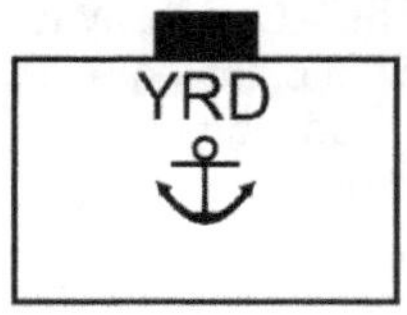 Maritime yard facility

SECTOR 2 MODIFIERS FOR ACTIVITIES AND INSTALLATIONS

3-11. Table 3-8 (on pages 105 through 106) shows sector 2 modifiers (field A). Sector 2 modifiers further identify affiliation, capability, special characteristic, or specialty.

Table 3-8. Sector 2 modifiers for individuals and organizations

Function	*Symbol*	*Symbol Usage Construct Example and Translation*
coerced recruitment—The process of adding new individuals by force or threat. ***Note.*** This modifier symbol is interchangeable. Symbol may be used as a sector 1 or sector 2 modifier with the same meaning.	CR	 Organized coerced terrorist recruitment activity
data—A collection of discrete or continuous values that convey information, describing the quantity, quality, fact, statistics, other basic units of meaning, or simply sequences of symbols that may be further interpreted formally.	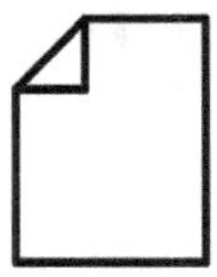	 Network data security facility

Table 3-8. Sector 2 modifiers for individuals and organizations (*continued*)

Function	*Symbol*	*Symbol Usage Construct Example and Translation*
leader—A person who has commanding authority or influence. ***Note.*** This modifier symbol is interchangeable. Symbol may be used as a sector 1 or sector 2 modifier with the same meaning.	**LDR**	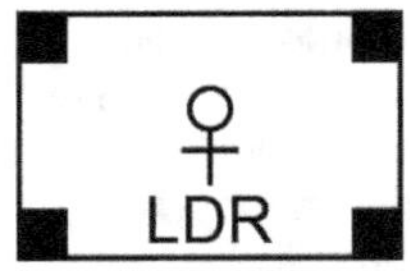 Friendly individual leader activity
operations—Capability to execute the principal planning and operating functions of an organization	**OPS**	 Wireless network operations facility
security—Measures taken by a military unit, activity, or installation to protect itself against all acts designed to, or which may, impair its effectiveness. (JP 3-10)	**SEC**	 Network cyber security facility
supply—The process of providing all items necessary to equip, maintain, and operate a military command. (FM 1-02.1)		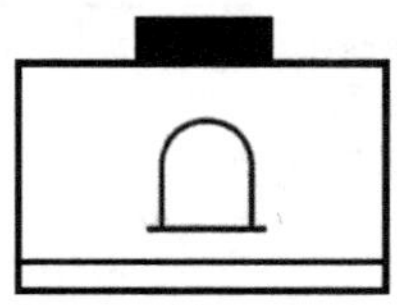 Ammunition cache
willing recruitment—The process of adding new individuals without coercion. ***Note.*** This modifier symbol is interchangeable. Symbol may be used as a sector 1 or sector 2 modifier with the same meaning.	**WR**	 Organized group willing terrorist recruitment activity

Chapter 4

Equipment Symbols

This chapter discusses Main function symbols, sector 1 modifiers, and mobility indicator amplifiers for equipment. Equipment is nonexpendable items needed to outfit or equip an individual or organization. This section includes the lists of main and modifier symbols and amplifiers for building land equipment symbols.

FRAMED AND UNFRAMED EQUIPMENT SYMBOLS

4-1. The equipment symbol construct standard permits the option to depict the symbol with frame or unframed. As discussed in chapter 1, the frame shape is what indicates the standard identity (friendly, enemy, neutral, and unknown) of a symbol. The unframed equipment symbol constructs must use colors (including blue, red, green, and yellow) in order to indicate the standard identity depiction of friendly, enemy, neutral, or unknown items. (See paragraph 1-9 for more information on standard identity colors.)

EQUIPMENT SYMBOL FRAME SHAPES

4-2. The symbol frame shape is used to identify friendly, enemy, neutral, and unknown equipment affiliation in or supporting an area of interest or operations. Table 4-1 provides the standard identity frame shapes for units and organizations. The frame shape construct guidelines for main and modifier symbols and amplifier placement locations are provided in paragraph 4-4.

Table 4-1. Equipment standard identity frame shapes

Domain	*Friendly*	*Hostile*	*Neutral*	*Unknown*
Land and sea surface	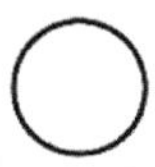			
	Assumed Friend	*Suspect*		*Pending*

Air (in flight)	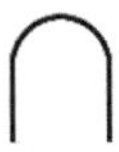		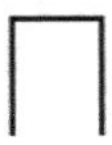	
	Assumed Friend	*Suspect*		*Unknown*

Space (in space)				
	Assumed Friend	*Suspect*		*Unknown*

MAIN AND MODIFIER SYMBOLS AND AMPLIFIERS FOR EQUIPMENT

4-3. The purpose of main and modifier symbols and amplifier fields is to standardize the display of optional alphanumeric information that graphically describes the equipment and provides additional information on capabilities, status, and location. Figure 4-1 shows the placement of land equipment symbol amplifiers around the equipment symbol using a friendly frame example for the purpose of reference location. Table 4-2 on pages 108 through 110 provides the descriptions and formats of each amplifier.

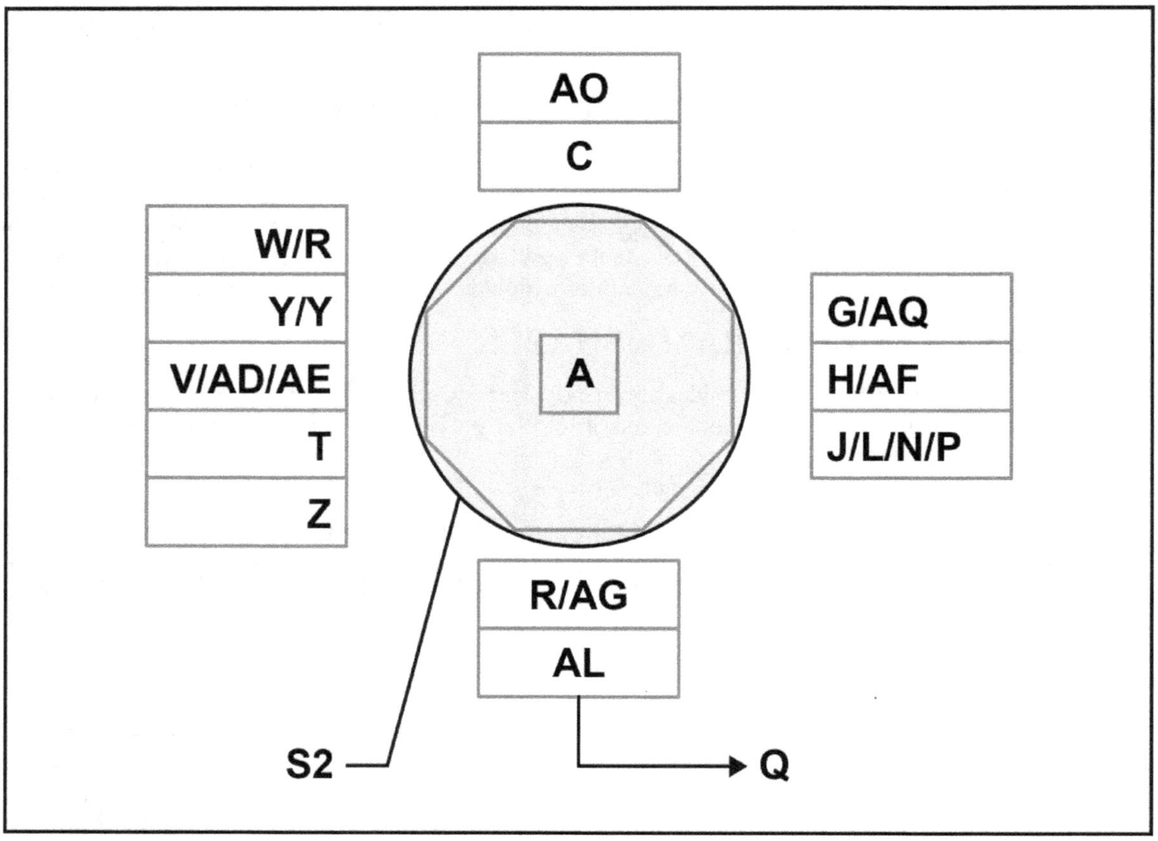

Figure 4-1. Placement of land equipment symbols main and modifier symbols and amplifiers

Table 4-2. Descriptions of main, modifier, and amplifier fields

Field	*Field Title*	*Description*
A	Main function and modifier symbols	The innermost part of a symbol that represents the main function and its capabilities (modifiers 1 and 2).
C	Quantity	A text amplifier identifies the number of items present. ***Note.*** This field allows a maximum of 9 characters.
G	Staff comments	A text amplifier content is implementation specific. ***Note.*** This field allows a maximum of 20 characters.
H	Additional information	A text amplifier content is implantation specific. ***Note.*** This field allows a maximum of 20 characters.

Table 4-2. Descriptions of main, modifier, and amplifier fields (*continued*)

Field	*Field Title*	*Description*
J	Evaluation rating	A text amplifier that consists of a one-letter reliability rating and a one-number credibility rating: **Reliability Ratings:** A-completely reliable. B-usually reliable. C-fairly reliable. D-not usually reliable. E-unreliable. F-reliability cannot be judged. **Credibility Ratings:** 1-confirmed by other sources. 2-probably true. 3-possibly true. 4-doubtfully true. 5-improbable. 6-truth cannot be judged. ***Note.*** This field allows a minimum of 2 characters.
L	Signature equipment	A text amplifier for hostile equipment; "!" indicates detectable electronic signatures. ***Note.*** This field allows a maximum of 1 characters.
N	Hostile (enemy)	A text amplifier for unframed equipment; letters "ENY" denote hostile symbols. ***Note.*** This field allows a maximum of 3 characters.
P	Identification, friend or foe Selective identification feature	A text amplifier displaying one or more identification, friend or foe or selective identification feature identification modes and codes. Display priority: Mode 5, Mode S, Mode 4, Mode 3, Mode 2. ***Note.*** This field allows a maximum of 15 characters.
Q	Direction of movement indicator	A graphic amplifier that identifies the direction of movement or intended movement of an object.
R	Mobility mode indicator	A graphic amplifier that depicts the mobility mode of transportation of an object.
S[2]	Offset location indicator	A graphic amplifier used to indicate the offset or precise location of a single point symbol.
T	Unique identifier	An amplifier field reserved for command and control systems that uniquely identifies a particular symbol with a track number. Prefix = TN: #####. Example: TN: 13579. ***Note.*** This field allows a maximum of 30 characters.
V	Type	A text amplifier for equipment that indicates types of equipment. ***Note.*** This field allows a maximum of 24 characters.
W	Date-time group	An alphanumeric designator for displaying a date-time group (DDHHMMSSZMONYYYY) or "O/O" for an order. The date-time group is composed of a group of six numeric digits with a time zone suffix and the standardized three-letter abbreviation for the month followed by four digits representing the year. The first pair of digits represents the day; the second pair, the hour; the third pair, the minutes. For automated systems, two digits may be added before the time zone suffix and after the minutes to designate seconds. ***Note.*** This field allows a maximum of 16 characters.

Table 4-2. Descriptions of main, modifier, and amplifier fields (*continued*)

Field	*Field Title*	*Description*
X	Altitude or depth	A text amplifier that displays either altitude, flight level, depth for submerged objects; or height of equipment or structures on the ground. Measurement units shall be displayed in the string. Examples: 1500MSL. FL150. ***Note.*** This field allows a maximum of 14 characters.
Y	Location	A text amplifier that displays a symbol's location in degrees, minutes, and decimal minutes (or in military grid reference system, global area reference system, or other applicable display formats). Examples: military grid reference system: 28SMB2649083145 global area reference system: 3317.0921N 04412.6332E ***Note.*** This field allows a maximum of 22 characters.
Z	Speed	A text amplifier that displays velocity. ***Note.*** This field allows a maximum of 8 characters.
AD	Platform type	Electronic intelligence notation or communications intelligence notation. ***Note.*** This field allows a maximum of 6 characters.
AE	Equipment teardown time	Equipment teardown time in minutes. ***Note.*** This field allows a maximum of 3 characters.
AF	Common identifier	A text amplifier used for placement of common system name or model type name. Example: "Hawk" for Hawk surface-to-air missile system. ***Note.*** This field allows a maximum of 12 characters.
AG	Auxiliary equipment indicator	Towed sonar array indicator: A graphic modifier for equipment that indicates the presence of a towed sonar array.
AL	Operational condition	A graphic amplifier that indicates operational condition or capacity. If used, it shall be comprised of only one color. Example: Aircraft: Red—destroyed, Green—fully capable. Example: Missile: Red—imminent threat, Green—no threat.
AO	Engagement bar	A graphic amplifier placed immediately atop the symbol. May denote 1) local/remote status, 2) engagement status, and 3) weapon type. Format: A:BBB-CC, where A = remote/local BBB = engagement status CC = weapon asset
AQ	Guarded unit	During ballistic missile defense, some tracks are designated as guarded by a particular unit. ***Note.*** This field allows a maximum of 2 characters.
AR	Special designator	Special track designators such as non-real time and tactically significant tracks are denoted here. ***Note.*** This field allows a maximum of 3 characters.

4-4. **Equipment direction of movement amplifier (Field Q).** The equipment direction of movement amplifier is an arrow or staff identifying the direction of movement or intended movement of an object. For equipment symbols, the amplifier is an angled arrow extending downward from the bottom center of the frame or symbol and pointing in the direction of movement. Figure 4-2 on page 111 provides a direction of movement example for an armored high mobility vehicle with medium gun system.

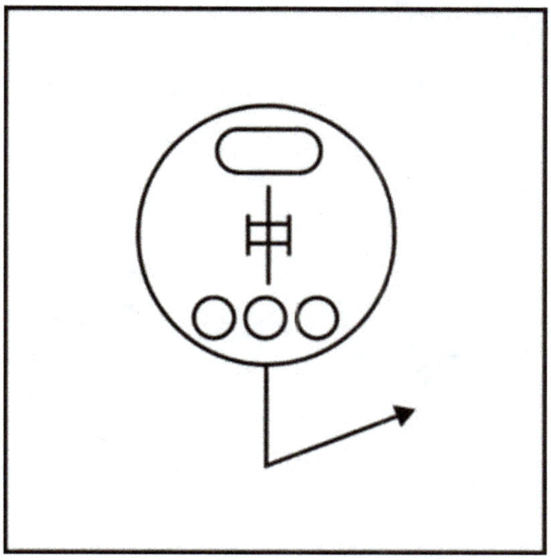

Figure 4-2. Equipment direction of movement usage construct example of an armored high mobility vehicle with medium gun system

4-5. **Engagement bar amplifier (Field AO).** The engagement amplifier bar may be used to designate engagements and to indicate targets. Both may be done in conjunction where depicted targets contain engagement information. The engagement bar may contain information on 1) remote or local engagement; 2) stage of the engagement (for example, assign, cover, engage, hold fire, cease fire, cease engage, break engagement, or missile in flight); and 3) type of weapon assignment (for example, missile, gun, or torpedo). Engagement bars use four colors: red, white, and orange for hostile targets, and blue for friendly participating. (See table 4-3.) Figure 4-3 depicts an example of a friendly self-propelled (tracked) long range surface to air missile launcher engaging an enemy attack rotary aircraft. (See MIL-STD 2525E for more detailed technical information concerning expanded usage of the engagement bar.)

Table 4-3. Engagement bar designation colors

Hostile target	
Hostile non-target	
Hostile expired target	
Friendly participating	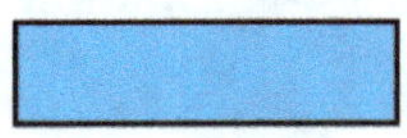

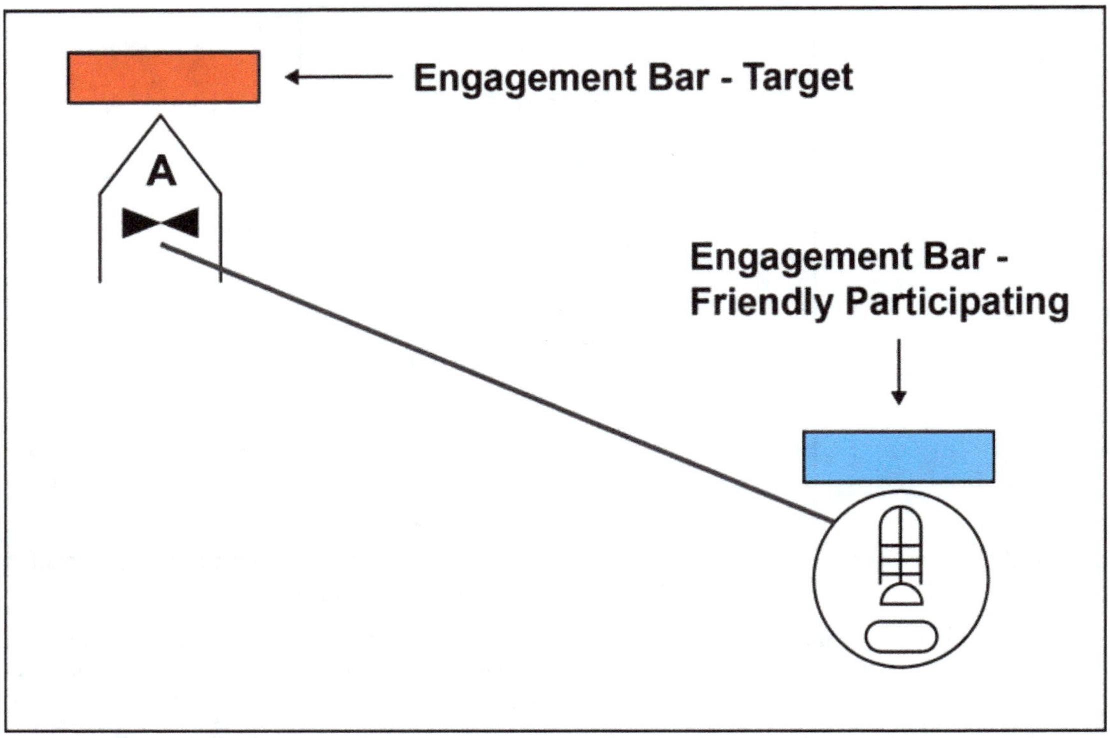

Figure 4-3. Example of armored self-propelled (tracked) long-range surface to air missile launcher engaging an enemy attack rotary aircraft

4-6. **Mobility (transportation) mode indicator (Field R).** The mobility (transportation) mode indicator is only used to depict the mode of transport of equipment. For example, a symbol for an armored self-propelled howitzer transported by train would include a railway mobility mode indicator in Field R (see figure 4-4.) Table 4-4 on page 113 provides a list of mobility (transportation) mode indicators with construct examples of the transporting of a howitzer.

Figure 4-4. Example of armored self-propelled howitzer moving by train

Table 4-4. Equipment mobility (transportation) mode indicators (Field R)

Description	*Mobility Symbol*	*Construct examples with howitzer*				
		Unframed	*Unknown*	*Friend*	*Neutral*	*Hostile*
Wheeled (limited to improved roads)		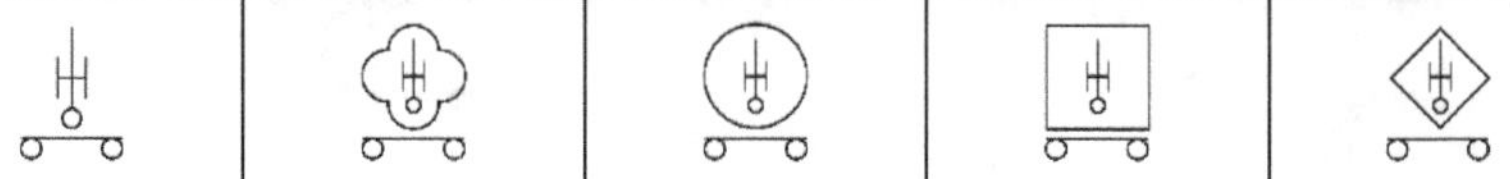				
Wheeled (cross-country)		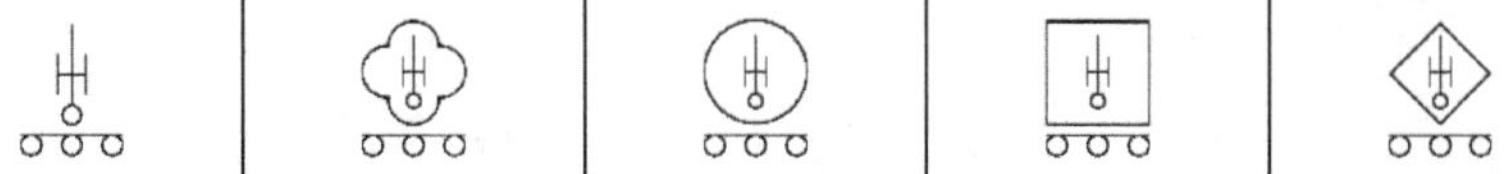				
Tracked	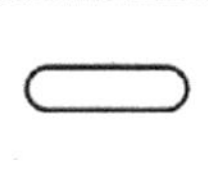	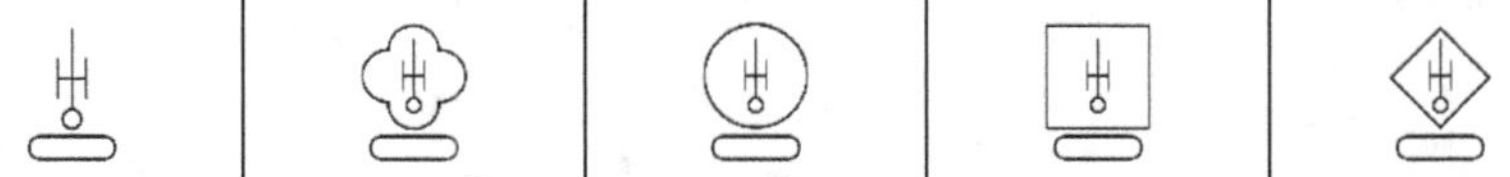				
Wheeled and tracked combination	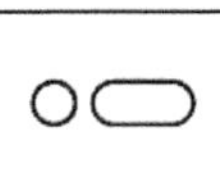	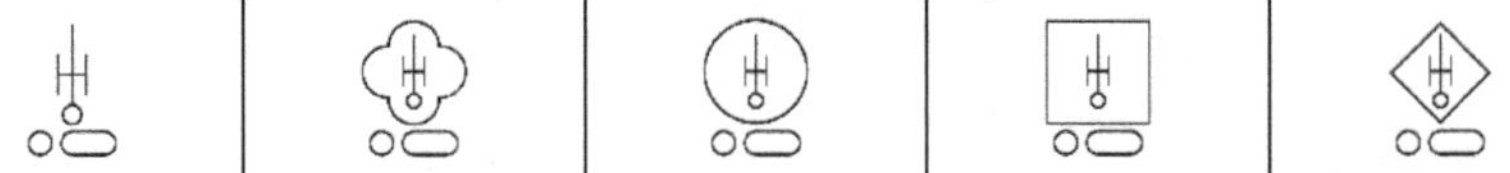				
Towed	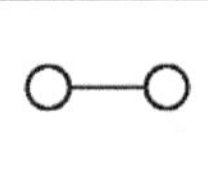	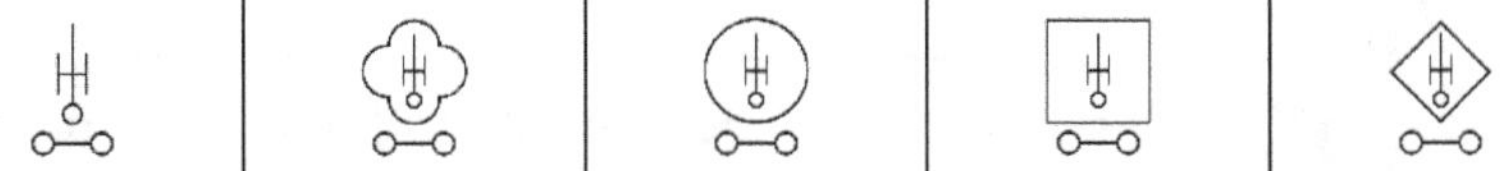				
Railway	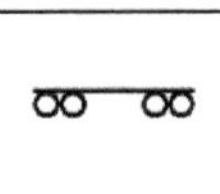					
Over-snow (prime mover)	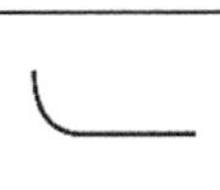	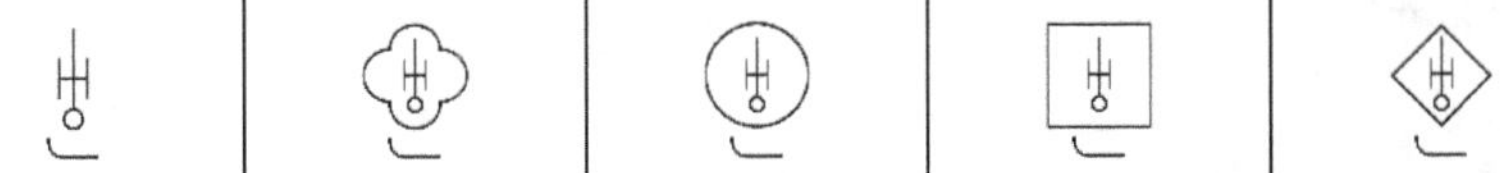				
Sled	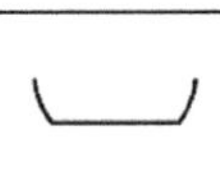	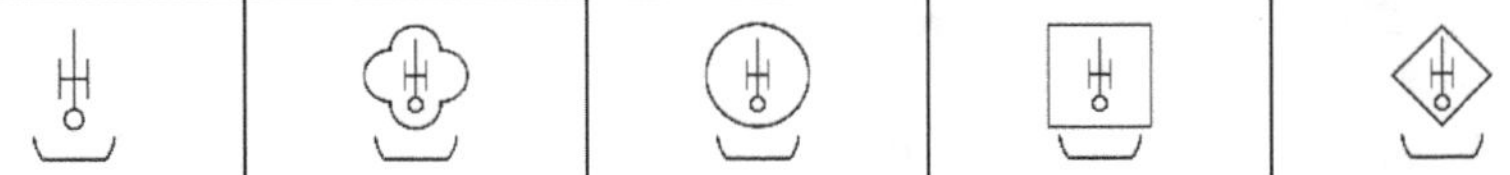				
Pack animals		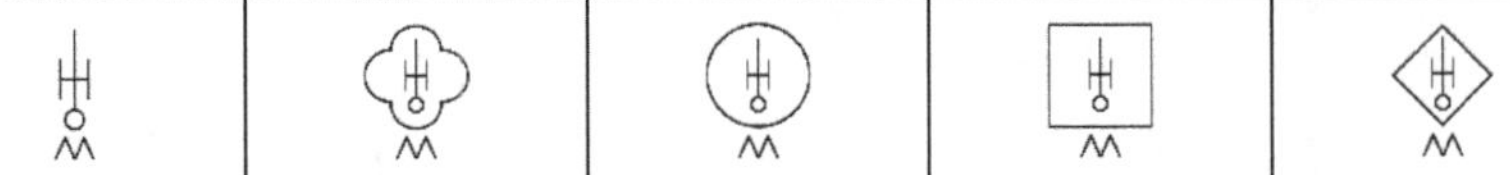				
Barge		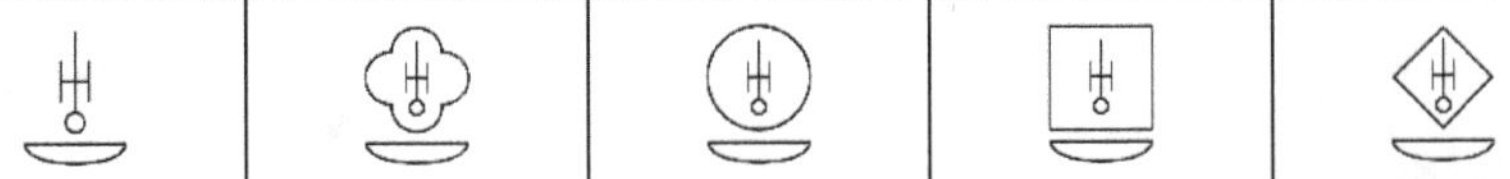				
Amphibious	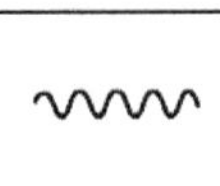	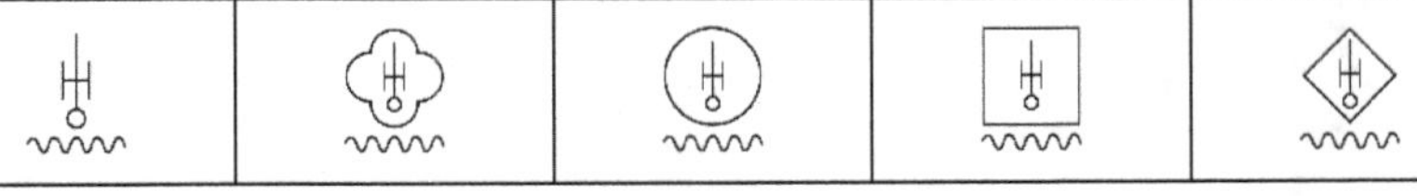				

4-7. **Speed (Field Z).** This field is used to display equipment velocity. (See figure 4-5 on page 114.) The first part of this eight-character (its maximum length) amplifier shall be a numeric value (quantity), and the second part shall be the speed or velocity unit of measure. Legal entries for this portion of the amplifier shall be "kph" (kilometers per hour), "mps" (meters per second), "kts" (knots per hour), or "mph" (miles per hour). Examples include—

- 220kph = 220 kilometers per hour.
- 974.5mps = 974.5 meters per second.
- 18.75kts = 18.75 knots per hour.
- 5mph = 5 miles per hour.

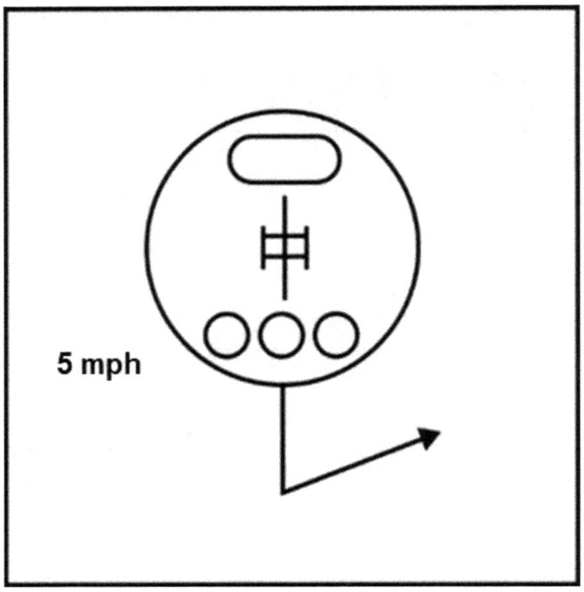

Figure 4-5. Speed usage construct example with direction of movement of an armored high mobility vehicle with medium gun system

4-8. Operational condition amplifiers (Field AL). This field allows the display of the level of operational condition of the equipment. Table 4-5 on page 115 shows operational condition amplifiers and construct examples.

Table 4-5. Operational condition amplifiers and construct examples

Function	***Amplifier***	***Example of Amplifier Construct Usage***
Fully operational		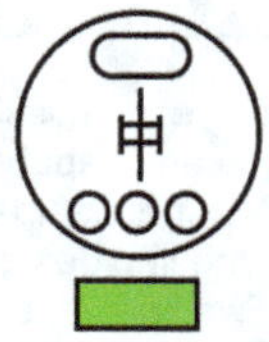 Fully operational armored high mobility wheeled vehicle with medium direct fire gun system
Damaged but substantially operational		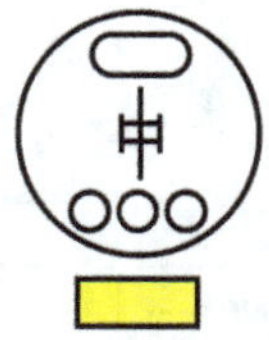 Damaged armored high mobility vehicle wheeled with medium direct fire gun system
Destroyed		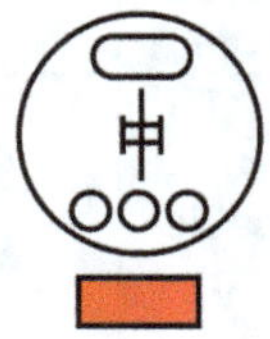 Destroyed armored high mobility wheeled vehicle with medium direct fire gun system
Full to capacity		 Full to capacity medium tractor trailer

MAIN FUNCTION SYMBOLS FOR EQUIPMENT

4-9. Main function symbols (Field A) reflect the main function of the symbol. Table 4-6 (on pages 116 through 138) shows the Main function symbols for equipment.

Table 4-6. MAIN FUNCTION SYMBOLS for equipment

Function	*Symbol*	*Symbol Usage Construct Example*	
		Friendly	*Enemy*
Weapons Systems			
Notes. 1. Weapons systems, missile launchers, and nonlethal weapons use a unique method for indicating size, altitude, or range. 2. Weapon size and capability is indicated by a horizontal line or lines perpendicular to the weapon symbol. a. No line: basic equipment symbol (generic). b. One line designates light, low altitude, or short-range. c. Two lines designates medium, medium altitude, or medium-range. d. Three lines designates heavy, high altitude, or long-range. 3. Land equipment symbols can be displayed without a frame, and color may be used to differentiate friend (blue), enemy (red), and neutral (green).			
unspecified weapon			
flame thrower—A device that expels from a nozzle a burning stream of liquid or semiliquid fuel under pressure.			
grenade launcher—A weapon capable of projecting a small missile that contains an explosive or a chemical agent (such as tear gas, a flame producer, or a smoke producer).	 Generic		
	 Short-range		
	 Medium-range		
	 Long-range		

Table 4-6. MAIN FUNCTION SYMBOLS for equipment (*continued*)

Function	*Symbol*	*Symbol Usage Construct Example*	
		Friendly	*Enemy*
Weapons Systems			
Guns			
air defense gun—A weapon designed to prevent or minimize the effects of attack by enemy aircraft or guided missiles. ***Note.*** The use of the air defense dome similar to the unit symbol at the base of the shaft indicates that it is primarily an air defense weapon.	 Generic		
	 Short-range		

		Self-propelled (tracked) short-range air defense gun	
	 Medium-range	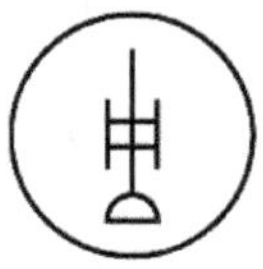	
	 Long-range	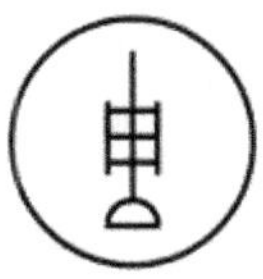	

Table 4-6. MAIN FUNCTION SYMBOLS for equipment (*continued*)

Function	***Symbol***	***Symbol Usage Construct Example***	
		Friendly	***Enemy***
		Guns	
anti-tank gun—A weapon designed to destroy armored vehicles. ***Note.*** The use of the inverted V similar to the unit symbol at the base of the shaft indicates that it is primarily an anti-tank weapon.	 Generic		

		Armored high mobility wheeled vehicle with anti-tank gun (generic)	
	 Light		
	 Medium		
	 Heavy		
direct fire gun—A weapon designed to engage gunfire by direct aiming on a visible target.	 Generic		
	 Light		

Table 4-6. MAIN FUNCTION SYMBOLS for equipment (*continued*)

Function	*Symbol*	*Symbol Usage Construct Example*	
		Friendly	***Enemy***
Guns			
direct fire gun (*continued*)			
		Armored high mobility wheeled vehicle with light direct fire gun	
	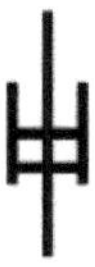 Medium		
	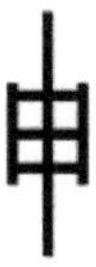 Heavy		
howitzer—A short cannon used to fire projectiles at medium muzzle velocities and with relatively high trajectories. ***Note.*** The use of the circle similar to the unit symbol for field artillery at the base of the shaft indicates that it is primarily a high trajectory.	 Generic		

		Howitzer (generic) towed	
	 120 millimeters or less		

Table 4-6. MAIN FUNCTION SYMBOLS for equipment (*continued*)

Function	*Symbol*	*Symbol Usage Construct Example*	
		Friendly	*Enemy*
		Guns	
howitzer (*continued*)	 Greater than 120 millimeters but less than 160 millimeters		
	 Greater than 160 millimeters but less than 210 millimeters		

		Howitzer (>160mm <210 mm) armored self-propelled (tracked)	
mortar—A portable muzzle-loading weapon having a tube short in relation to its caliber that is used to project rounds at high angles.	 Generic		
	 60 millimeters or less	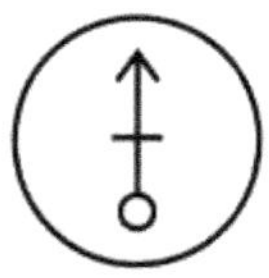	

Table 4-6. MAIN FUNCTION SYMBOLS for equipment (*continued*)

Function	*Symbol*	*Symbol Usage Construct Example*	
		Friendly	***Enemy***
Guns			
mortar (*continued*)	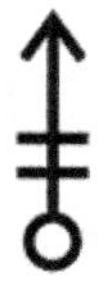 Greater than 60 millimeters but less than 107 millimeters		

		Armored high mobility wheeled vehicle with mortar (>60mm <107mm)	
	 Greater than 107 millimeters		
recoilless gun—A weapon with venting expanding propellant gas before recoil is produced.	 Generic		
	 Light		

Table 4-6. MAIN FUNCTION SYMBOLS for equipment (*continued*)

Function	*Symbol*	*Symbol Usage Construct Example*	
		Friendly	*Enemy*
Guns			
recoilless gun (*continued*)	 Medium		

		High mobility wheeled vehicle with medium recoilless gun	
	 Heavy		
rifle—A shoulder weapon with a rifled bore.	 Generic		
	 Single		
	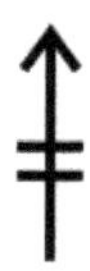 Semi-automatic		
	 Automatic		

Table 4-6. MAIN FUNCTION SYMBOLS for equipment (*continued*)

Function	***Symbol***	***Symbol Usage Construct Example***	
		Friendly	***Enemy***
		Guns	
machine gun—An automatic weapon for sustaining rapid fire that uses bullets.	 Generic		
	 Light		
	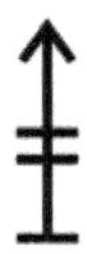 Medium		

		High mobility wheeled vehicle with medium machine gun	
	 Heavy	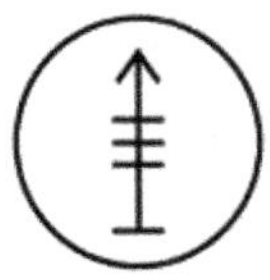	
		Missiles	
missile launcher—A weapon capable of launching a projectile to strike something at a distance. ***Note.*** The use of the dome covering most or the entire shaft similar to the unit symbol indicates that it is a missile launcher.	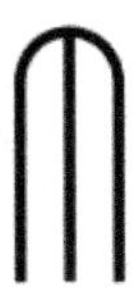 Generic		

		Armored high mobility wheeled vehicle with generic missile launcher	

Table 4-6. MAIN FUNCTION SYMBOLS for equipment (*continued*)

Function	*Symbol*	*Symbol Usage Construct Example*	
		Friendly	*Enemy*
Missiles			
missile launcher (*continued*)	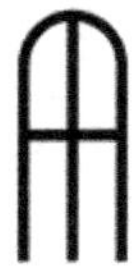 Short-range		
	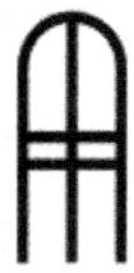 Medium-range		
	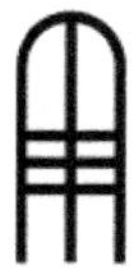 Long-range		
air defense missile launcher or surface-to-air missile launcher—A weapon designed to use missiles to prevent or minimize the effects of attack by enemy aircraft or guided missiles.	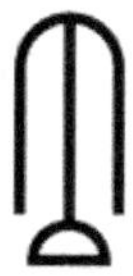 Generic		
	 Short-range		

		Armored high mobility wheeled vehicle with short-range air defense surface to air missile launcher	

Table 4-6. MAIN FUNCTION SYMBOLS for equipment (*continued*)

Function	*Symbol*	*Symbol Usage Construct Example*	
		Friendly	*Enemy*
Missiles			
air defense missile launcher or surface-to-air missile launcher (*continued*)	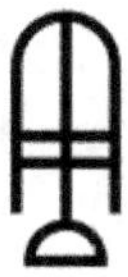 Medium-range		
	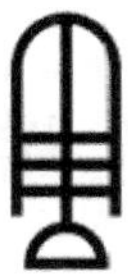 Long-range		

		Armored high mobility wheeled vehicle with long-range missile launcher	
anti-tank missile launcher—A weapon designed to use missiles to destroy armored vehicles.	 Generic		
	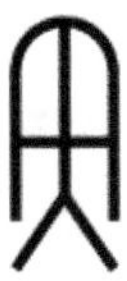 Short-range		

Table 4-6. MAIN FUNCTION SYMBOLS for equipment (*continued*)

Function	***Symbol***	***Symbol Usage Construct Example***	
		Friendly	***Enemy***
		Missiles	
anti-tank missile launcher (*continued*)	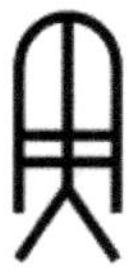 Medium-range		

		Armored high mobility wheeled vehicle with medium-range anti-tank missile launcher	
	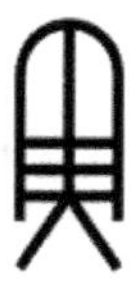 Long-range		
Surface-to-surface missile launcher—A weapon designed to use missiles to engage targets situated, transported, or employed on the earth's surface.	 Generic		

		High mobility wheeled vehicle with generic surface-to-surface missile launcher	
	 Short-range		

Table 4-6. MAIN FUNCTION SYMBOLS for equipment (*continued*)

Function	*Symbol*	*Symbol Usage Construct Example*	
		Friendly	*Enemy*
Missiles			
surface-to-surface missile launcher (*continued*)	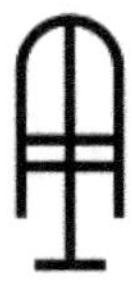 Medium-range		
	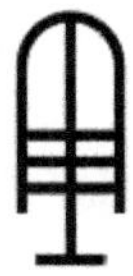 Long-range		
Rockets			
anti-tank rocket launcher—A weapon designed to use rocket-propelled explosive devices to destroy armored vehicles.	 Generic		
	 Short-range		

Table 4-6. MAIN FUNCTION SYMBOLS for equipment (*continued*)

Function	*Symbol*	*Symbol Usage Construct Example*	
		Friendly	*Enemy*
Rockets			
anti-tank rocket launcher (*continued*)	 Medium-range		

		Armored tracked vehicle with medium- range anti-tank rocket launcher	
	 Long-range		
single rocket launcher—A weapon designed to use a rocket-propelled explosive device to destroy a target. ***Note.*** The use of the double inverted Vs similar to the multiple rocket launcher unit symbol indicates that it is a rocket launcher.	 Generic		
	 Short-range		
	 Medium-range		
	 Long-range		

Table 4-6. MAIN FUNCTION SYMBOLS for equipment (*continued*)

Function	*Symbol*	*Symbol Usage Construct Example*	
		Friendly	*Enemy*
Rockets			
multiple rocket launcher—A weapon designed to use multiple rocket-propelled explosive device to destroy a target.	 Generic		
	 Short-range		
	 Medium-range		
	 Long-range		

		Armored tracked vehicle with heavy multiple rocket launcher	
Nonlethal Weapons			
nonlethal weapon—A weapon designed to not cause death.			
stun gun—A devise that fires electrified darts to stun and immobilize a person.			

Table 4-6. MAIN FUNCTION SYMBOLS for equipment (*continued*)

Function	***Symbol***	***Symbol Usage Construct Example***	
		Friendly	***Enemy***
		Vehicles	
water cannon—A device that shoots a high-velocity stream of water.			

		Armored low mobility wheeled vehicle with water cannon	
		Armored	
armored fighting vehicle—An armed combat vehicle protected by armor, generally combining operational mobility with offensive and defensive capabilities.			
armored personnel carrier—A type of vehicle designed to transport personnel and equipment in combat zones.	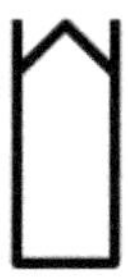		

Table 4-6. MAIN FUNCTION SYMBOLS for equipment (*continued*)

Function	***Symbol***	***Symbol Usage Construct Example***	
		Friendly	***Enemy***
		Armored	
armored protected—Vehicle designed and equipped with harden protective covering. ***Notes.*** 1. This symbol represents armored protected and requires a sector 2 wheeled vehicle modifier (provided in table 4-8 on page 143) to complete a non-tracked vehicle type capability. 2. If used without a sector 2 wheeled vehicle modifier present, symbol represents an armored tracked vehicle.		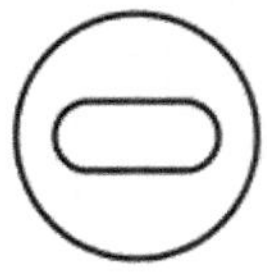	
tank—An enclosed heavily armed and armored combat vehicle that moves on tracks. ***Notes.*** 1. Vehicle symbol construct indicates size by using the same method as weapons systems symbols. 2. Size is indicated by a vertical line or lines within the symbol. a. No line: basic equipment symbol (generic). b. One line designates light. c. Two lines designates medium. d. Three lines designates heavy. 3. Land equipment symbols can be displayed without a frame, and color may be used to differentiate friend (blue), enemy (red), and neutral (green).	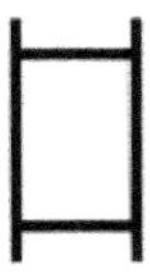 Generic		
	 Light		
	 Medium		
	 Heavy		

Table 4-6. MAIN FUNCTION SYMBOLS for equipment (*continued*)

Function	*Symbol*	*Symbol Usage Construct Example*	
		Friendly	*Enemy*
Vehicle Platforms			
Notes. 1. Land equipment symbols can be displayed without a frame, and color may be used to differentiate friend (blue), enemy (red), and neutral (green). 2. Vehicle platforms require a mobility modifier to complete the vehicle type symbol construct.			
engine—A machine with moving parts that converts power into motion.	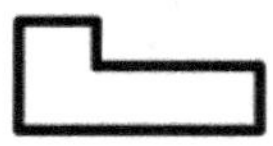		
		Engine (locomotive) railway	

		Standard Mobility Semi-trailer truck	
flatbed—A solid cargo bed, usually of wooden planks with no roof and no fixed sides. Loads are secured by manually tie downs.			
		Flatbed railway car	

		Flatbed tractor trailer	
vehicle—A piece of equipment designed to transport people or cargo.			
		Railway boxcar	

		High mobility wheeled petroleum, oils, and lubricants vehicle	

Table 4-6. MAIN FUNCTION SYMBOLS for equipment (*continued*)

Function	*Symbol*	*Symbol Usage Construct Example*	
		Friendly	*Enemy*
Vehicle Platforms			
vehicle (***continued***)			
		Standard mobility wheeled armored vehicle (armored car)	

		Medium tractor trailer	
Engineer Equipment			
bridge—A structure carrying a road, path, railroad, or canal across a river, ravine, road, railroad, or other obstacle.			

		Rail bridge	
drill—An instrument with an edged or pointed end for making holes in hard substances by revolving or by a succession of blows.			

		Medium drill	

Table 4-6. MAIN FUNCTION SYMBOLS for equipment (*continued*)

Function	*Symbol*	*Symbol Usage Construct Example*	
		Friendly	*Enemy*
Engineer Equipment			
mine clearing—Equipment designed to disarm and remove explosive mines from an area.	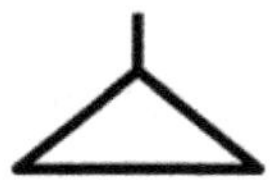		

		Armored (tracked) robotic mine clearing vehicle	
mine laying—Equipment designed to place explosive mines.			

		Tracked mine laying vehicle	
Aircraft			
Note. For all aircraft, an unframed symbol denotes the equipment is in flight—in addition to the in flight construct provided.			
fixed wing—A heavier-than-air flying machine, which is capable of flight using wings that generate lift caused by the aircraft's forward airspeed and the shape of the wings.			
		Fixed-wing aircraft on ground	

		Fixed-wing aircraft in flight	

Table 4-6. MAIN FUNCTION SYMBOLS for equipment (*continued*)

Function	*Symbol*	*Symbol Usage Construct Example*	
		Friendly	*Enemy*
Aircraft			
rotary-wing (helicopter)—A lifting rotor or wing which spins to provide aerodynamic lift.			
		Rotary-wing aircraft on ground	

		Rotary-wing aircraft in flight	
unmanned aircraft—(DOD) An aircraft that does not carry a human operator and is capable of flight with or without human remote control. (JP 3-30)			
		Unmanned aircraft on ground	

		Unmanned aircraft in flight	
Maritime			
destroyer—A warship equipped to conduct anti-aircraft warfare, tactical land strikes, anti-submarine warfare, and anti-surface warfare.	**DD**		
		Military Destroyer ship	

Table 4-6. MAIN FUNCTION SYMBOLS for equipment (*continued*)

Function	*Symbol*	*Symbol Usage Construct Example*	
		Friendly	*Enemy*
		Maritime	
military noncombatant ship (generic)—A military maritime vessel that does not engage in combat.			
		Military noncombatant ship (generic)	

		Military noncombatant hospital ship	

		Military noncombatant cargo ship	
civilian merchant ship (generic)—A non-military maritime vessel used for transporting trade goods.			
		Civilian merchant ship (generic)	

		Civilian merchant cargo ship	
		Other Equipment	
antenna—A device for radiating or receiving radio waves.			

		High mobility vehicle with mounted antenna	

Table 4-6. MAIN FUNCTION SYMBOLS for equipment (*continued*)

Function	*Symbol*	*Symbol Usage Construct Example*	
		Friendly	*Enemy*
Other Equipment			
chemical, biological, radiological, or nuclear (CBRN)			
		CBRN armored high mobility wheeled vehicle	
communications satellite—A manufactured object or vehicle intended to orbit the earth, the moon, or another celestial body for the purpose of transmitting or exchanging information.			
		Space communications satellite	
computer—A programmable electronic device that can store, retrieve, and process data.			
cyber-server—Computer in a network used to provide services to other computers in the network.	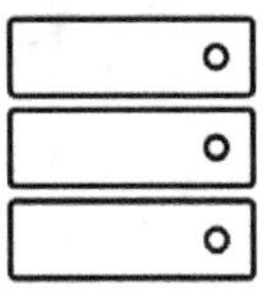		
		Command and control cyber-server	
directed energy—An umbrella term covering technologies that relate to the production of a beam of concentrated electromagnetic energy or atomic or subatomic particles. Also called DE. (JP 3-85)			

		Directed energy mounted on high mobility wheeled vehicle	

Table 4-6. MAIN FUNCTION SYMBOLS for equipment (*continued*)

Function	*Symbol*	*Symbol Usage Construct Example*	
		Friendly	*Enemy*
Other Equipment			
psychological operations—A robust military influence capability. (See FM 3-53 for more information on psychological operations forces and their operations.)			None applicable
			None applicable
		Psychological operations armored high mobility wheeled vehicle	
radar—A radiolocation system that uses radio waves to determine the distance (ranging), angle (azimuth), and radial velocity of objects relative to the site.			

		High mobility wheeled radar vehicle	
sensor—A device that responds to a physical stimulus and transmits a resulting impulse.			

Sector 1 Modifiers for Equipment

4-10. Table 4-7 (on pages 139 through 144) shows sector 1 modifiers (Field A) for equipment.

Table 4-7. Sector 1 modifiers for equipment

Function	*Symbol*	*Symbol Usage Construct Example and Translation*	
		Friendly	*Enemy*
armored (protection)—A vehicle hull equipped or protected with armor. ***Notes.*** 1. As a sector 1 modifier, this symbol represents armored protected and requires a sector 2 wheeled vehicle modifier to complete a non-tracked vehicle type capability. 2. If this sector 1 is used without a sector 2 wheeled vehicle modifier present, it means the unit has armored tracked capability.			
		Psychological operations armored high mobility wheeled vehicle	
attack—Designed, planned, or used for carrying out a military attack.	**A**		
		Attack helicopter in flight	
battalion (echelon of support)—Provides support to a battalion. (See ATP 3-96.1 for more information on battalion echelon of support.)	**II**		
		High mobility vehicle supporting battalion (combat trains)	

		High mobility vehicle supporting battalion (field trains)	

Table 4-7. Sector 1 modifiers for equipment (*continued*)

Function	*Symbol*	*Symbol Usage Construct Example and Translation*	
		Friendly	*Enemy*
cargo—The goods or merchandise temporarily stored or conveyed in a mode of transport.			
		Cargo helicopter in flight	

		High mobility wheeled cargo vehicle	
command and control—The exercise of authority and direction by a properly designated commander over assigned and attached forces in the accomplishment of the mission. Also called C2. (JP 1, Volume 2)	**C2**	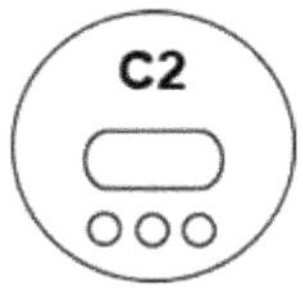	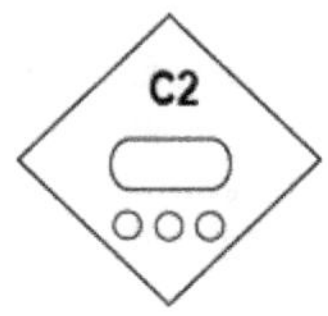
		Command and control armored high mobility wheeled vehicle	
heavy—Greater capability than the average of its kind or class. ***Note.*** This modifier symbol is interchangeable. Symbol may be used as a sector 1 or sector 2 modifier with the same meaning.	**H**		
		Heavy helicopter in flight	

Table 4-7. Sector 1 modifiers for equipment (*continued*)

Function	*Symbol*	*Symbol Usage Construct Example and Translation*	
		Friendly	*Enemy*
light—Lesser capability than the average of its kind or class. ***Note.*** This modifier symbol is interchangeable. Symbol may be used as a sector 1 or sector 2 modifier with the same meaning.	**L**		
		Light helicopter in flight	

		Light high mobility wheeled vehicle	
medium—Average capability of its kind or class. ***Note.*** This modifier symbol is interchangeable. Symbol may be used as a sector 1 or sector 2 modifier with the same meaning.	**M**		
		Medium helicopter in flight	

		Medium standard mobility wheeled vehicle	
medical evacuation—Capability to provide movement of wounded, injured, or ill to and between medical treatment facilities with en route medical care.			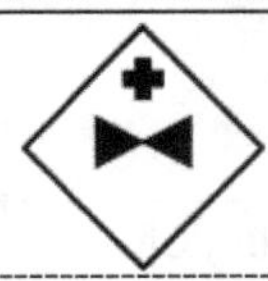
		Medical evacuation helicopter on ground	

		Medical evacuation helicopter in flight	

		Medical evacuation fixed wing in flight	

Table 4-7. Sector 1 modifiers for equipment (*continued*)

Function	*Symbol*	*Symbol Usage Construct Example and Translation*	
		Friendly	*Enemy*
medical evacuation (*continued*)		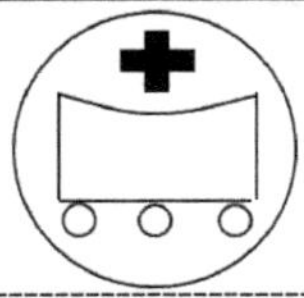	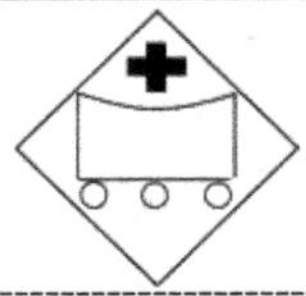
		Medical evacuation wheeled high mobility vehicle	
		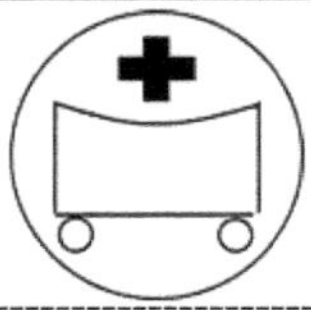	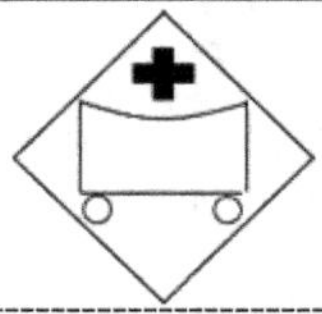
		Medical evacuation wheeled low mobility vehicle	
		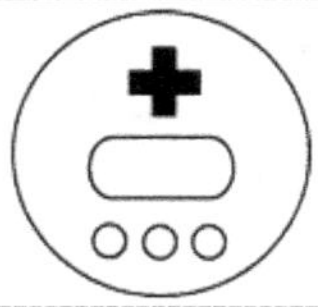	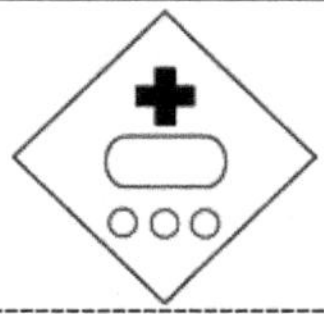
		Medical evacuation armored wheeled high mobility vehicle	
		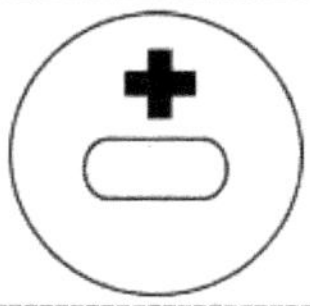	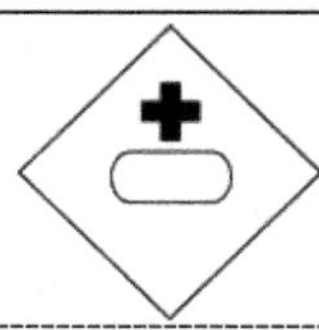
		Medical evacuation armored tracked vehicle	
multifunctional—Capable of performing more than one function.	**MF**		
		Multifunctional high mobility wheeled vehicle	
petroleum, oils, and lubricants (POL)—Petroleum based products used for equipment and vehicles.			
		POL transport high mobility wheeled vehicle	
			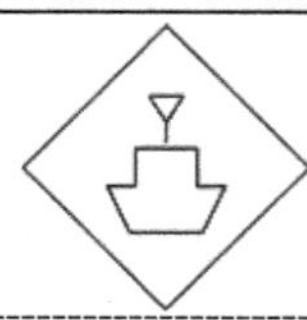
		Civilian merchant POL ship	

Table 4-7. Sector 1 modifiers for equipment (*continued*)

Function	*Symbol*	*Symbol Usage Construct Example and Translation*	
		Friendly	*Enemy*
recovery and maintenance—Capability of recovering and providing maintenance.			
		High mobility wheeled recovery and maintenance vehicle	

		Armored tracked heavy recovery and maintenance vehicle	
Robotic (guided and autonomous)—Equipment that has robot guided or autonomous control mechanism characteristic capability.			
		Robotic armored tracked vehicle with heavy anti-tank gun	

		Robotic rotary wing aircraft in flight	
utility—Designed or adapted for general use.	**U**		
		Utility helicopter on ground	

		High mobility wheeled utility vehicle	

Table 4-7. Sector 1 modifiers for equipment (*continued*)

Function	*Symbol*	*Symbol Usage Construct Example and Translation*	
		Friendly	*Enemy*
water—Capability of providing portable water.	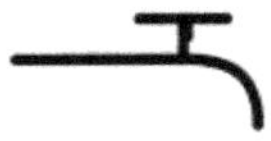		
		Water transport high mobility wheeled vehicle	

SECTOR 2 MODIFIERS FOR EQUIPMENT

4-11. Table 4-8 (on pages 144 through 146) shows sector 2 modifiers (Field A) for equipment.

Table 4-8. Sector 2 modifiers for equipment

Function	*Symbol*	*Symbol Usage Construct Example and Translation*	
		Friendly	*Enemy*
armored tracked—An armored combat vehicle that moves on tracks.			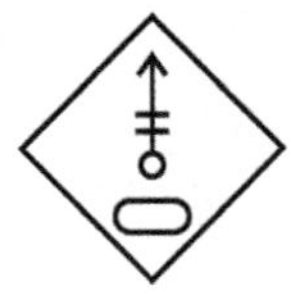
		Armored tracked vehicle with heavy mortar	
amphibious—Capability to move over land and water.			
		Amphibious armored tracked vehicle	
heavy—Greater capability than the average of its kind or class. ***Note.*** This modifier symbol is interchangeable or symbol may be used as a sector 1 or sector 2 modifier with the same meaning.	**H**		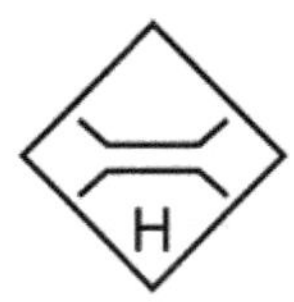
		Heavy bridge	
launcher—Capability to launch objects.	**—**		
		Mine laying launcher	

Table 4-8. Sector 2 modifiers for equipment (*continued*)

<table>
<tr><th rowspan="2">Function</th><th rowspan="2">Symbol</th><th colspan="2">Symbol Usage Construct Example and Translation</th></tr>
<tr><th>Friendly</th><th>Enemy</th></tr>
<tr><td rowspan="2">light—Lesser capability than the average of its kind or class.
Note. This modifier symbol is interchangeable or symbol may be used as a sector 1 or sector 2 modifier with the same meaning.</td><td rowspan="2">L</td><td></td><td></td></tr>
<tr><td colspan="2">Light bridge</td></tr>
<tr><td rowspan="2">medium—Average capability of its kind or class.
Note. This modifier symbol is interchangeable or symbol may be used as a sector 1 or sector 2 modifier with the same meaning.</td><td rowspan="2">M</td><td></td><td></td></tr>
<tr><td colspan="2">Medium cargo helicopter on ground</td></tr>
<tr><td rowspan="2">pack animal—Animals used as a mode of transport.</td><td rowspan="2"></td><td></td><td></td></tr>
<tr><td colspan="2">Cargo pack animal</td></tr>
<tr><td rowspan="2">rail—Capability to move on a line of rails fixed to ties and laid on a roadbed and providing a track for cars or equipment drawn by locomotives or propelled by self-contained motors.</td><td rowspan="2"></td><td></td><td></td></tr>
<tr><td colspan="2">Rail boxcar</td></tr>
<tr><td rowspan="2">tractor trailer—A large truck with a long trailer attached to the back of it.</td><td rowspan="2"></td><td></td><td></td></tr>
<tr><td colspan="2">Heavy tractor trailer vehicle</td></tr>
<tr><td rowspan="2">wheeled high mobility (cross-country)—Capable of traversing all terrain (off road).</td><td rowspan="2">OOO</td><td></td><td></td></tr>
<tr><td colspan="2">Armored high mobility wheeled vehicle with missile launcher</td></tr>
</table>

Table 4-8. Sector 2 modifiers for equipment (*continued*)

Function	*Symbol*	*Symbol Usage Construct Example and Translation*	
		Friendly	*Enemy*
wheeled standard mobility—Capable of traveling on surface roads designed for vehicles.			
		Medium limited mobility vehicle	

Chapter 5

Control Measures and Operation Planning Symbols

This chapter discusses military symbols used to control, coordinate, and identify military and related actions in the conduct of land operations.

FUNDAMENTALS

5-1. **Control Measures Symbols.** A *control measure* is the means of regulating forces or warfighting functions (ADP 6-0). Control measure symbols use a combination of military symbols constructs to depict offense and defense operational capabilities and to de-conflict areas, space, and time to regulate forces or warfighting functions. Control measures provide the means to visualize and understand current combat capabilities when combined with other symbols and amplifiers to depict operational information on maps and displays. Figure 5-1 provides an example of the use of a unit symbol construct with a control measure to depict a controlled offensive maneuver.

5-2. **Operations Planning Symbols.** Operation planning symbols identify planned actions, events, affects, built-up areas (building, fortified area, obstacles), and gap crossings. Some operation planning symbols are related to offensive and defensive operations but are not control measures. They may be used to depict planned actions conducted during offense and defense operations to enhance the operational picture of an operation.

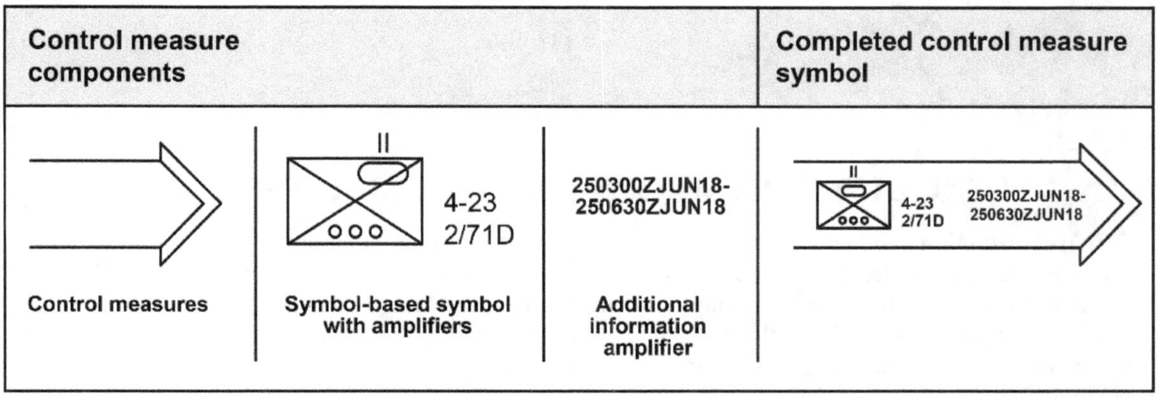

Figure 5-1. Composition of control measure symbol

STANDARD IDENTITY COLORING

5-3. Friendly graphic control measures use black or blue when drawn manually or on a color computer-generated display. Hostile graphic control measures use red. If red is not available, they are drawn in black with the abbreviation "ENY" placed on the graphic in at least two places. Obstacles as shown in this chapter are drawn using green. If green is not available, obstacles should be drawn using black. The hatching for CBRN contaminated areas uses the color yellow.

ACRONYMS AND ABBREVIATIONS USAGE

5-4. The acronyms and abbreviations in this chapter are considered symbols that are part of the military symbol construct for use with Army control measure symbols. No acronyms or abbreviations other than those provided in this publication may be used. When acronyms or abbreviations are approved for use with military symbols, they become part of the military symbol lexicon.

LABELING CONTROL MEASURES

5-5. Symbol makers make all text labeling in upper case letters. The reader should be able to read the labels for all text labels of modifier or amplifier fields for control measure symbols when the bottom of the overlay is closest to the reader. Labeling written on an angle should be readable to viewers horizontally.

5-6. Status refers to whether a control measure exists at the location identified (status is "present") or will in the future reside at that location (status is "planned" or "suspected"). In general, linear control measures shall be a solid line when indicating present status and a dashed line when indicating suspected or planned status, as depicted in table 5-1. There are certain control measures such as counterattack which are drawn in the "present" status with dashed lines.

Table 5-1. Present and planned status for control measure symbols

	Point Graphics	***Boundary Line Graphics***	***Area Graphics***
Present position	 Isolated personnel pickup point	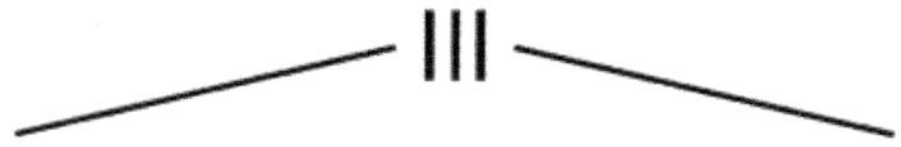 Regiment boundary	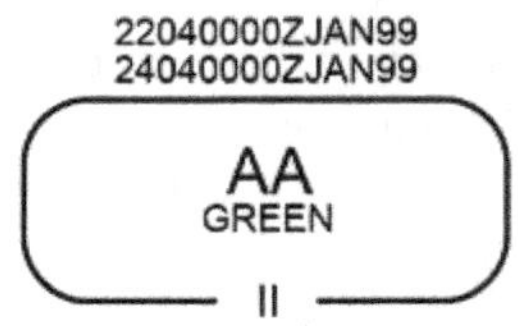 Assembly area green
Planned or suspected		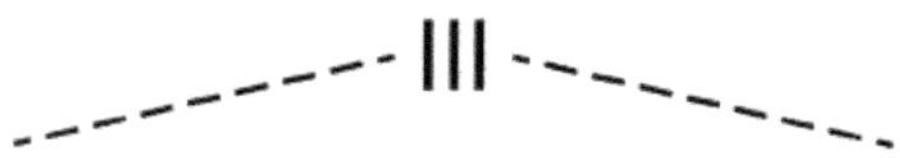	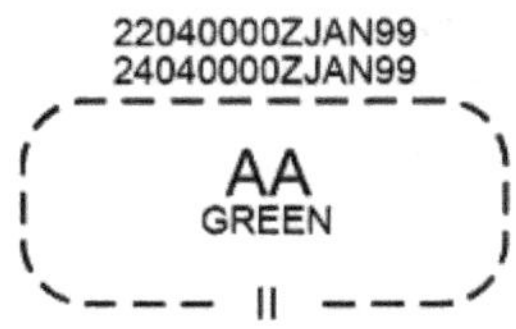

MAIN AND MODIFIER SYMBOLS AND AMPLIFIERS

5-7. **Main function symbols.** Field A is the placement area for control measure Main function symbols and provides the ability to depict the main or supporting function within the construct composition. Main function symbols for control measures are framed or unframed symbol constructs, or they are unique approved symbols that can be embedded within the symbol to effectively translate the intent or function. Not all control measures have this placement field, and the control measure template will indicate if the construct composition provides the capability to add a Main function symbol to the symbol.

5-8. **Sector 1 modifier symbols.** Only limited access areas and minefields use modifiers in their symbol construct. Each of these constructs have unique sector 1 modifier placement templates that are displayed in their military symbol construct sections. (See table 5-4 [on page 155] for limited access area modifier usage constructs, and tables 5-20 [on page 209] and 5-21 [on page 212] for minefield modifier usage constructs.)

5-9. **Amplifiers.** An amplifier provides optional additional information about a tactical symbol. The field identification, field title, description, and maximum allowable display lengths of tactical symbol amplifiers are presented in table 5-2 on pages 149 through 150. An example of each amplifier (including both text and graphic indicators) is included in figure 5-2 on page 150. Amplifiers can be defined as either static or dynamic:

- Static amplifiers are amplifiers whose size and placement are based on the attributes of an object, and they can change as these attributes and the scale of the background change.
- Dynamic amplifiers are amplifiers whose size and placement are based on the attributes of an object and change as these attributes change.

Table 5-2. Amplifier descriptions for control measure symbols

Field	*Field Title*	*Description*
A or sector 1 modifier	Main function symbol or modifier	The part of a symbol that represents main function, capability, type, or classification.
B	Echelon	A unit symbol that identifies command level.
H	Additional information	Content is implementation specific. ***Note.*** This field allows a maximum of 20 characters.
N	Hostile (enemy)	The letters "ENY" denote hostile control measure symbols. ***Note.*** This field allows a maximum of 3 characters.
Q	Direction of movement indicator	Identifies the direction of movement or intended movement of an object.
S	Offset location indicator	Used for points and chemical, biological, and radiological events to display a symbol away from its position while retaining its actual location.
T, T1	Unique identifier	A text amplifier used to differentiate a symbol by numbering, lettering, or a combination of both, and/or may be used to include the unit designation. ***Notes.*** 1. In some cases, this tactical symbol may require multiple instances of a "T" amplifier to fully create or represent an object. 2. "T1" maybe be used if field used displayed more than once in a tactical symbol. 3. This field allows a maximum of 30 characters.
W, W1	Date-time group (DTG)	"W" identifies the start DTG and can be displayed alone or in conjunction with "W1" to identify the projected DTG end date. The "W" represents an alphanumeric designator for displaying a date-time group (DDHHMMSSZMONYYYY) or "O/O" for an order. When "W" and "W1" are used in conjunction, they identify the time control measure in effect. The date-time group is composed of a group of six numeric digits with a time zone suffix and the standardized three-letter abbreviation for the month followed by four digits. The first pair of digits represents the day; the second pair, the hour; the third pair, the minutes. The last four digits after the month are the year. For automated systems, two digits may be added before the time zone suffix and after the minutes to designate seconds. ***Note.*** This field allows a maximum of 16 characters.
X	Altitude or depth	Displays the minimum, maximum, or specific altitude (in feet or meters in relation to a reference datum), flight level, or depth (for submerged objects in feet below sea level). ***Note.*** This field allows a maximum of 15 characters.
Y	Location	Displays a symbol's location in degrees, minutes, and decimal minutes. ***Note.*** This field allows a maximum of 22 characters.
AM	Distance	A numeric amplifier that displays a minimum, maximum, or specific distance (range, radius, width, or length) in meters or feet. ***Note.*** This field allows a maximum of 7 characters.
AN	Azimuth	A numeric amplifier that displays an angle measured from true north to any other line in degrees. ***Note.*** This field allows a maximum of 3 characters.

Table 5-2. Amplifier descriptions for control measure symbols (*continued*)

Field	*Field Title*	*Description*
AP	Target number	A six-character text modifier used in fire support operations to uniquely designate targets where characters 1 and 2 are alphabetic and characters 3–6 are numeric (for example, AANNNN). ***Note.*** This field allows a maximum of 6 characters.
AP1	Target number extension	A *target number extension* is a sequentially assigned number identifying the individual elements in a target (MIL-STD 6017), where character 1 is a dash and characters 2 and 3 are numeric, from 1 through 15. It is applicable only to the "point or single target" symbol, is conditional upon the presence of the target designator amplifier, and is visually displayed appended to the target number amplifier. ***Note.*** This field allows a maximum of 3 characters.
AS	Country	A three-letter code representing geographical entity. ***Note.*** This field allows a maximum of 3 characters.

ECHELON INDICATOR (FIELD B)

5-10. The echelon indicator provides a graphic representation of command level and is used to show the element echelon on lines and areas. The indicator is positioned as shown in figure 5-2 and Field B is defined in table 5-1 on page 148.

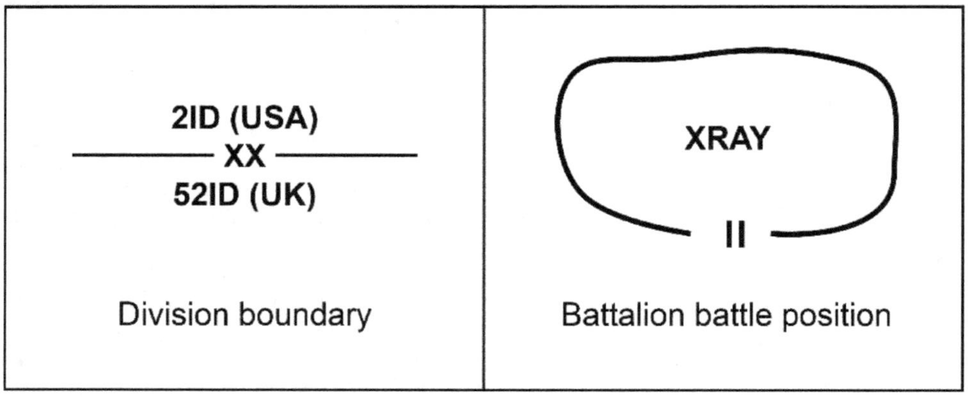

Figure 5-2. Echelon indicator usage construct examples

DIRECTION OF MOVEMENT INDICATOR (FIELD Q)

5-11. The direction of movement indicator is an arrow identifying the direction of movement of events. The arrow extends downward from the center of the symbol and points in the direction of movement. The indicator is positioned as shown in figure 5-3 on page 151 and Field Q is defined in table 5-1 on page 148.

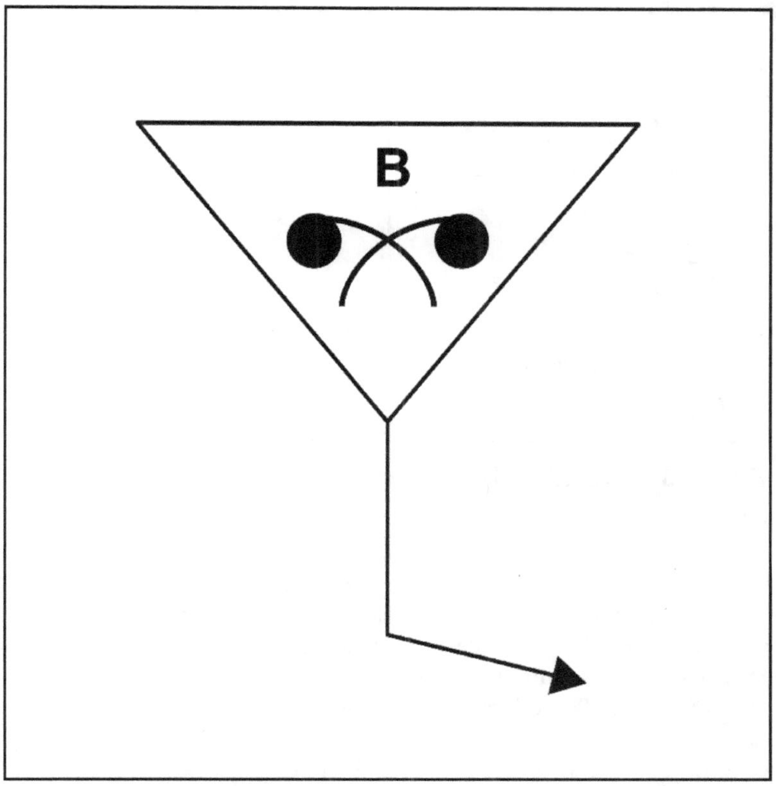

Figure 5-3. Direction of movement usage construct example

OFFSET LOCATION INDICATOR (FIELD S)

5-12. The offset location indicator is used when placing an object away from its actual location. The indicator is a line extending downward from an appropriate anchor point on a symbol. This amplifier permits including the actual location in latitude and longitude between the anchor point of the symbol and the line extending downward. The indicator is positioned as shown in figure 5-4 and Field S is defined in table 5-1 on page 148.

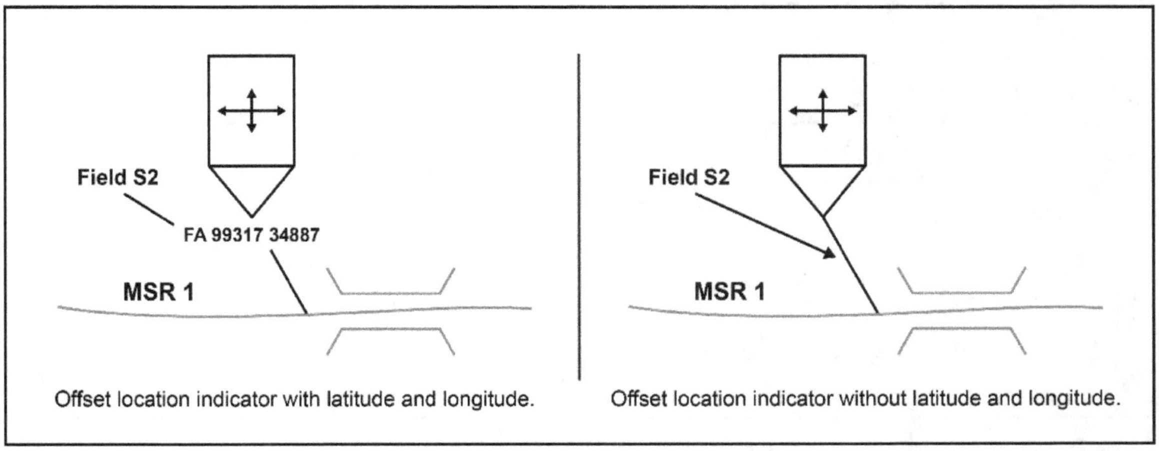

Figure 5-4. Offset location indicator usage construct examples.

ADDITIONAL INFORMATION AMPLIFIER (FIELD H)

5-13. The additional information amplifier field is used to add content that is implementation specific, provided it does not exceed the allowed 20 maximum number of characters. This amplifier is positioned differently based on the symbol used. Figure 5-5 provides 2 different position variation constructs for this field.

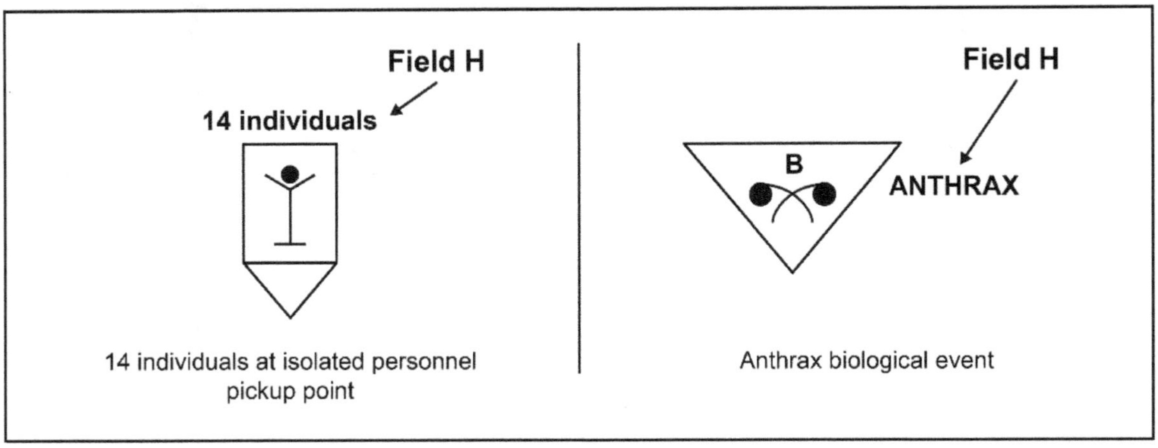

Figure 5-5. Additional information usage construct example

BOUNDARIES

5-14. A *boundary* is a line that delineates surface areas for the purpose of facilitating coordination and deconfliction of operations between adjacent units, formations, or areas (JP 3-0). Figure 5-6 provides a boundary composition template that includes orientation of optional field amplifier labels for horizontal (east—west) and vertical (north—south) boundaries. (See table 5-3 on pages 153 through 154 for depictions of boundary control measures. The following are optional amplifiers that may be used with this control measure construct:

- Country code Field AS to identify the units' three-letter geographical entity country code.
- Echelon indication Field B to identify the boundary echelon.
- Hostile (enemy) Field N to identify enemy boundaries.
- Unique identifier Field T to identify designations of adjacent units.

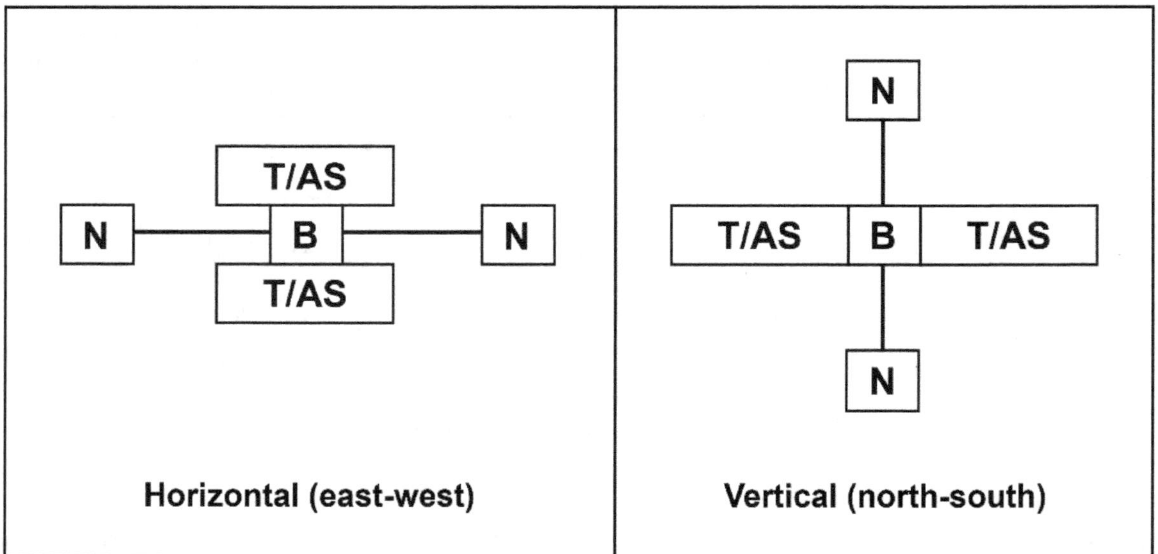

Boundary symbol amplifiers

AS: Country code field identifies the units' three-letter graphical entity country code.
B: Echelon indication field identifies enemy boundaries.
N: Hostile (enemy) field identifies designations of adjacent units.
T: Unique identifier field identifies designation of adjacent units.

Figure 5-6. Boundary composition template

Table 5-3. Boundaries

Function	*Template*	*Construct Example*
Friendly present boundary	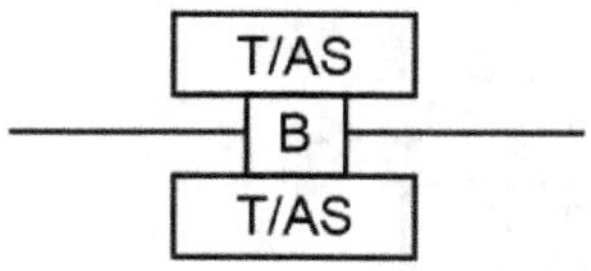	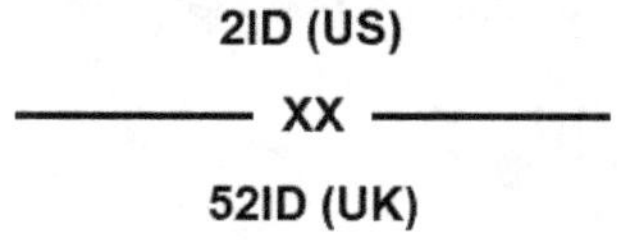 Division boundary between United States 2nd Infantry Division and United Kingdom 52nd Infantry Division
	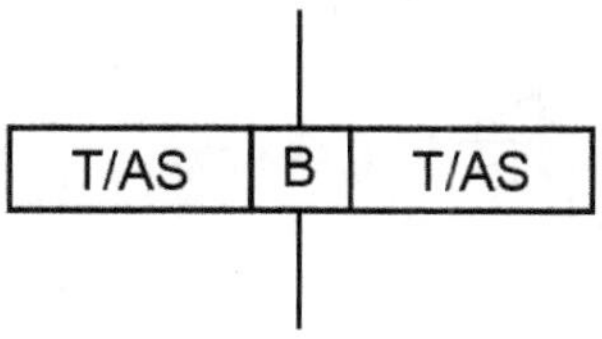	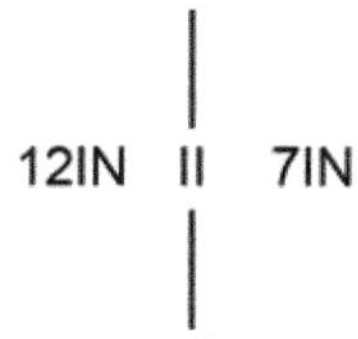 Battalion boundary between 12th Infantry battalion and 7th Infantry battalion
Friendly planned boundary	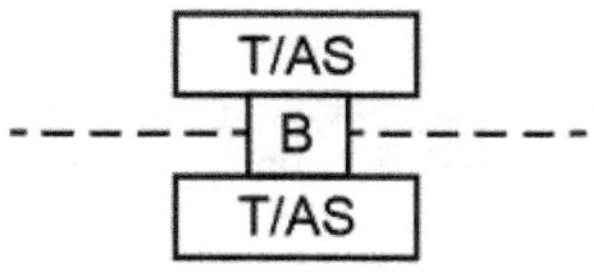	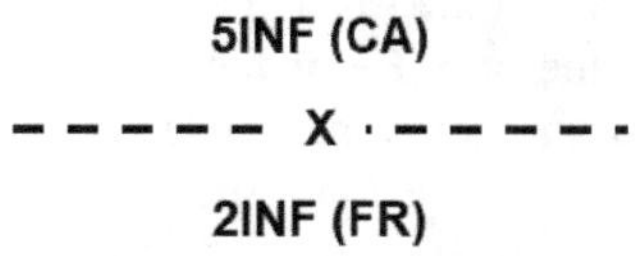 Brigade boundary between Canada 5th Infantry brigade and France 2nd Infantry brigade
Enemy known boundary	Monochrome 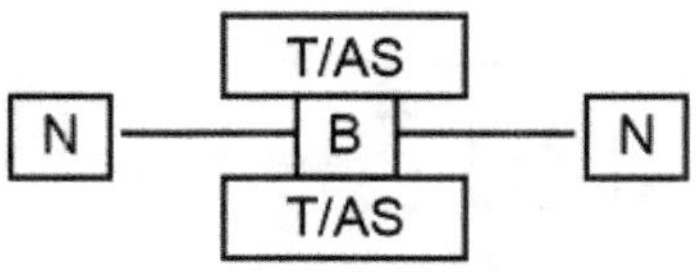	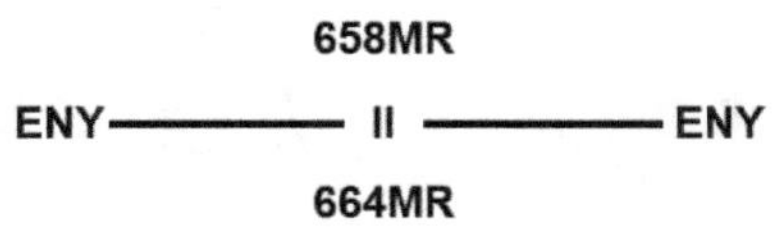
	Color 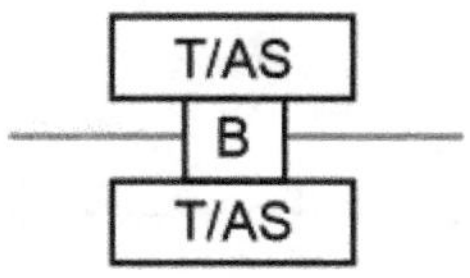	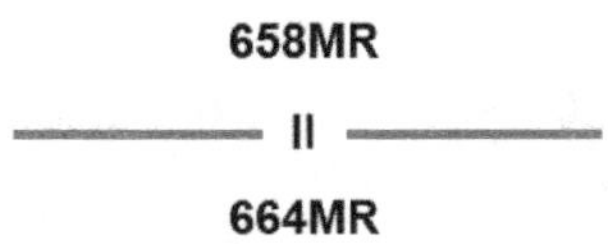
	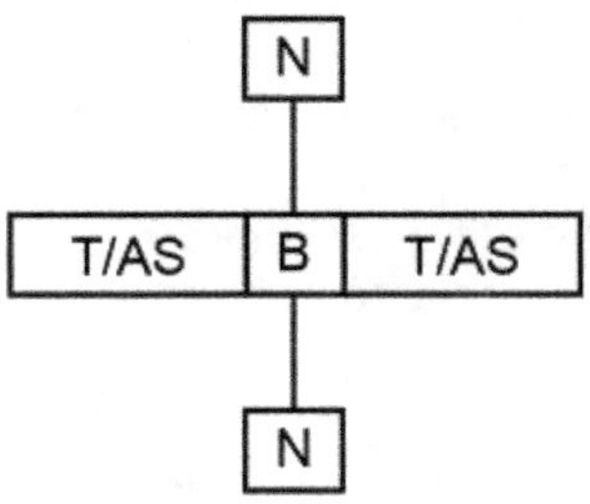	

Table 5-3. Boundaries (*continued*)

Function	*Template*	*Construct Example*
Enemy suspected boundary	Monochrome 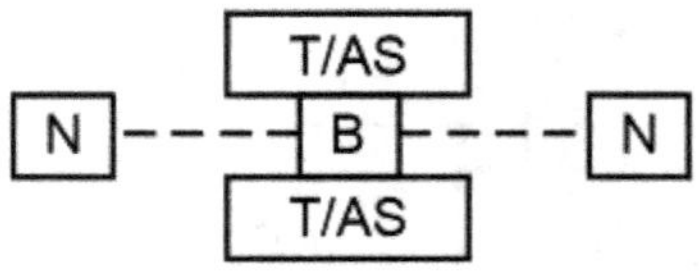	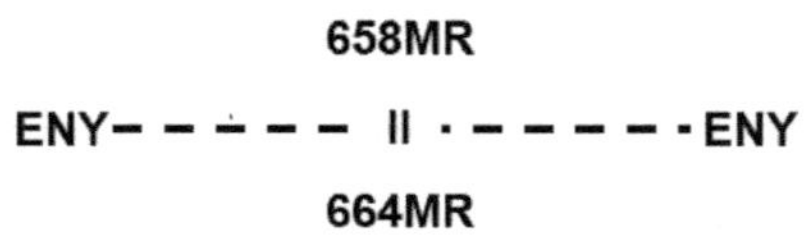
	Color 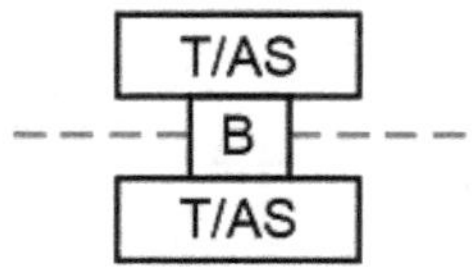	

AREAS

5-15. An area is a specified geographic surface included within a delineated set of lines (boundaries) used for the purpose of facilitating coordination and deconfliction between adjacent units, formations, or other specific geographical surfaces. Figure 5-7 provides the dynamic figure template for areas that provide the option to use 7 amplifier fields with the operations area Main function symbols listed in table 5-4 (on pages 155 to 159). Table 5-5 (on pages 160 through 163) lists battle positions and operations area templates with unique drawing constructs.

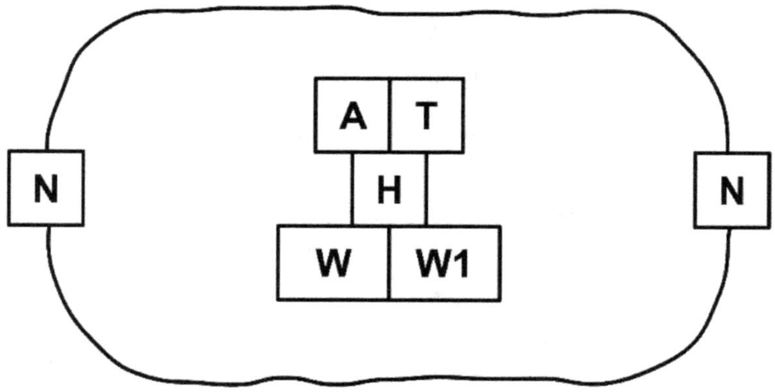

A: Symbol that identifies the main function.
H: Additional information (content is implementation specific)
N: Enemy: Letters "ENY" denote hostile.
T: Identifies the unique text or numerical designation.
W/W1: Identifies the date-time group associated with area.

NOTES.
1. Not all fields are required for each area as some areas may use only one field, while others will use several.
2. The information fields are moveable and scalable as a block in the area.
3. The figure is dynamic.

Figure 5-7. Template for area control measure symbols

5-16. Table 5-4 on pages 155 through 159, lists area control measure Main function symbols that follow the specific format as shown in figure 5-7.

Table 5-4. Area control measures MAIN FUNCTION SYMBOLS

Function	*Main function symbol (Field A)*	*Construct Example*
	Areas	
airfield—An area prepared for the accommodation (including any buildings, installations, and equipment), landing, and takeoff of aircraft. (JP 3-36)	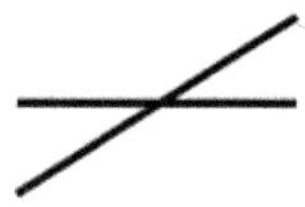	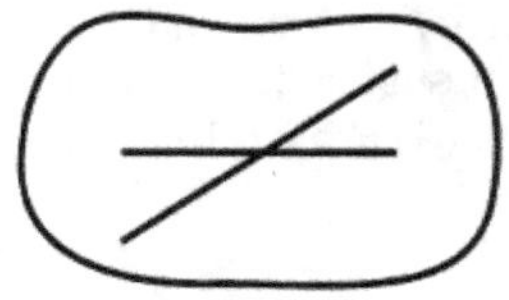
area of operations—An operational area defined by a commander for the land or maritime force commander to accomplish their missions and protect their forces. (JP 3-0)	**AO**	 Area of operations BUFFALO
assembly area—An area a unit occupies to prepare for an operation. (FM 3-90)	**AA**	 Assembly area BLUE
		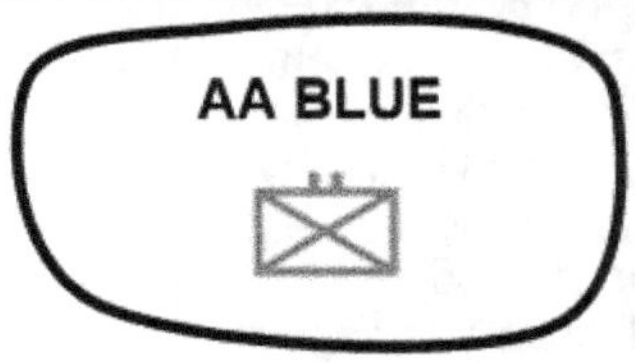 Occupied
		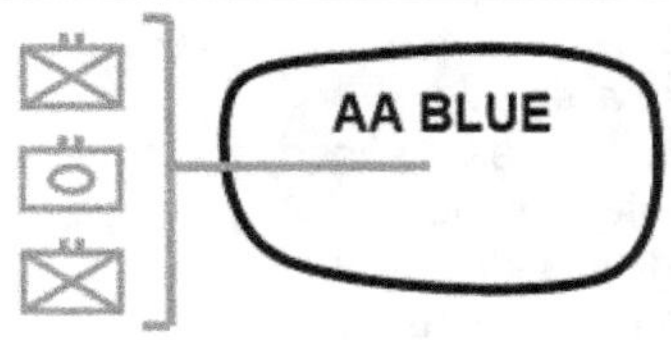 Occupied using offset for units
		 Planned

Table 5-4. Area control measures MAIN FUNCTION SYMBOLS (*continued*)

Function	*Main function symbol (Field A)*	*Construct Example*
	Areas	
base camp—An evolving military facility that supports the military operations of a deployed unit and provides the necessary support and services for sustained operations. (ATP 3-37.10)	**BC**	 Base camp SOTO
engagement area—An area where the commander masses effects to contain and destroy an enemy force. (FM 3-90)	**EA**	 Engagement area ROCK with three company battle positions covering fields of fire
guerrilla base—A temporary site where guerrilla installations, headquarters, and some guerrilla units are located. (ATP 3-05.1)	**GB**	 Guerrilla base BOOGEYMAN
named area of interest—A geospatial area or systems node or link against which information that will satisfy a specific information requirement can be collected, usually to capture indications of adversary courses of action. (JP 2-0)	**NAI**	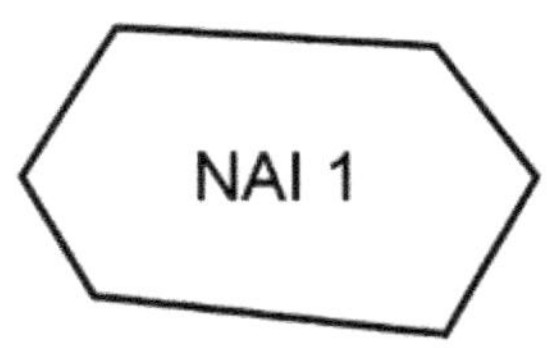 Named area of interest 1
objective area—A geographical area, defined by competent authority, within which is located an objective to be captured or reached by military forces. (JP 3-06)	**OBJ**	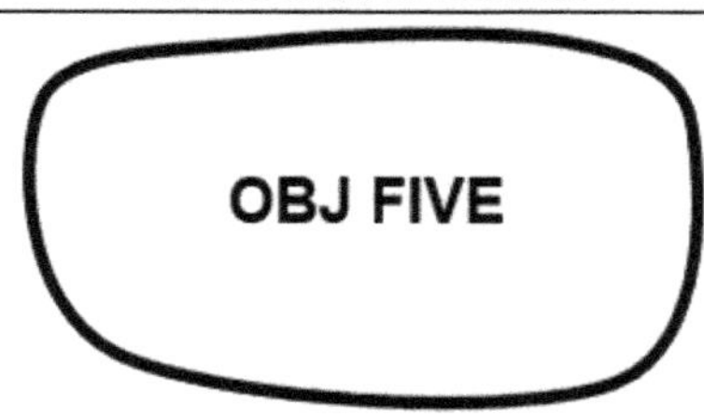 Objective area FIVE

Table 5-4. Area control measures MAIN FUNCTION SYMBOLS (*continued*)

Function	*Main function symbol (Field A)*	*Construct Example*
	Areas	
target area of interest—The geographical area where high-value targets can be acquired and engaged by friendly forces. (JP 2-0)	**TAI**	 Target area of interest ALY
	Positions	
assault position—A covered and concealed position short of the objective from which final preparations are made to assault the objective. (ADP 3-90)	**ASLT**	 Assault position DANUB
attack position—(Army) The last position an attacking force occupies or passes through before crossing the line of departure. (ADP 3-90)	**ATK**	 Attack position NILE
		 Combined arms unit in attack position AMAZON
	Note. The attack position area is superimposed over the unit location. Offset indicator may also be used.	
	Holding Areas	
detainee holding area—A facility or other location where detainees are administratively processed and provided custodial care pending disposition and subsequent release, transfer, or movement to a theater detention facility. (JP 3-31)	**DHA**	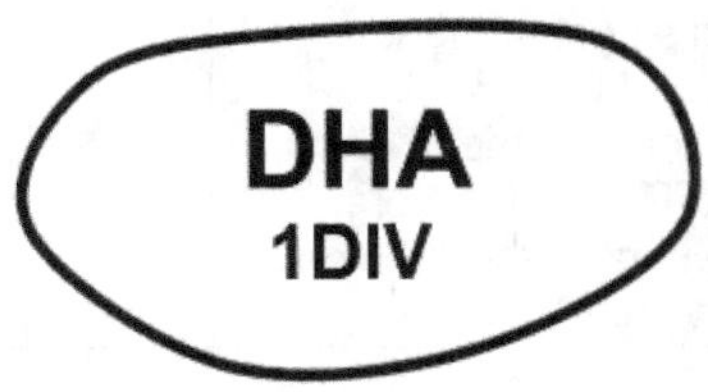 1st Division holding area

Table 5-4. Area control measures MAIN FUNCTION SYMBOLS (*continued*)

Function	*Main function symbol (Field A)*	*Construct Example*
	Holding Areas	
refugee holding area ***Note.*** The use of "refugee holding area" acronym is permitted to conserve space.	**REFUGEE HOLDING AREA**	 15th Military Police refugee holding area

	Support Area The portion of the commander's area of interest that is designated to facilitate the positioning, employment, and protection of base sustainment assets required to sustain, enable, and control operations. (ADP 3-0)	
forward arming and refueling point—A temporary facility, organized, equipped, and deployed, to provide fuel and ammunition necessary for the employment of aviation maneuver units in combat. (JP 3-09.3)	**FARP**	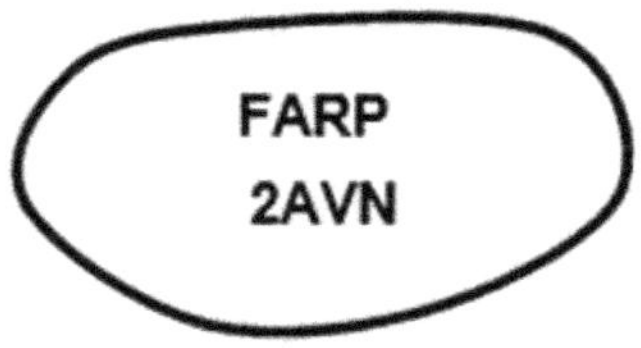 2nd Aviation forward arming and refueling point
brigade support area—A designated area in which sustainment elements locate to provide support to a brigade. (See ATP 4-90 for more information on brigade support area.)	**BSA**	 Brigade support area CAPA
division support area—A designated area within the division commander's area of operations that provides a location to base the division's sustainment assets and provide sustainment to the division. (See ATP 3-91 for more information on the division support area.)	**DSA**	 Division support area CRUZ

Table 5-4. Area control measures MAIN FUNCTION SYMBOLS (*continued*)

Function	*Main function symbol (Field A)*	*Construct Example*
	Support Area	
corps support area—Established by the corps headquarters. (See FM 3-0 for more information on the corps support area.)	**CSA**	 Corps support area MARIA
	Zone Areas	
drop zone—A specific area upon which airborne troops, equipment, or supplies are airdropped. (JP 3-36) (See FM 3-99 for more information on drop zone.)	**DZ**	 Drop zone HAWK
landing zone—Any specified zone used for the landing of aircraft. (JP 3-36) (See FM 3-99 for more information on landing zone)	**LZ**	 Landing zone SILVER
kill zone—The location where fires are concentrated in an ambush. (FM 3-90)	**Kill ZONE**	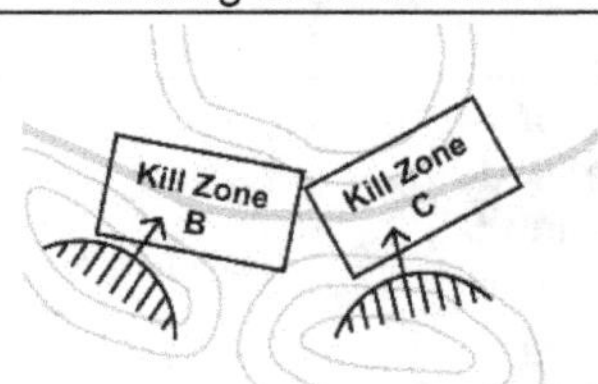
pickup zone—A geographic area used to pick up troops or equipment by helicopter. (See ATP 3-04.1 for more information on pickup zone.)	**PZ**	 Pickup zone WOLF

5-17. Table 5-5 on pages 160 through 163, lists functions, templates, and examples for battle positions and operations area control measures that have unique usage constructs. A battle position is a defensive location oriented on a likely enemy avenue of approach. (See ADP 3-90 for more information on battle positions.) These control measures must follow the design and construct guidelines of this table.

Table 5-5. Battle position and unique operation area templates with examples

Battle Position A defensive location oriented on a likely enemy avenue of approach. (ADP 3-90)		
Function	***Template***	***Example***
battle position	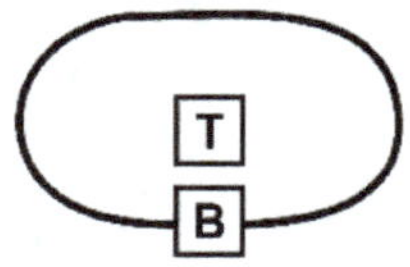	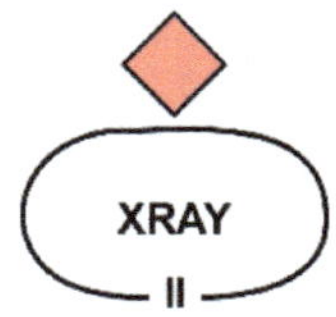 Battle position XRAY
battle position planned but not prepared	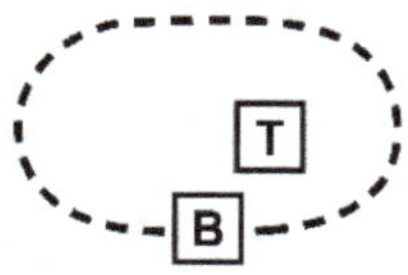	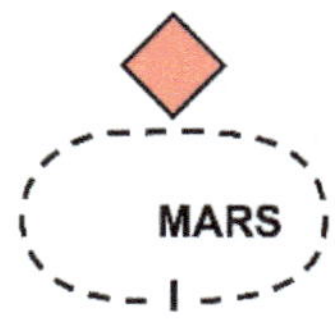 Company battle position MARS planned but not prepared
battle position prepared but not occupied	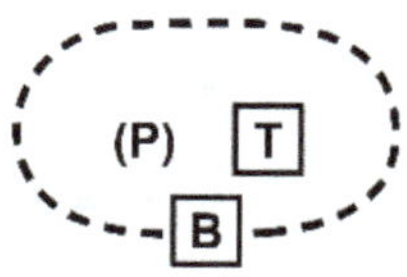	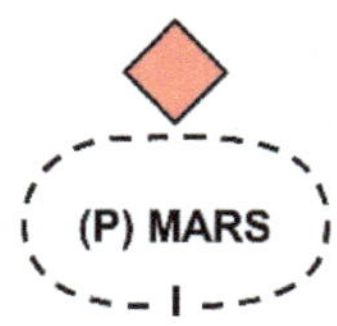 Company battle position MARS prepared but not occupied
strong point—A heavily fortified battle position tied to a natural or reinforcing obstacle to create an anchor for the defense or to deny the enemy decisive or key terrain. (ADP 3-90)	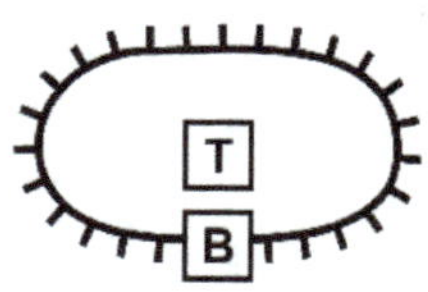	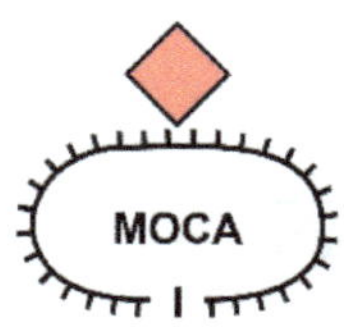 Company strong point MOCA
Unique Area Templates with Examples		
fortified area		 Fortified area TANGO

BODY-172

Table 5-5. Battle position and unique operation area templates with examples (*continued*)

Function	*Template*	*Example*
encirclement—Where one force loses its freedom of maneuver because an opposing force is able to isolate it by controlling all ground lines of communications and reinforcement. (FM 3-90)		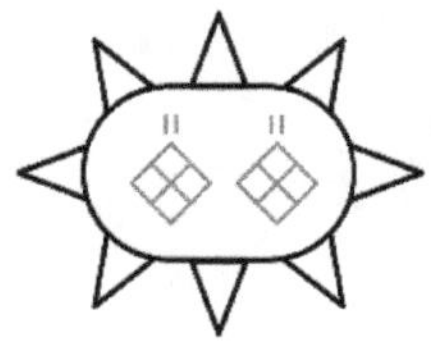 Friendly encirclement operations
		 Enemy encirclement operations
	Note. Numerous unit symbols can be included in the area.	
airhead line—A line denoting the limits of the objective area for an airborne assault. (JP 3-18)		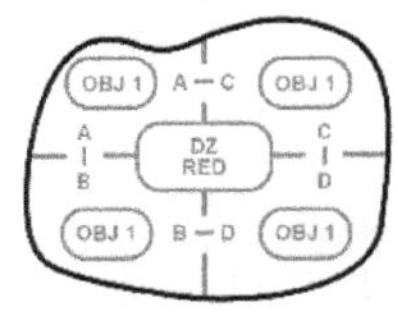

Table 5-5. Battle position and unique operation area templates with examples (*continued*)

Function	*Template*	*Example*
unexploded explosive ordnance (UXO) area—An area identified to have explosive ordnance that has been primed, fused, armed or otherwise prepared for action, and that has been fired, dropped, launched, projected, or placed in such a manner as to constitute a hazard to operations, installations, personnel, or material and remains unexploded either by malfunction or design or for any other cause. (See JP 3-42 for more information on UXO)	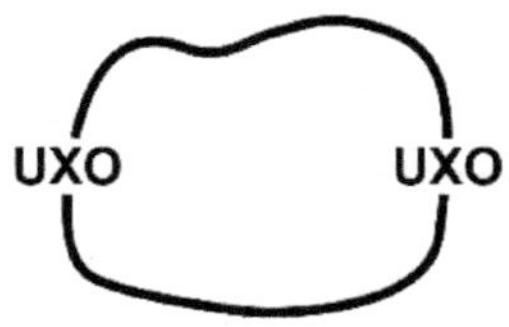	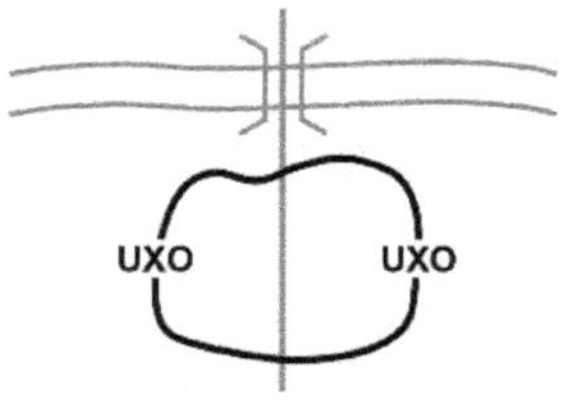
limited access area ***Note.*** The limited access area point symbol will be oriented upright, as shown in the template and example. The listed mobility sector 1 modifiers that can be used with this control measure.	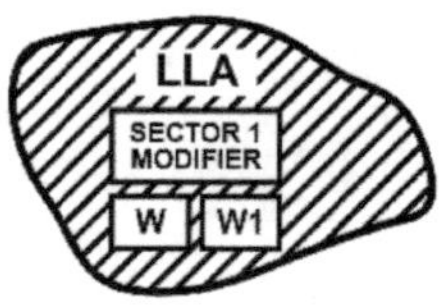	
	Mobility Sector 1 Modifiers	
	Wheeled (limited mobility)	
	Wheeled (high mobility)	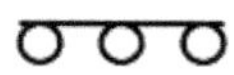
	Wheeled and tracked combination	
	Tracked	
	Towed	

Table 5-5. Battle position and unique operation area templates with examples (*continued*)

Function	*Template*	*Example*
limited access area (*continued*)	**Mobility Sector 1 Modifiers (*continued*)**	
	Railway	
	Over-snow (prime mover)	
	Sled	
	Pack animal	
	Barge	
	Amphibious	
	No vehicles	**ALL**

POINTS

5-18. Points are specified geographic positions used for identifying and regulating movement and maneuver, consolidation, de-confliction, distribution, and reducing hazards within an area of operation. They also can be special interest locations used to coordinate actions or locations of special interest to the mission.

5-19. **Point templates.** Figure 5-8 provides the templates used for point control measure Main function symbols listed in table 5-6 (pages 164 through 177) that include contact, coordination, decision, targets, sustainment, special supply distribution, and CBRN decontamination points. Supply points follow a slightly different format by using the unit sector 2 supply symbol in table 2-4 (on page 19) to modify the frame toward the bottom of the vertical rectangle and make it supply unique. Figure 5-8 demonstrates this supply unique modification that is only used for supply point control measures.

5-20. The point template includes amplifier fields (see table 5-2 on page 148) that can be used for additional information. As a minimum, a point shall include the function (Field A) information, and all other fields are optional information. Point symbols cannot be rotated and therefore text will not be written on an angle. Point symbol fields include—

- Field A (required) can use acronyms or graphics in to identify the point name or function.
- Field H (optional) used for additional information (content is implementation specific).
- Field N (optional) used to identify an enemy point using "ENY."
- Field T, T1 (optional) text amplifiers used to differentiate a symbol by numbering, lettering, or a combination of both, and/or unit designations.
- Field S2 (optional) used to indicate the offset or precise location of a single point symbol.
- Field W, W1 (optional) used to identify the date-time group associated with point.

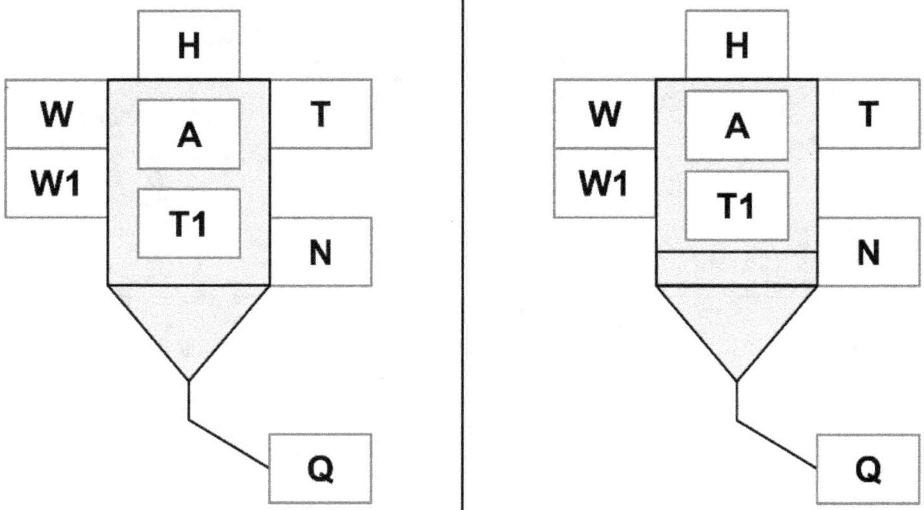

General purpose points | **Supply distribution points**

Point symbol fields

A (required) can use acronyms or graphics in to identify the point name or function.
H (optional) used for additional information (content is implementation specific).
N (optional) used to identify an enemy point using "ENY."
T, T1 (optional) for text amplifiers used to differentiate a symbol by numbering, lettering or a combination of both, and or unit designations.
S2 (optional) used to indicate the offset or precise location of a single point symbol.
W, W1 (optional) used to identify the date-time group associated with point.

Figure 5-8. Template for points (left) and supply distribution points (right) control measure symbols

5-21. **Point control measure.** Table 5-6 on page 164 through 176 lists point control measure Main function symbols that follow figure 5-8 template construct formats. A properly designated commander uses these points to exercise authority and direction over assigned and attached forces to the accomplishment of the mission.

Table 5-6. Point control measure Main function symbols

Function	*Main function symbol (Field A)*	*Construct Example and Symbol Translation*
	Movement and Maneuver	
checkpoint—A predetermined point on the ground used to control movement, tactical maneuver, and orientation. (See FM 3-90 for more information on checkpoints.)	**CP**	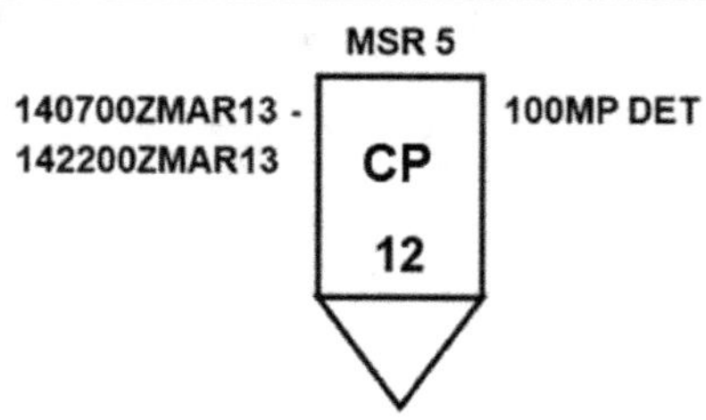 Checkpoint 12 on main supply route 5, controlled by 100th Military Police Detachment, operational from 0700 Zulu hour 14 MAR 2013 to 2200 Zulu hour 14 March 2013

Table 5-6. Point control measure Main function symbols (*continued*)

Function	*Main function symbol (Field A)*	*Construct Example and Symbol Translation*
	Movement and Maneuver	
checkpoint (*continued*)	**CKP** Alternate MIL-STD 2525E and North Atlantic Treaty Organization (NATO) APP 6(E) symbol usage	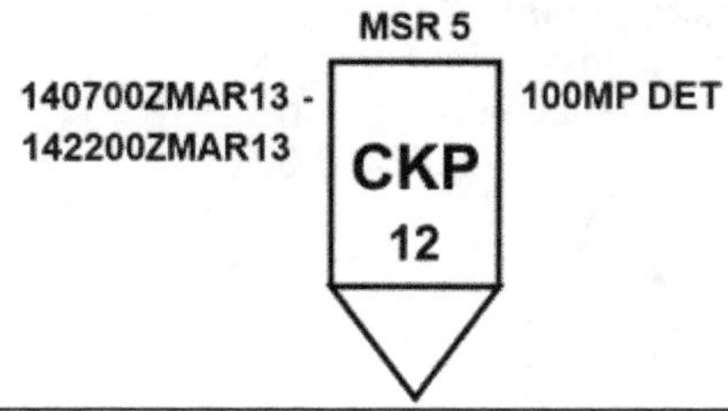
control—An action taken to eliminate a hazard or reduce its risk. (ATP 5-19) ***Note.*** This symbol demonstrates the capability and authority to exercise restraining or directing influence (regulating) over a specific function.	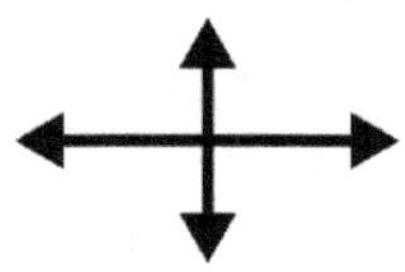	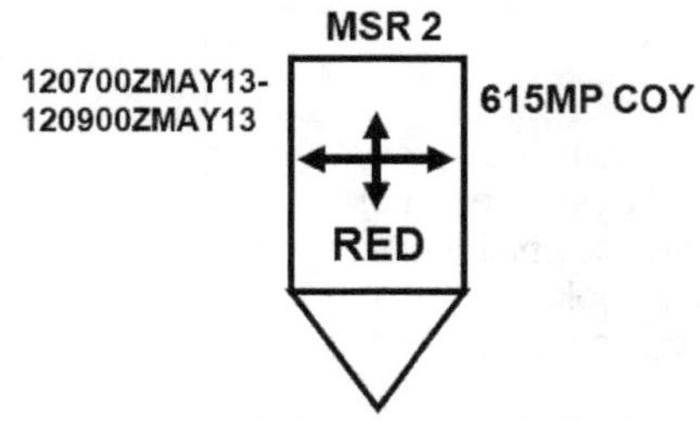 Control point RED on main supply route 2, controlled by 615th Military Police Company, operational from 0700 Zulu hour 12 May 2013 to 0900 Zulu hour 12 May 2013
engineer regulating point—A location used to ensure that vehicles do not exceed the capacity of the crossing means and to give drivers final instructions on site-specific procedures and information. Also called ERP. (ATP 3-90.4)	**ERP**	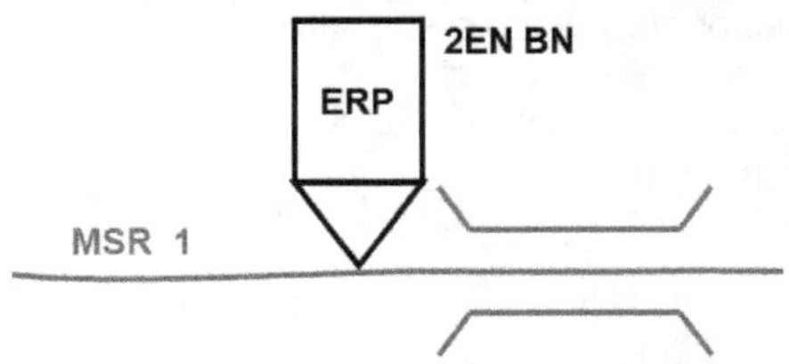 Engineer regulating point on main supply route 1, controlled by 2nd Engineer Battalion
linkup point—A designated place where two forces are scheduled to meet. (FM 3-90)	**LU**	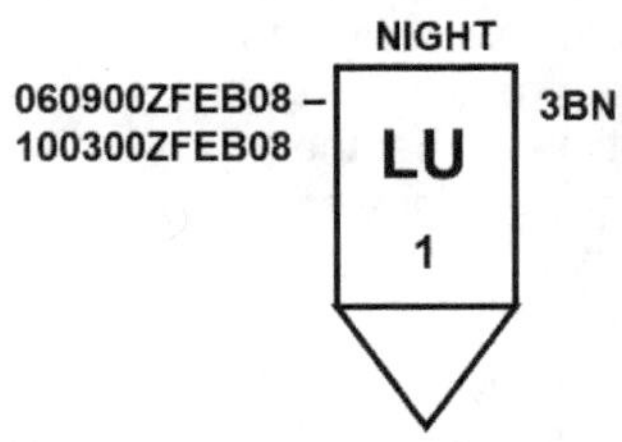 Night linkup point 1, controlled by 3rd Battalion, operational from 0900 Zulu hour 6 FEB 2008 to 0300 Zulu hour 10 FEB 2008

Table 5-6. Point control measure Main function symbols (*continued*)

Function	*Main function symbol (Field A)*	*Construct Example and Symbol Translation*
	Movement and Maneuver	
passage point—A designated place where passing units pass through the stationary unit. (FM 3-90)	**PP**	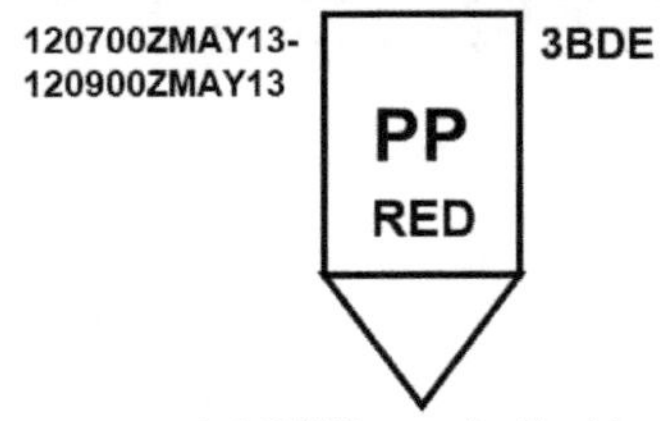 Passage point RED, controlled by 3rd Brigade, operational from 0700 Zulu hour 12 May 2013 to 0900 Zulu hour 12 May 2013
rally point—An easily identifiable point on the ground at which units can reassemble and reorganize if they become dispersed. (FM 3-90)	**RLY**	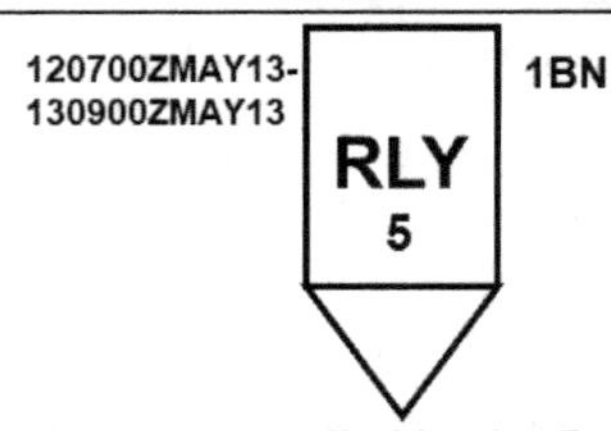 Rally point 5, controlled by 1st Battalion, operational from 0700 Zulu hour 12 MAY 2013 to 0900 Zulu hour 13 May 2013
release point—A designated place on a route where elements are released from centralized control. (FM 3-90)	**RP**	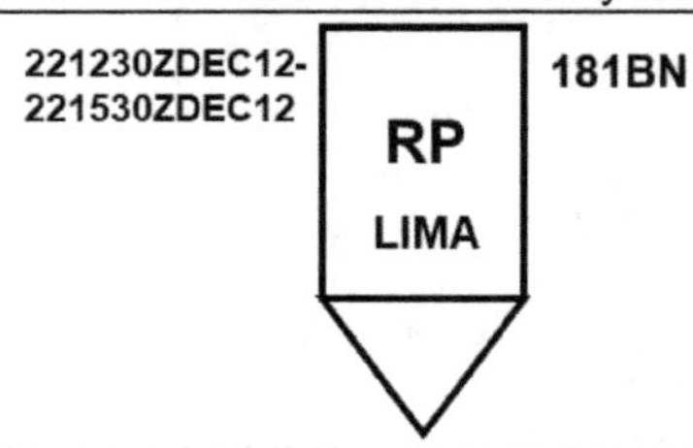 Release point LIMA, controlled by 181st Battalion, operational from 1230 Zulu hour 22 DEC 2012 to 1530 Zulu hour 22 DEC 2012
start point—A designated place on a route where elements fall under the control of a designated march commander. (FM 3-90)	**SP**	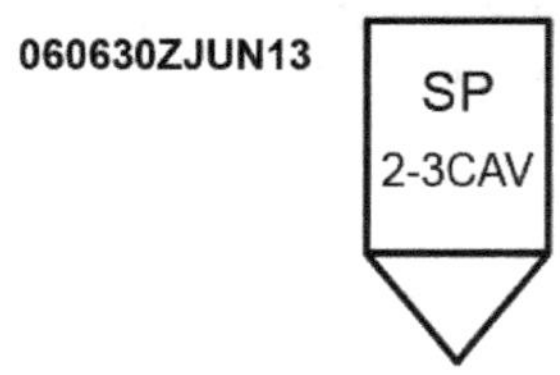 Start point, controlled by 2nd Troop, 3rd Cavalry Squadron, operational at 0630 Zulu hour 6 JUN 2013

Table 5-6. Point control measure Main function symbols (*continued*)

Function	*Main function symbol (Field A)*	*Construct Example and Symbol Translation*
	Movement and Maneuver	
point of departure—The point where the unit crosses the line of departure and begins moving along a direction of attack. (ADP 3-90)	**PD**	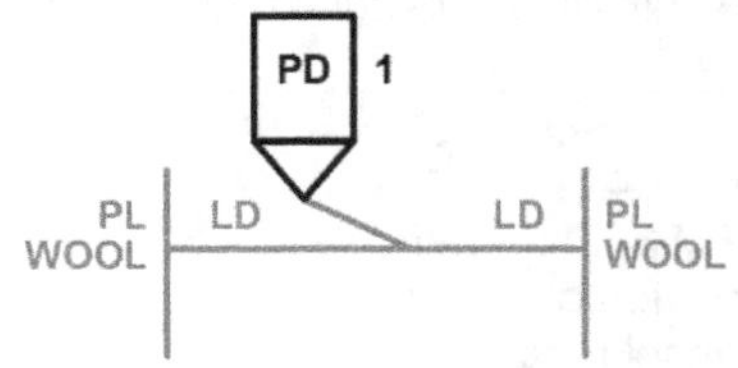 Point of departure 1. ***Note.*** The offset indicator is used in the example to allow the viewer to better see the line of departure. It is not required.
	Consolidation	
amnesty point—A location where individuals can return ammunition and explosives inadvertently kept, found, or stolen without the fear of prosecution. (See DA PAM 385-64 for more information)	**AMN**	 United Nations weapons amnesty point, controlled by New Zealand, operational from 0700 Zulu hour 14 MAR 2013 to 0700 Zulu hour 12 MAY 2013
civilian collection point—A specific location where civilians are assembled to be transported to another location.	**CIV**	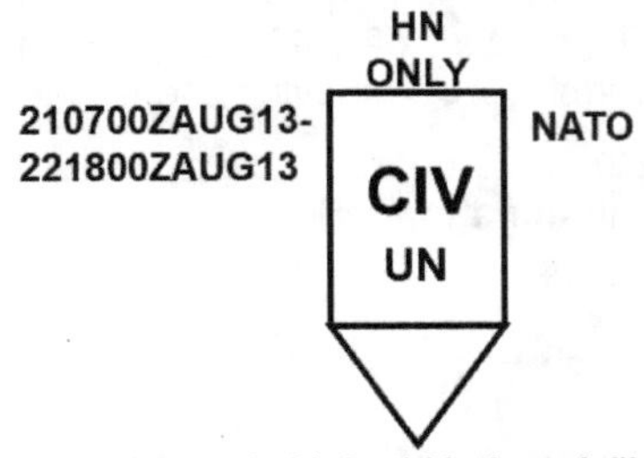 Host nation only United Nation civilian collection point, controlled by North Atlantic Treaty Organization, operational from 0700 Zulu hour 21 AUG 2013 to 0800 Zulu hour 22 AUG 2013
detainee collection point—A specific location where detainees are assembled until transportation becomes available for relocation to detainee holding area or theater detention facility. (See JP 3-31 and FM 3-63 for more information on detainee collection point.)	**DET**	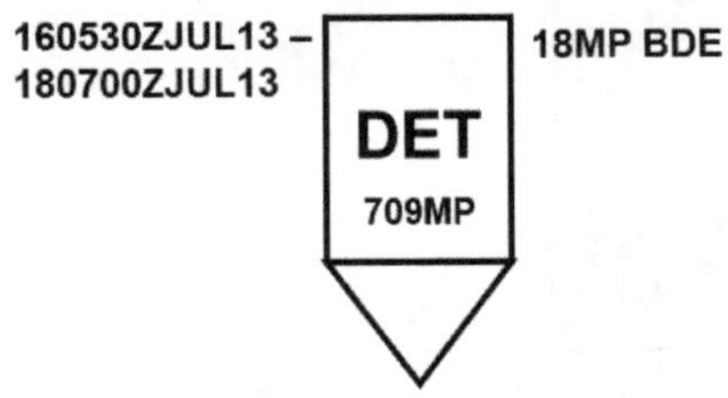 Detainee collection point, controlled by 709th Military Police of the 18 Military Police Brigade, operational from 0530 Zulu hour 16 JUL 2013 to 0700 Zulu hour 18 JUL 2013

Table 5-6. Point control measure Main function symbols (*continued*)

Function	*Main function symbol (Field A)*	*Construct Example and Symbol Translation*
Personnel Recovery		
The sum of military, diplomatic, and civil efforts to prepare for and execute the recovery and reintegration of isolated personnel. Also called PR. (JP 3-50)		
isolated personnel recovery—(See JP 3-50 and FM 3-50 for more information on isolated personnel recovery.)		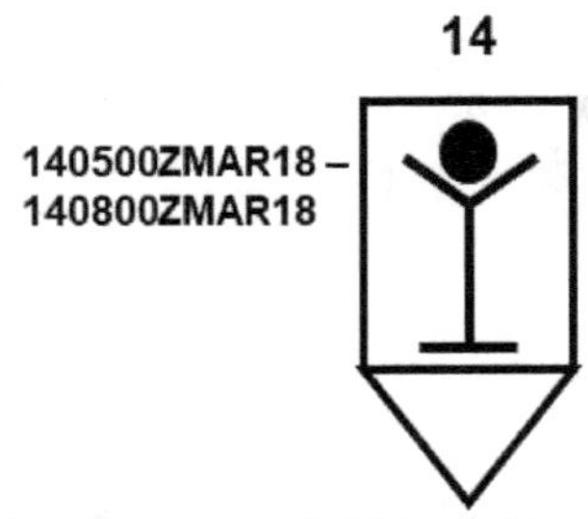 Isolated personnel (14 individuals) recovery point between 0500 Zulu hour 14 MAR 2018 and 0800 Zulu hour 14 MAR 2018
search and rescue point—A predesignated specific location, relative to which isolated personnel provide their position to recovery forces. (JP 3-50)	**SAR**	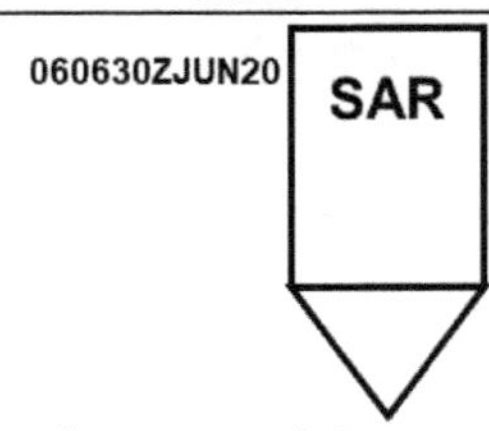 Search and rescue point, operational at 0630 Zulu hour 6 JUN 2020
Decontamination		
The process of making any person, object, or area safe by absorbing, destroying, neutralizing, making harmless, or removing chemical or biological agents, or by removing radioactive material clinging to or around it. (JP 3-11)		
decontamination point	**DCN**	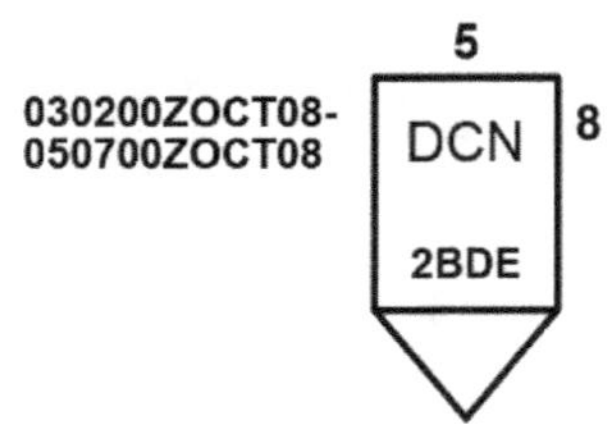
alternate decontamination point	**DCN ALT**	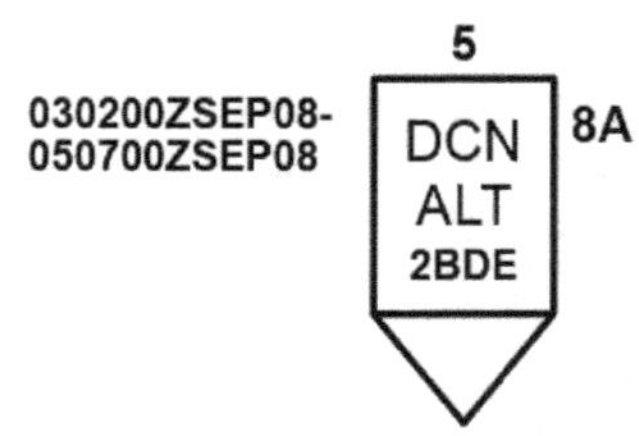

Table 5-6. Point control measure Main function symbols (*continued*)

Function	*Main function symbol (Field A)*	*Construct Example and Symbol Translation*
Decontamination		
equipment decontamination point	**DCN E**	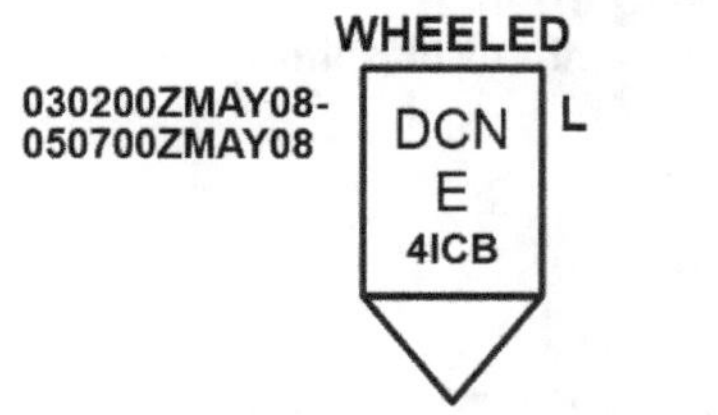
troop decontamination point	**DCN T**	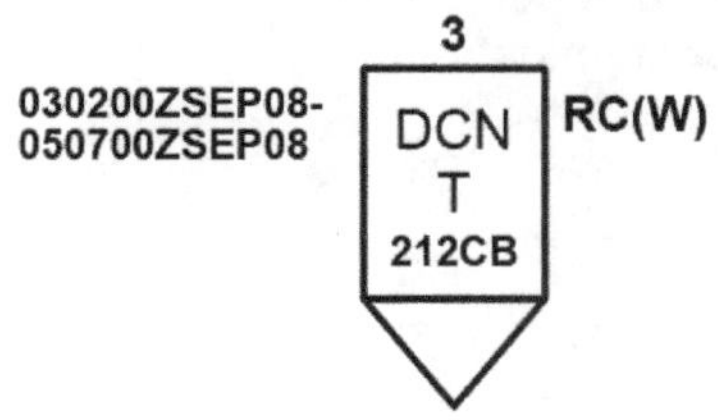
equipment or troop decontamination point	**DCN E/T**	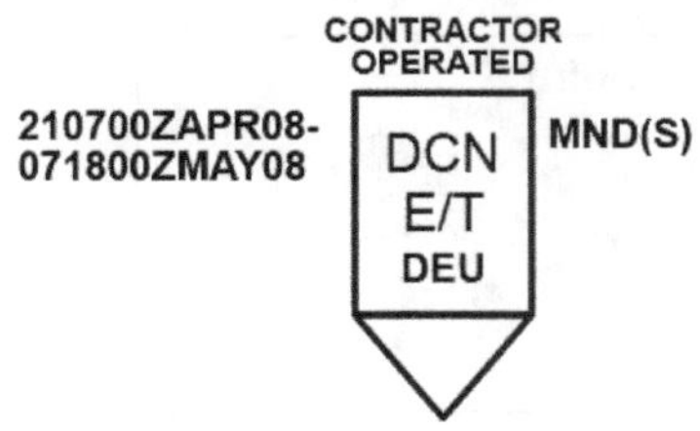
operational decontamination point	**DCN O**	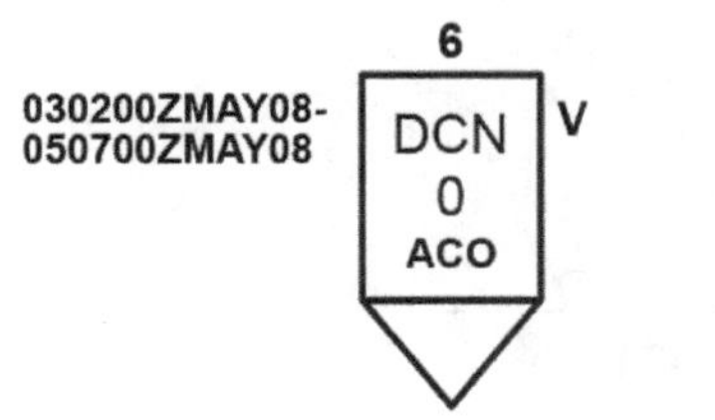
thorough decontamination point	**DCN TH**	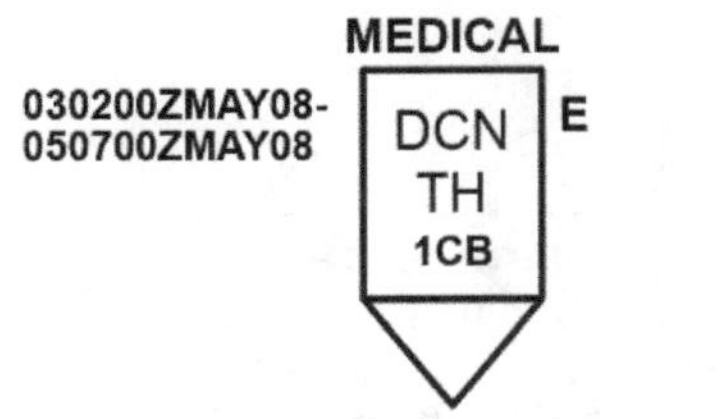
main equipment decontamination point	**DCN (M) E**	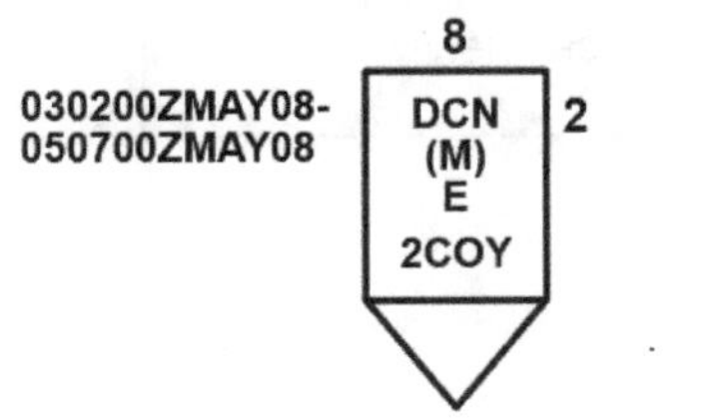

Table 5-6. Point control measure Main function symbols (*continued*)

Function	*Main function symbol (Field A)*	*Construct Example and Symbol Translation*
Decontamination		
forward troop decontamination point	**DCN (F) T**	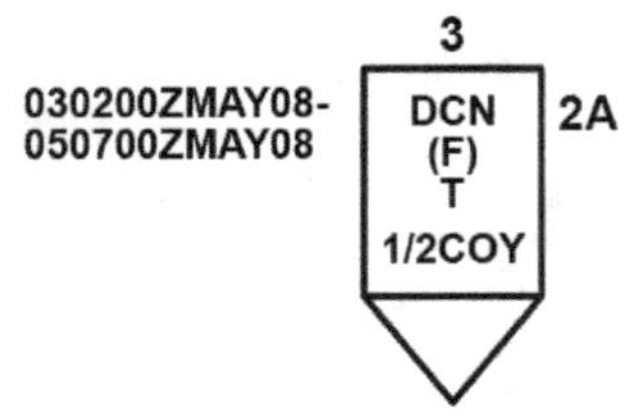
wounded personnel decontamination point	**DCN W**	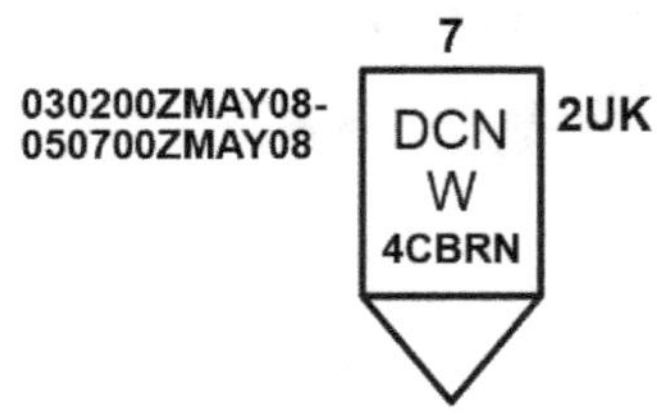
Field Artillery		
firing point	**FP**	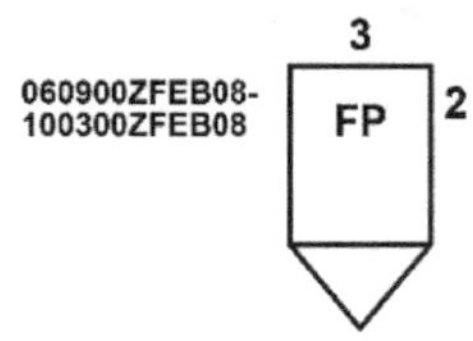
launch point	**LP**	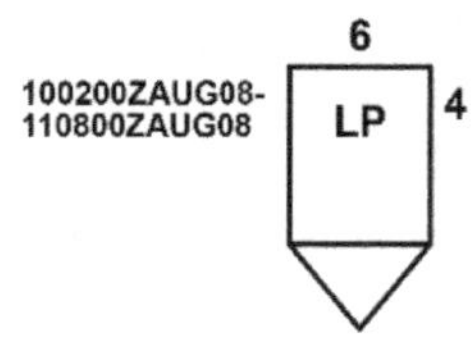
reload point	**RLP**	
survey control point—Initializes and updates data to ensure accuracy levels are maintained. (See ATP 3-09.02 for more information on logistics release point usage.)	**SCP**	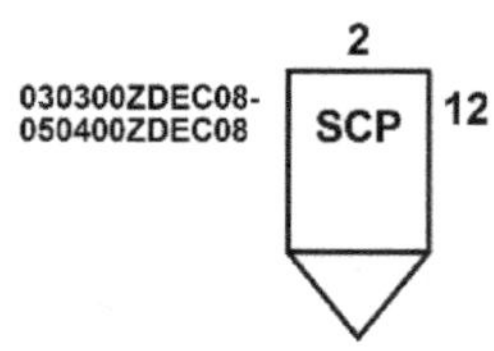

Table 5-6. Point control measure Main function symbols (*continued*)

Function	*Main function symbol (Field A)*	*Construct Example and Symbol Translation*
	Sustainment	
ambulance exchange point—A location where a patient is transferred from one ambulance to another en route to a medical treatment facility. (ATP 4-02.2)	**AXP**	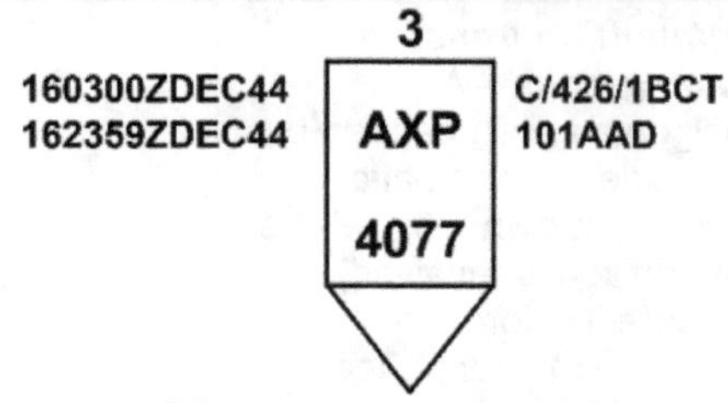
ambulance control point—A manned traffic regulating, often stationed at a crossroad or road junction, where ambulances are directed to one of two or more directions to reach loading points and medical treatment facilities. (ATP 4-02.2)	**ACP**	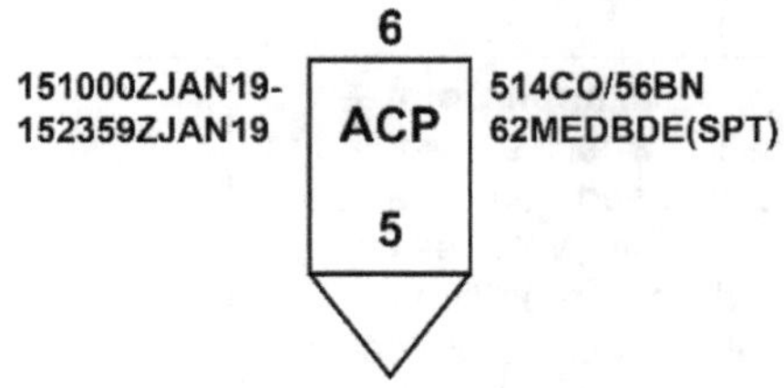
ambulance loading point—This is the point in the shuttle system where one or more ambulances are stationed ready to receive patients for evacuation. (ATP 4-02.2)	**ALP**	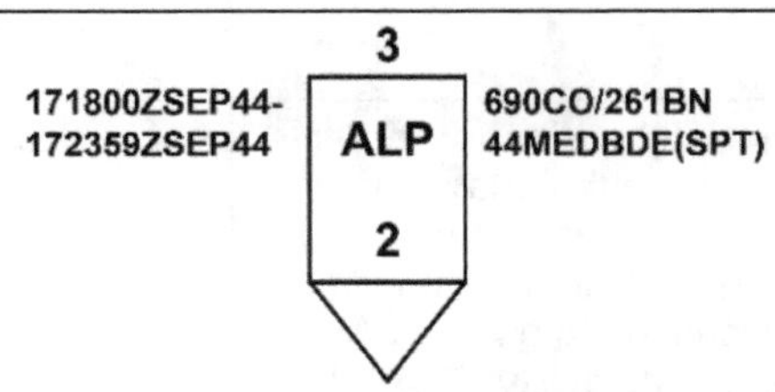
ambulance relay point—A point in the shuttle system where one or more empty ambulances are stationed to advance to a loading point or to the next relay post to replace departed ambulances. (ATP 4-02.2)	**ARP**	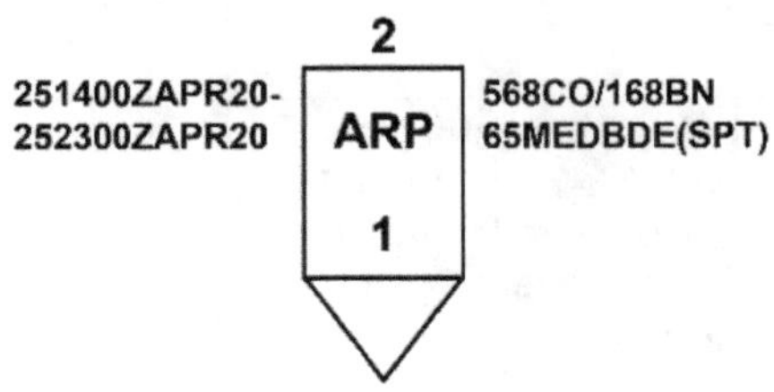
ammunition supply point—A retail ammunition support activity where ammunition is received, stored, issued, and accounted for. (ATP 4-35)	**ASP**	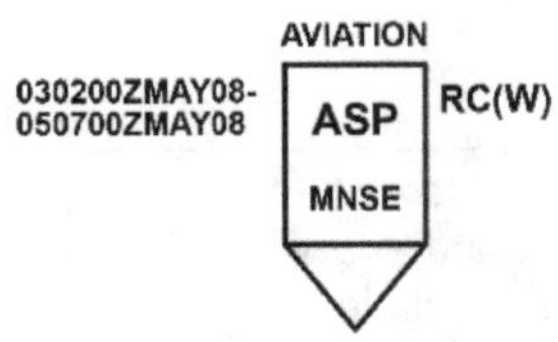

Table 5-6. Point control measure Main function symbols (*continued*)

Function	*Main function symbol (Field A)*	*Construct Example and Symbol Translation*
	Sustainment	
ammunition transfer holding point—A designated site operated by a brigade support battalion distribution company, where ammunition is received, temporarily stored, or transferred to supported units within a functional or multifunctional brigade. (ATP 4-35)	**ATHP**	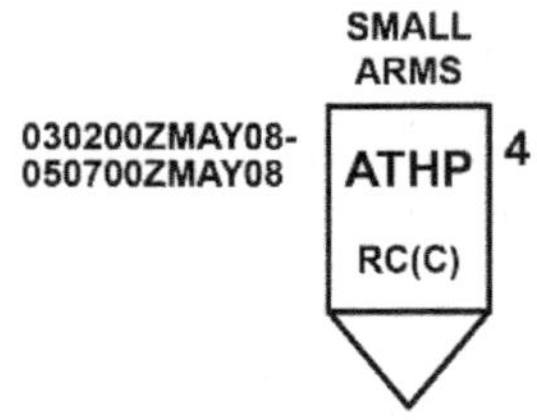
cannibalization point—A location where salvage parts are obtained from disabled machines or vehicles to use in building or repairing another machine or vehicle.	**CAN**	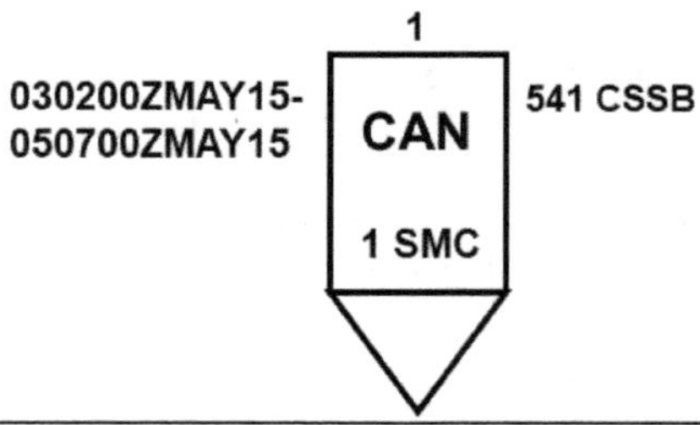
casualty collection point—A location that may or may not be staffed, where casualties are assembled for evacuation to a medical treatment facility. (ATP 4-02.2)	**CCP**	
logistics release point—(See ATP 3-21.11 and ATP 3-21.8 for more information on logistics release point usage.)	**LRP**	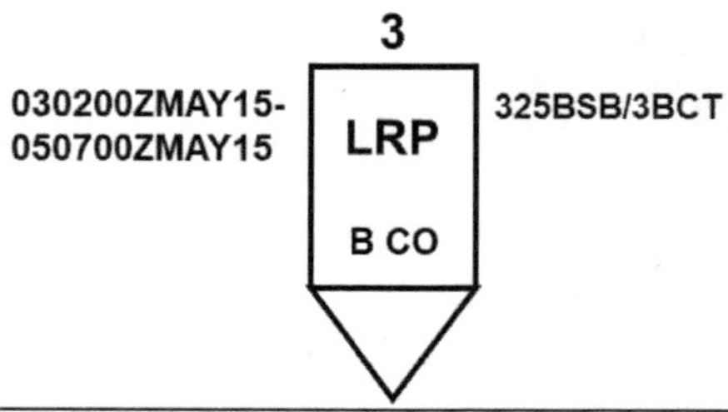
maintenance collection point—A temporary location established within the battalion echelon for the collection of equipment needing or undergoing field maintenance. (ATP 4-33)	**MCP**	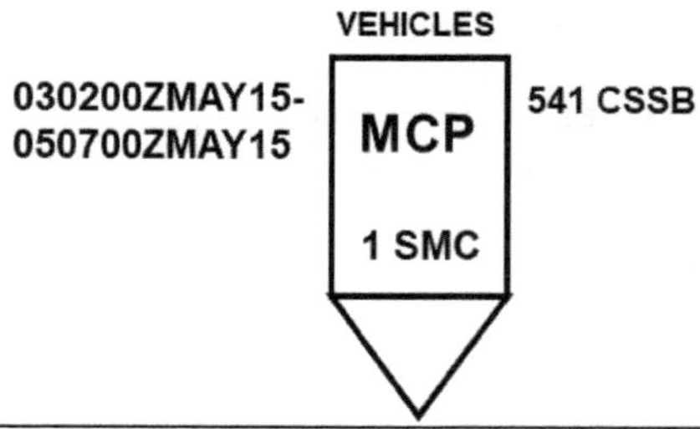

Table 5-6. Point control measure Main function symbols (*continued*)

Function	*Main function symbol (Field A)*	*Construct Example and Symbol Translation*
	Sustainment	
medical evacuation pickup point—A location staffed with medical personnel, where casualties are assembled for evacuation to a medical treatment facility on dedicated and properly marked medical platforms with en route care provided by medical personnel.		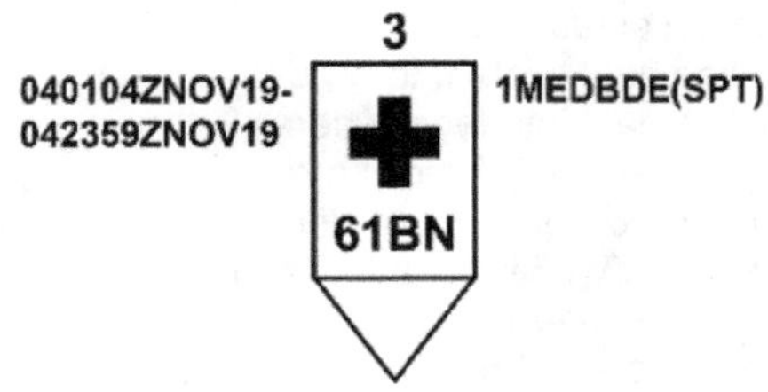
mortuary affairs collection point—The location for receiving, establishing chain of custody, and evacuating human remains and personal effects. (See ATP 4-46 for more information on mortuary affairs collection points.)	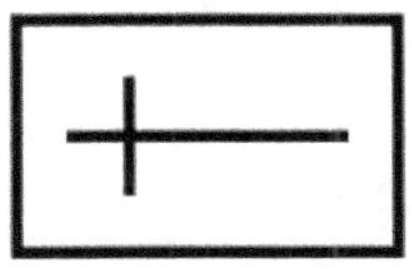	
rearm, refuel, and resupply point—A designated location through which a unit passes where it receives fuel, ammunition, and other necessary supplies to continue operations.	**R3P**	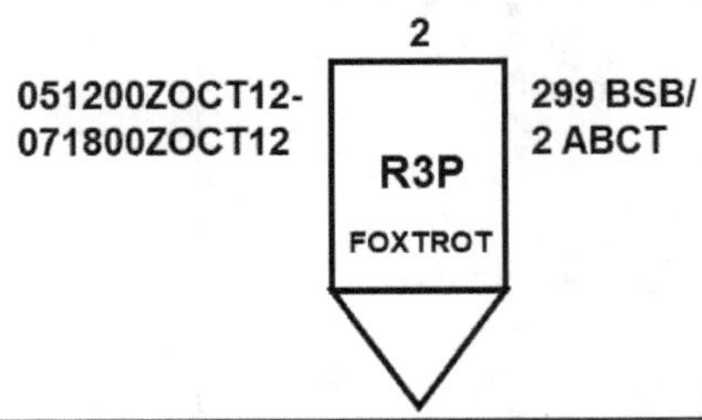
refuel on the move point—A location established to ensure that fuel tanks on combat and fuel servicing vehicles are full before they arrive in the unit's tactical assembly area.	**ROM**	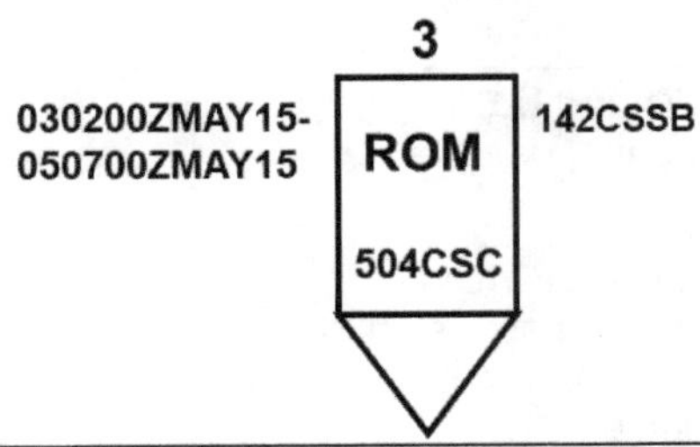
traffic control post—A manned post that is used to preclude the interruption of traffic flow or movement along a designated route. (FM 3-39)	**TCP**	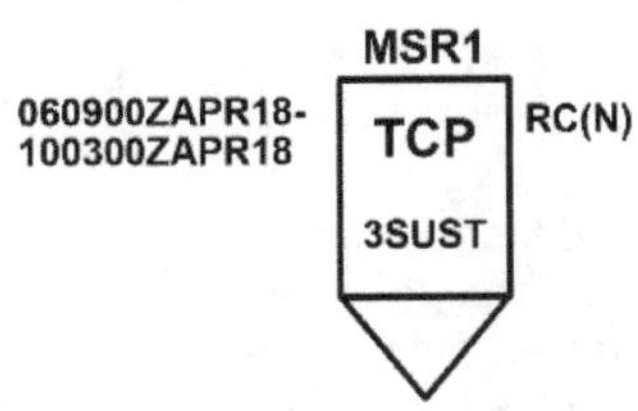

Table 5-6. Point control measure Main function symbols (*continued*)

Function	*Main function symbol (Field A)*	*Construct Example and Symbol Translation*
	Sustainment	
trailer transfer point—A location established along the line haul system to divide the line haul into legs, where semitrailers or flat racks are transferred from one carrier to another while en route. (See ATP 4-11 for more information on trailer transfer points.)	**TTP**	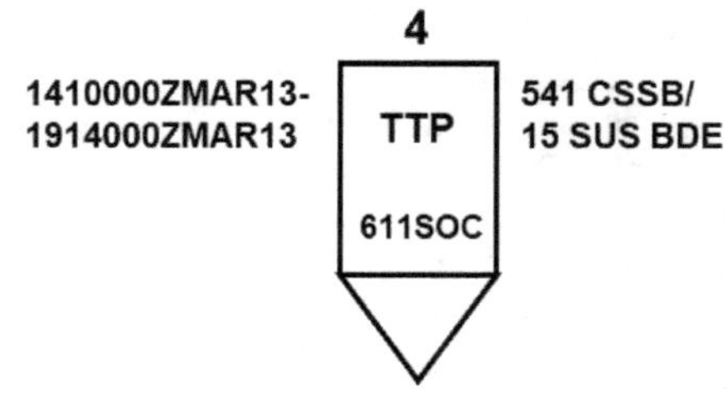
	Sustainment Distribution	
North Atlantic Treaty Organization (NATO) **Class I** Those items which are consumed by personnel or animals at the approximately uniform rate, irrespective of local changes in combat or terrain conditions.	**I**	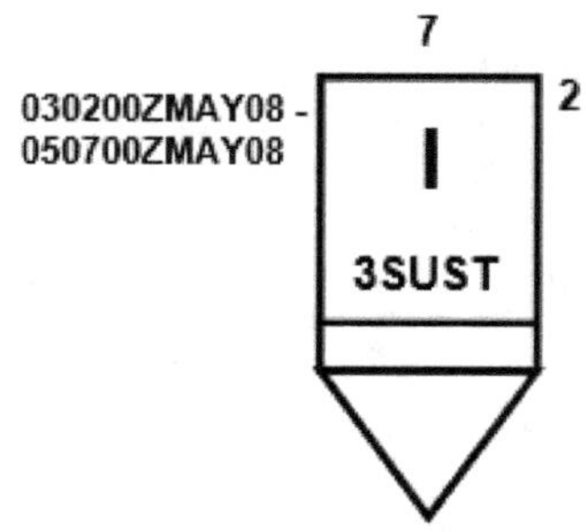
U.S. **Class I** (subsistence items)		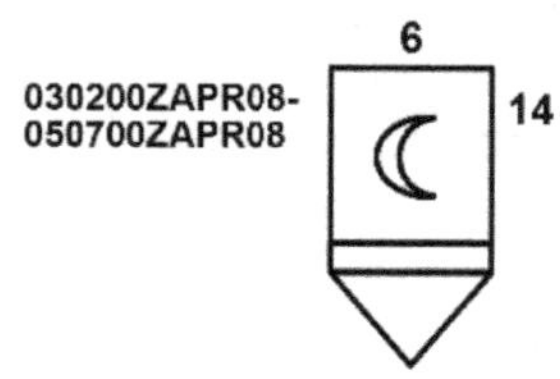
NATO and U.S. **Class II** Supplies for which allowances are established by tables of organization and equipment.	**I I**	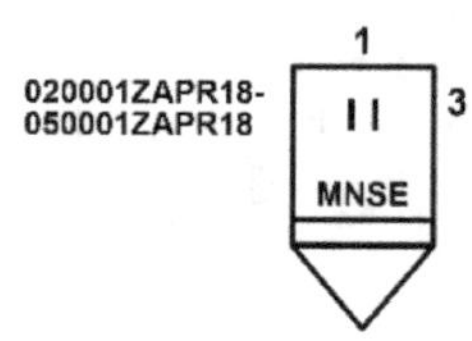
NATO **Class III** Fuels and lubricants for all purposes, except for operating aircraft or for use in weapons such as flame throwers.	**I I I**	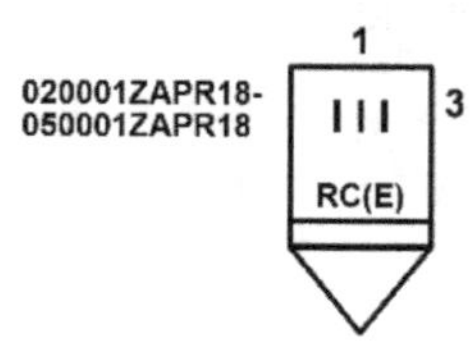

Table 5-6. Point control measure Main function symbols (*continued*)

Function	*Main function symbol (Field A)*	*Construct Example and Symbol Translation*
Sustainment Distribution		
U.S. **Class III** (petroleum, oils, and lubricants)		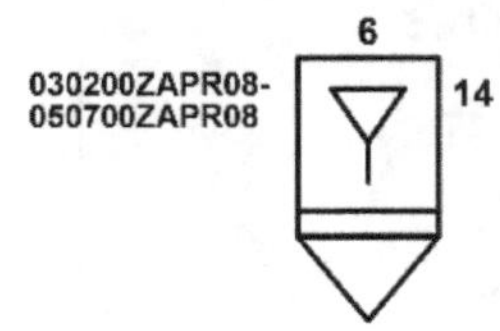
NATO **Class IV** Supplies for which initial issue allowances are not prescribed by approved issue tables.	**I V**	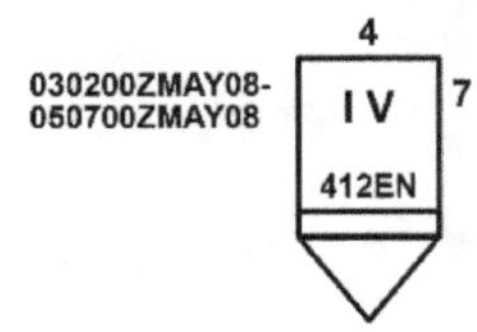
U.S. **Class IV** (construction and barrier materials)		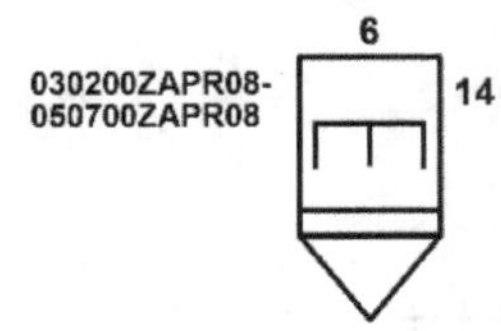
NATO and U.S. **Class V** Ammunition, explosives, and chemical agents of all types.		
U.S. **Class VI** (personal demand items)		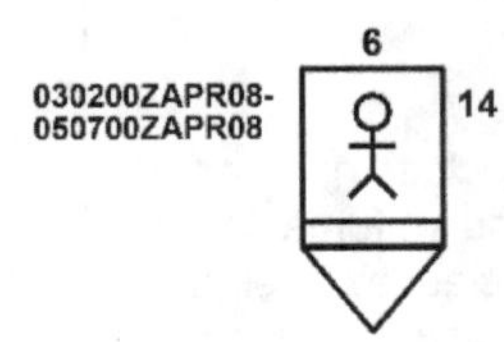
U.S. **Class VII** (major end items)		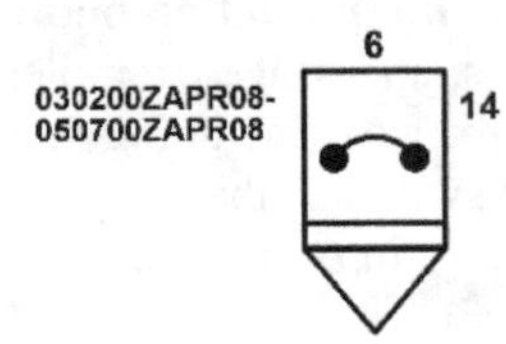

Table 5-6. Point control measure Main function symbols (*continued*)

Function	*Main function symbol (Field A)*	*Construct Example and Symbol Translation*
Sustainment Distribution		
NATO and U.S. **Class VIII** (medical)	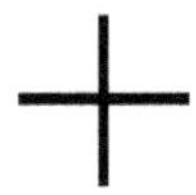	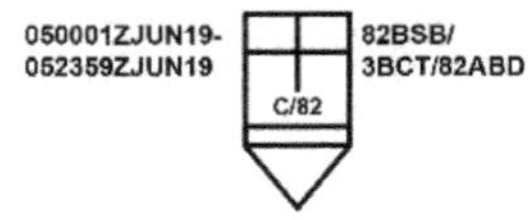 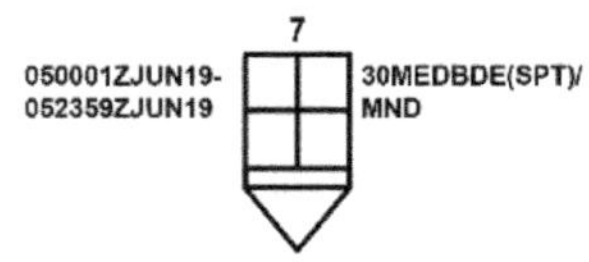
U.S. **Class IX** (repair parts)		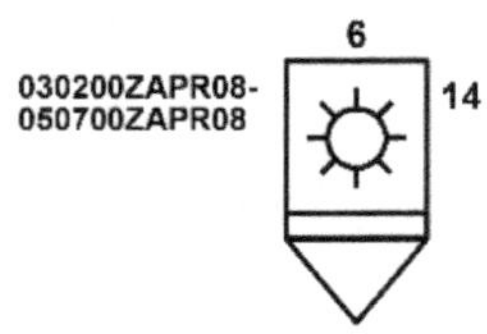
U.S. **Class X** (material to support nonmilitary programs)	**CA**	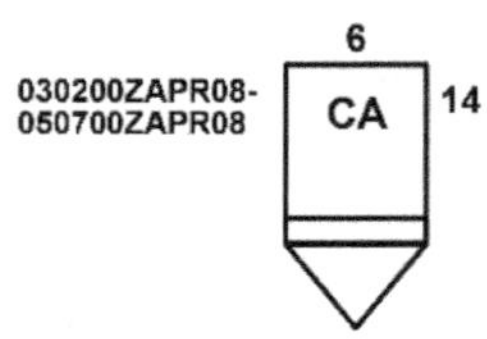
NATO multiple supply class point Use supply class numbers (I, II, III, IV, and V) for A field or ALL for all classes of supply.	**I/III/V**	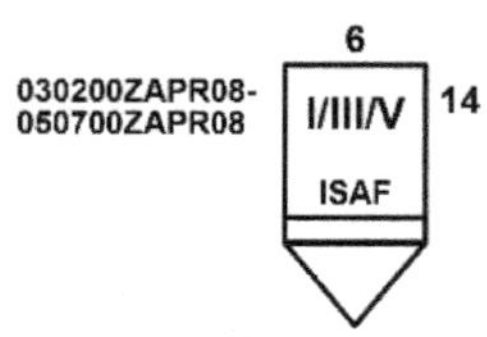

5-22. **Distinctive action point.** Table 5-7 on pages 177 through 178, lists distinctive action point functions, templates, and usage examples that must follow the guidelines of this table. Distinctive action points have unique formats (including square, circular, star, and cross) that do not follow a specific template and are unique symbols that represent specific actions (including contact, coordination, decision, targets, communication, and air).

Table 5-7. Distinctive action point control measure functions, templates, and examples

Function	***Template***	***Construct Example and Symbol Translation***
contact point—In land warfare, a point on the terrain, easily identifiable, where two or more units are required to make contact. (JP 3-50)	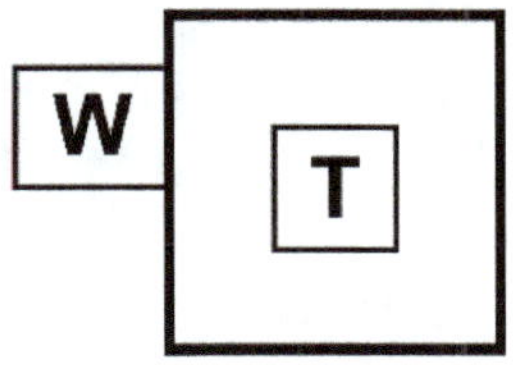	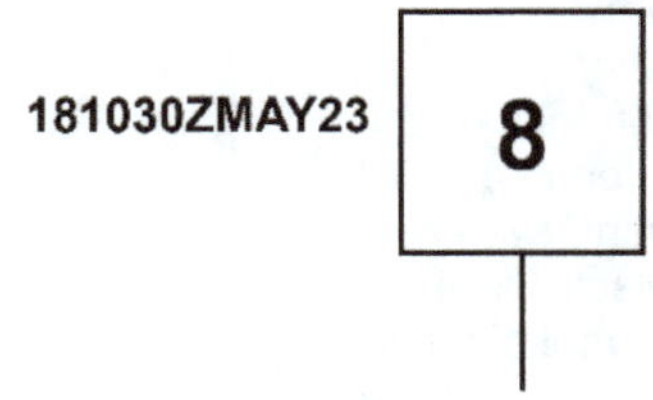 Contact point 8 at 1030 Zulu hour 18 MAY 2023
coordination point—A point that indicates a specific location for the coordination of tactical actions between adjacent units. (FM 3-90)		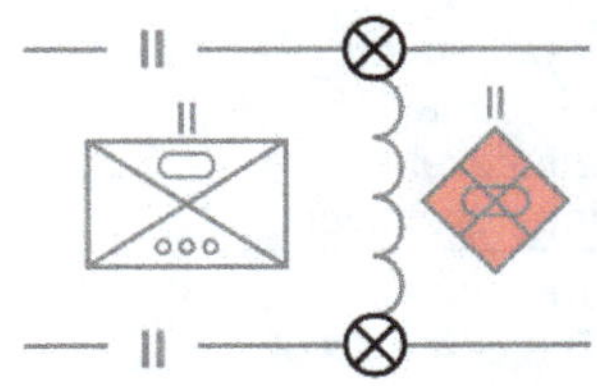
decision point—A point in space and time when the commander or staff anticipates making a key decision concerning a specific course of action. (JP 5-0)		 Decision point 3
isolated personnel initial location (See JP 3-50 and FM 3-50 for more information on isolated personnel initial locations.)	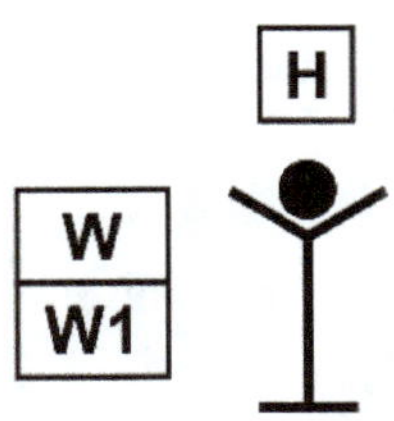	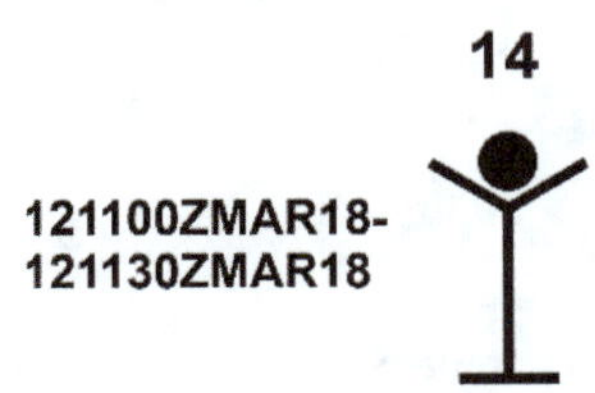 Isolated personnel (14 individuals) initial location between 1100 Zulu hour 12 MAR 2018 and 1130 Zulu hour 12 MAR 2018
key terrain—An identifiable characteristic whose seizure or retention affords a marked advantage to either combatant. (ADP 3-90)		 Key terrain 7

Table 5-7. Distinctive action point control measure functions, templates, and examples (*continued*)

Function	*Template*	*Construct Example and Symbol Translation*
waypoint—A designated point or series of points loaded and stored in a global positioning system or other electronic navigational aid system to facilitate movement.	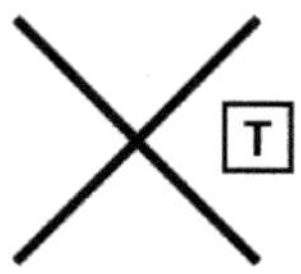	 Waypoint 8
target handover—Depicts a direction fire target handover mission is in progress and facilitates orientation of direct fires to maximize combat effectiveness and minimize fratricide.	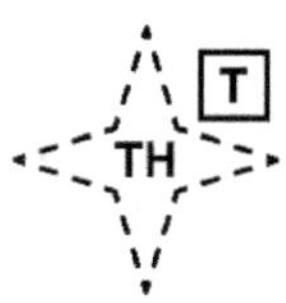	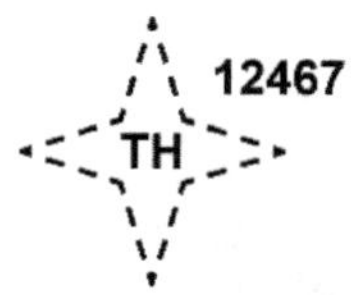 Target handover 12467
harbor entrance point—The entrance of a part of a body of water protected and deep enough to furnish anchorage.		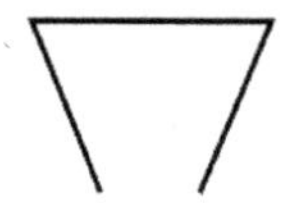 Without Text 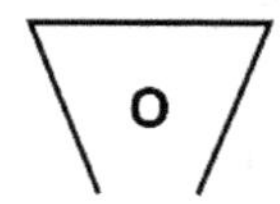 With Text

LINES

5-23. In table 5-8, on pages 179 through 185, there are line control measure symbols that follow a specific template format as shown in figure 5-9.

- Most lines are also names as a phase line for ease of reference in orders and during transmissions.
- Lines that have a specific purpose and are also named as phase lines (such as a restrictive fire line) should have the primary purpose in the Field T1 labeled on top of the line at both ends of the line inside the lateral boundary or as often as necessary for clarity.
- The designation of the controlling headquarters for fire support coordination measures is depicted in Field T2.
- The use of phase lines to mark line control measure symbols is not mandatory.

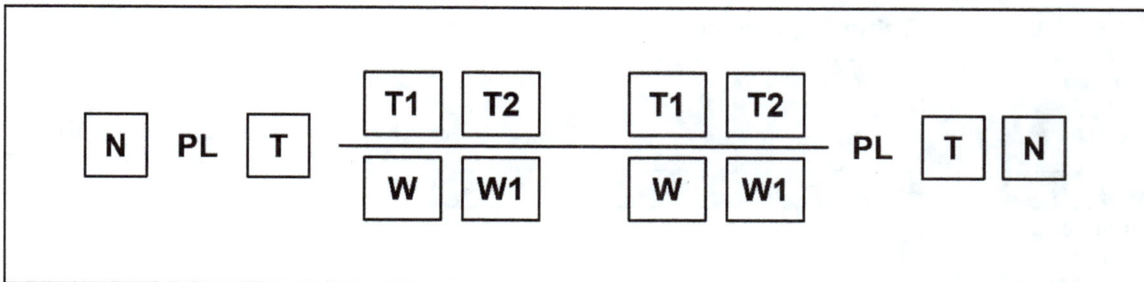

N	Field identifies hostile (enemy).
T	Text amplifier used to differentiate a symbol by numbering, lettering, or a combination of both.
T1	Text amplifier used to identify primary purpose (function).
T2	Text amplifier used to identify the controlling headquarters.
W-W1	Text amplifier for date and time group.

Figure 5-9. Line control measure template

Table 5-8. Line control measure symbols

Function	***Template***	***Construct Example*** **Note.** Symbols colored gray are used to help explain how the control measure is used and are not a part of the control measure.
	Lines	
forward line of own troops—A line which indicates the most forward position of friendly forces in any kind of military operation at a specific time. (FM 3-90) ***Note.*** The open side of the arc reflects the reported unit.		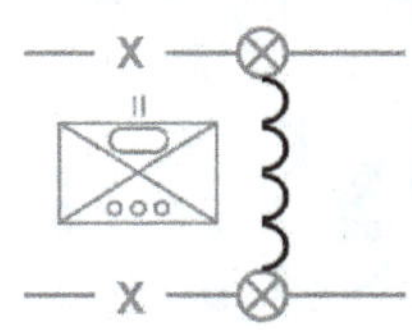 Friendly present
		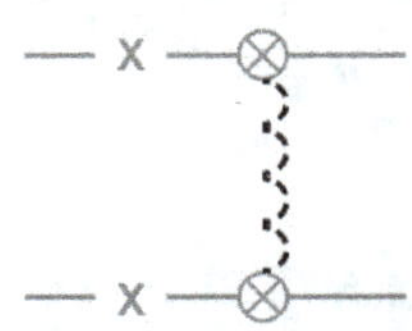 Friendly planned
		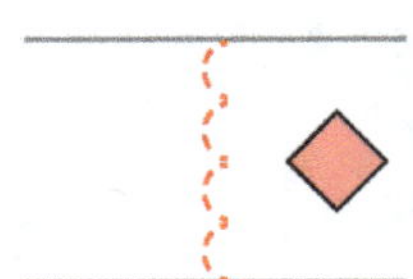 Enemy suspected

Table 5-8. Line control measure symbols (*continued*)

Function	*Template*	*Construct Example* ***Note.*** Symbols colored gray are used to help explain how the control measure is used and are not a part of the control measure.
	Lines	
line of contact—A general trace delineating the location where friendly and enemy forces are engaged. (FM 3-90) The line of contact symbol is created when both the friendly and enemy forward line of troops symbols are displayed. ***Note.*** The open side of the art reflects the reported unit.		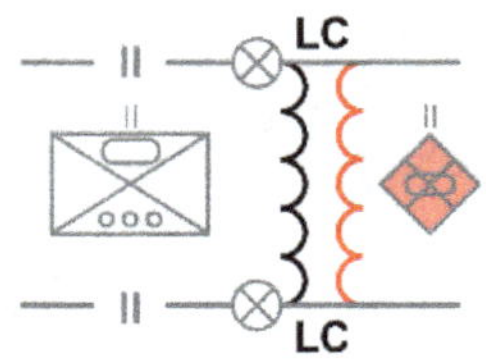
phase line—An easily identified feature in the operational area utilized for control and coordination of military operations. (JP 3-09)		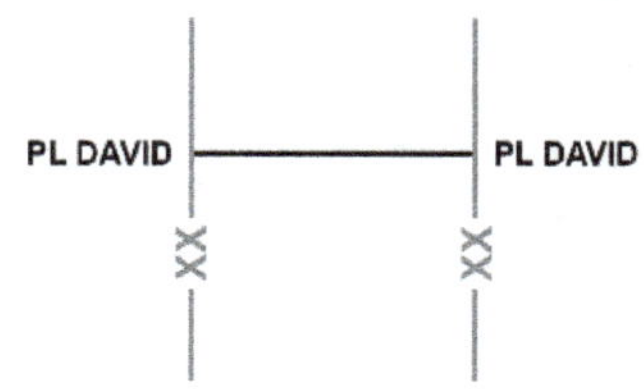
forward edge of the battle area—The foremost limits of a series of areas in which ground combat units are deployed to coordinate fire support, the positioning of forces, or the maneuver of units, excluding areas in which covering or screening forces are operating. (JP 3-09.3)		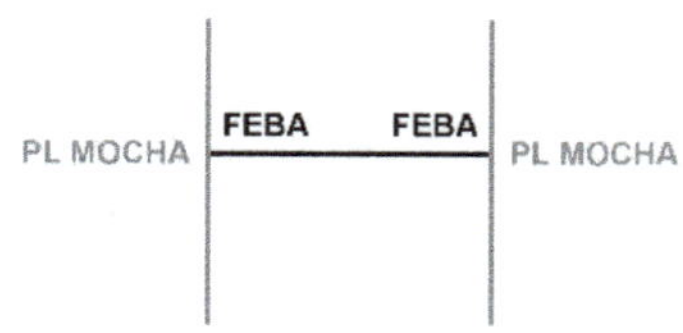

Table 5-8. Line control measure symbols (*continued*)

Function	*Template*	*Construct Example* ***Note.*** Symbols colored gray are used to help explain how the control measure is used and are not a part of the control measure.
	Lines	
release line—Phase line used in river crossing operations that delineates a change in the headquarters controlling movement. (See ATP 3-90.4 for more information on release line.)		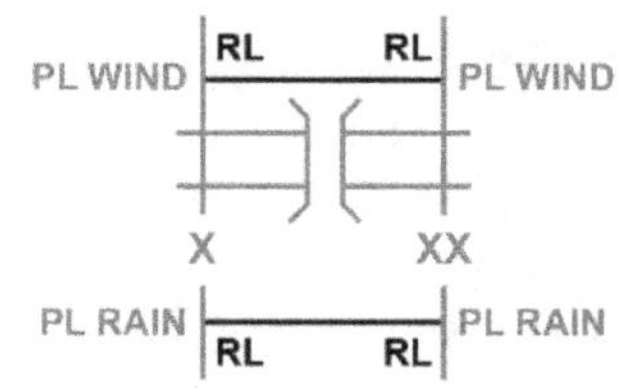
bridgehead line—The limit of the objective area in the development of the bridgehead. (ATP 3-90.4)		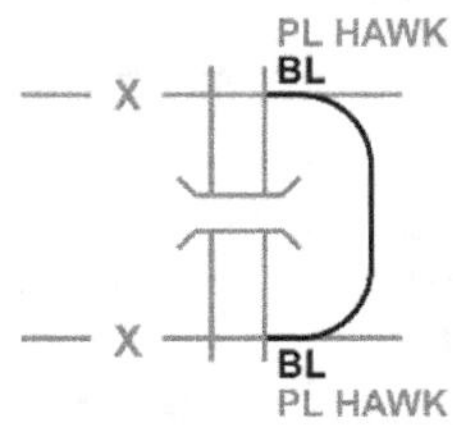
	Offensive Lines	
battlefield handover line—A designated phase line where responsibility transitions from the stationary force to the moving force and vice versa. (ADP 3-90)		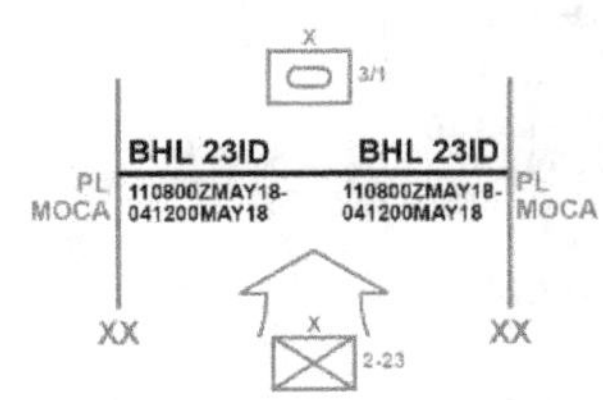
delay line—A phase line over which an enemy is not allowed to cross before a specific date and time or enemy condition. (FM 3-90)		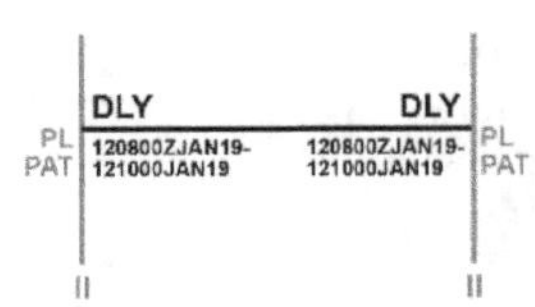

Table 5-8. Line control measure symbols (*continued*)

Function	***Template***	***Construct Example*** **Note.** Symbols colored gray are used to help explain how the control measure is used and are not a part of the control measure.
	Offensive Lines	
final coordination line—A phase line close to the enemy position used to coordinate the lifting or shifting of supporting fires with the final deployment of maneuver elements. (ADP 3-90)	**FCL FCL**	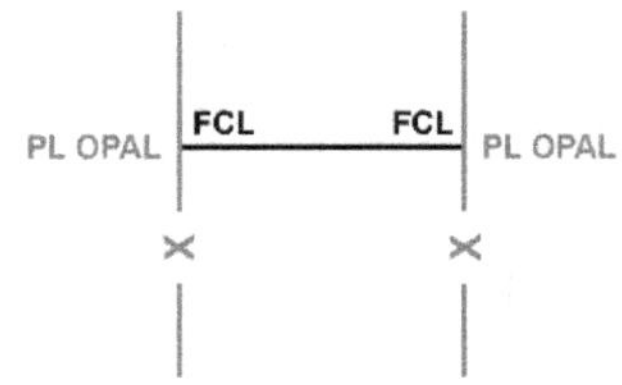
limit of advance—A phase line used to control forward progress of the attack. (ADP 3-90)	**LOA LOA**	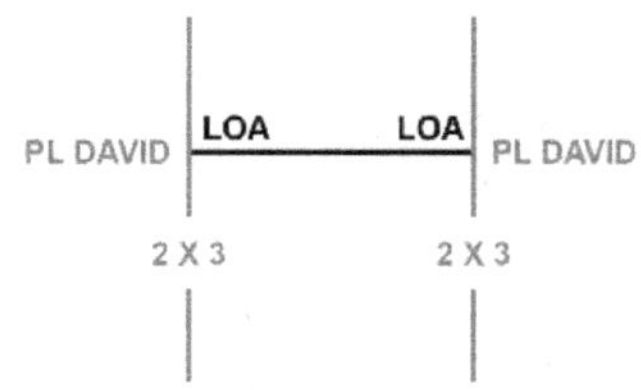
line of departure—In land warfare, a line designated to coordinate the departure of attack elements. (JP 3-31)	**LD LD**	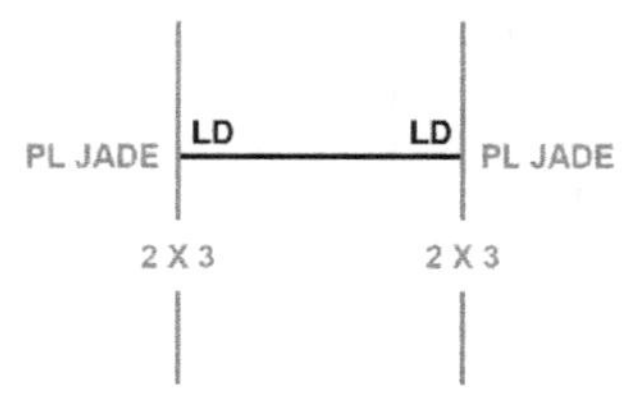
line of departure or line of contact—A general trace delineating the locations where friendly and enemy forces are engaged. (ADP 3-90)	**LD/LC LD/LC**	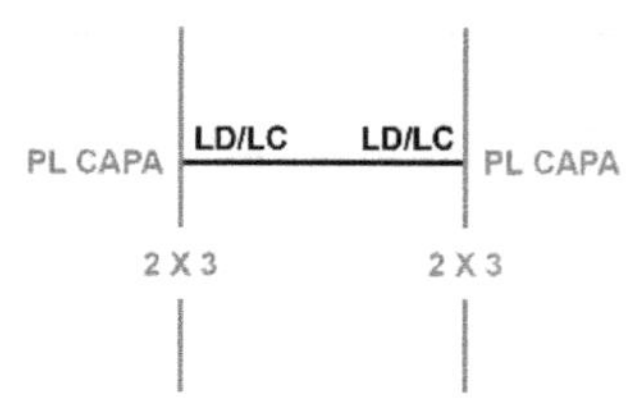

Table 5-8. Line control measure symbols (*continued*)

Function	*Template*	*Construct Example* ***Note.*** Symbols colored gray are used to help explain how the control measure is used and are not a part of the control measure.
Offensive Lines		
probable line of deployment—A phase line that designates the location where the commander intends to deploy the unit into assault formation before beginning the assault. (ADP 3-90) ***Note.*** The dashed lines in this symbol shall be displayed in present and anticipated status.		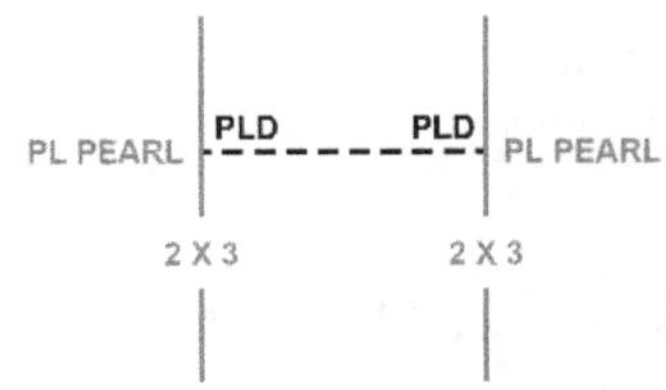
Fire Lines		
common sensor boundary—A line depicted by a series of grid coordinates, grid line, phase line or major terrain feature that divides target acquisition search areas into radar acquisition management areas. (FM 3-09)	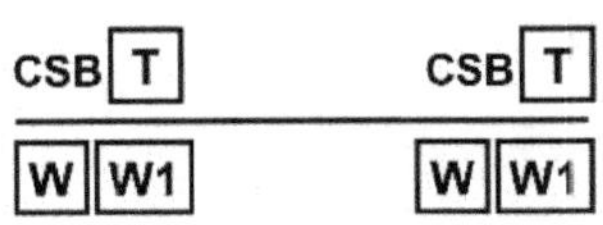	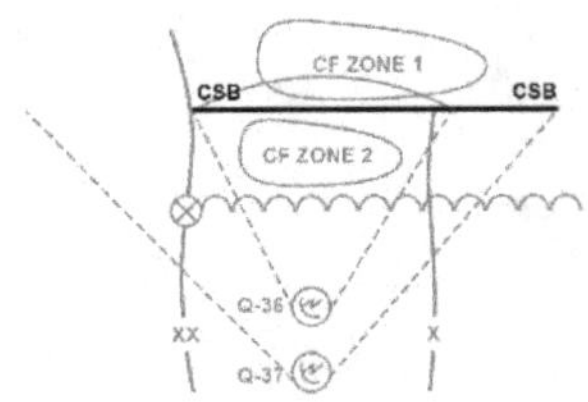

Table 5-8. Line control measure symbols (*continued*)

Function	*Template*	*Construct Example* **Note.** Symbols colored gray are used to help explain how the control measure is used and are not a part of the control measure.
Fire Lines		
coordinated fire line—A line beyond which conventional surface-to-surface direct fire and indirect fire support means may fire at any time within the boundaries of the establishing headquarters without additional coordination but does not eliminate the responsibility to coordinate the airspace required to conduct the mission. (JP 3-09)		
fire support coordination line—A fire support coordination measure established by the land or amphibious force commander to support common objectives within an area of operation; beyond which all fires must be coordinated with affected commanders prior to engagement, and short of the line, all fires must be coordinated with the establishing commander prior to engagement. (JP 3-09)	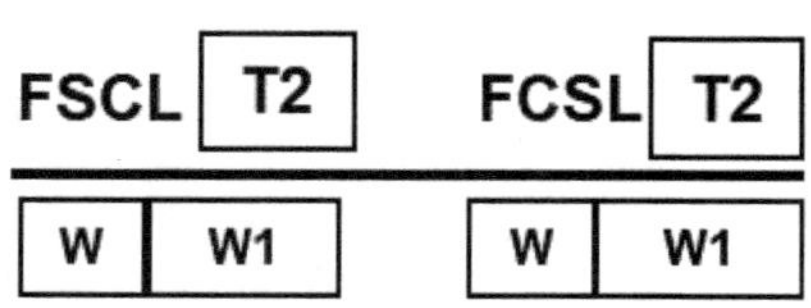	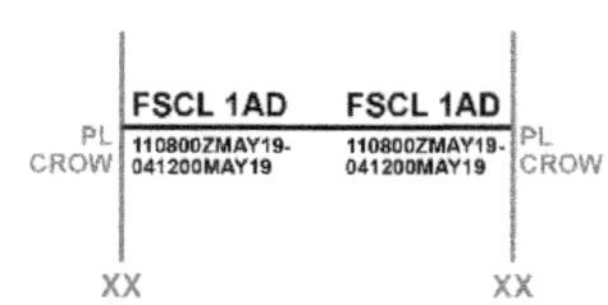

Table 5-8. Line control measure symbols (*continued*)

Function	***Template***	***Construct Example*** ***Note.*** Symbols colored gray are used to help explain how the control measure is used and are not a part of the control measure.
Fire Lines		
restrictive fire line—A specific boundary established between converging friendly surface forces that prohibits fires or their effects from crossing. (JP 3-09)	RFL T2 RFL T2 W W1 W W1	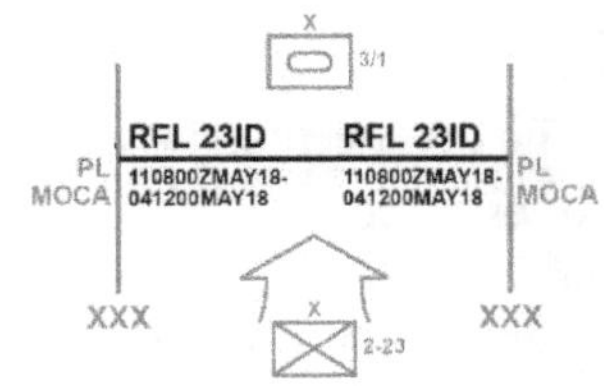 RFL 23ID RFL 23ID PL MOCA 110800ZMAY18-041200MAY18 110800ZMAY18-041200MAY18 PL MOCA XXX XXX
Engineer Line		
engineer work line—A coordinated boundary line used to compartmentalize an area of operations to indicate where specific engineer units have primary responsibility for the engineer effort. (FM 3-34) ***Note.*** Letters "EWL" are an integral part of the symbol, but they can rotate based on the direction of the line.	T/AS EWL EWL T/AS	EWL 326EN BN (USA) EWL 127EN BN (USA)
	EWL T/AS T/AS EWL	EWL 326EN BN (USA) 127EN BN (USA) EWL
Intelligence Line		
intelligence coordination line	ICL T ICL T W W1 W W1	ICL MOCANO ICL MOCANO PL BORICUA 140800ZMAY18-141200MAY18 140800ZMAY18-141200MAY18 PL BORICUA XXX XXX

MOVEMENT CONTROL MEASURES AND FORMS OF MANEUVER SYMBOLS

5-24. *Movement* is the positioning of combat power to establish the conditions for maneuver (ADP 3-90), and *maneuver* is the employment of forces in the operational area, through movement in combination with fires and information, to achieve a position of advantage in respect to the enemy (DOD Dictionary). (See FM 3-90 for additional information on movement and maneuver.)

5-25. Table 5-9, on pages 186 through 190 includes movement control measures and forms of maneuver symbols. *Forms of maneuver* are distinct tactical combinations of fire and movement with a unique set of doctrinal characteristics that differ primarily in the relationship between the maneuvering force and the enemy (ADP 3-90).

Table 5-9. Movement control measures and forms of maneuver symbols

Control Measure	*Template*	*Construct Example and Symbol Translation*
	Movement Control Measures	
	Axis of Advance The general area through which the bulk of a unit's combat power must move. (ADP 3-90)	
airborne or aviation axis of advance	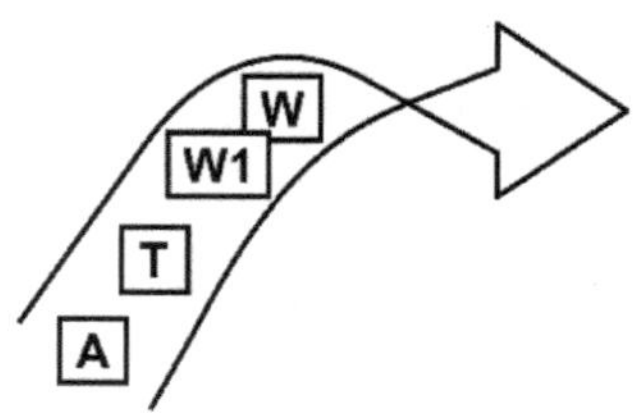	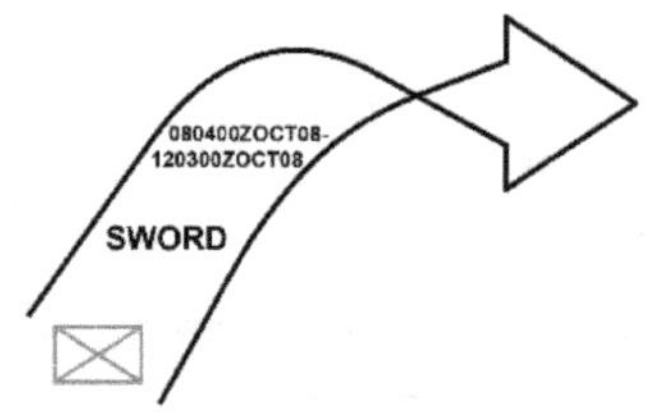 Airborne infantry unit moving on airborne axis of advance SWORD from 0400 Zulu hour 8 October 2008 to 0300 Zulu hour 12 October 2008
		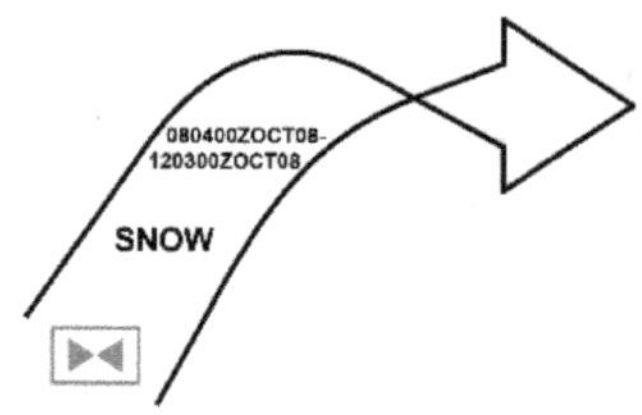 Aviation unit moving on aviation axis of advance SNOW from 0400 Zulu hour 8 October 2008 to 0300 Zulu hour 12 October 2008
attack helicopter axis of advance	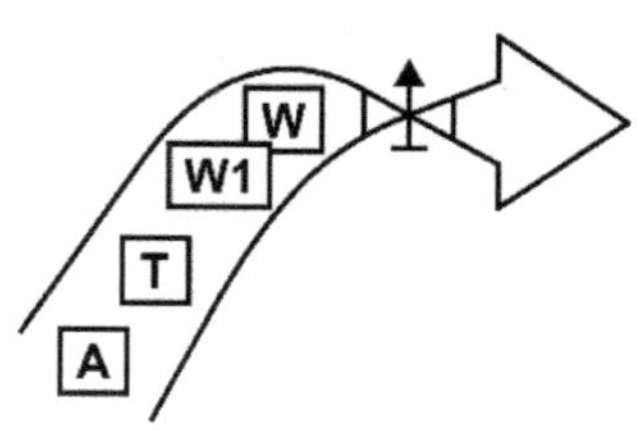	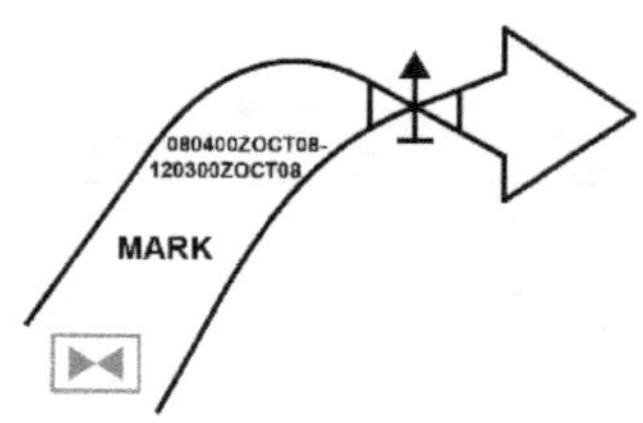 Attack aviation unit moving on attack helicopter axis of advance MARK from 0400 Zulu hour 8 October 2008 to 0300 Zulu hour 12 October 2008

Table 5-9. Movement control measures and forms of maneuver symbols (*continued*)

<table>
<tr><th>Control Measure</th><th>Template</th><th>Construct Example and Symbol Translation</th></tr>
<tr><td colspan="3">Axis of Advance (continued)</td></tr>
<tr><td>supporting axis of advance</td><td>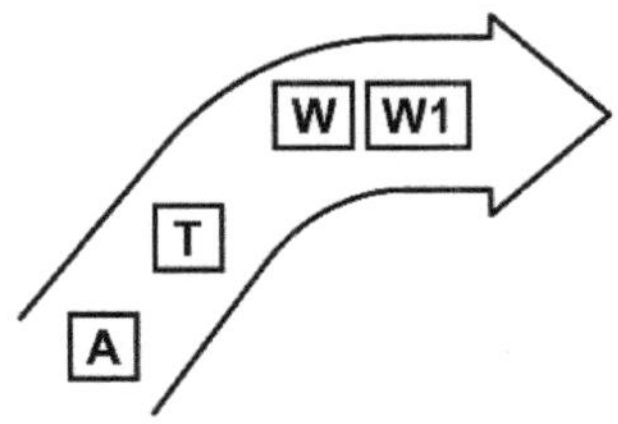</td><td>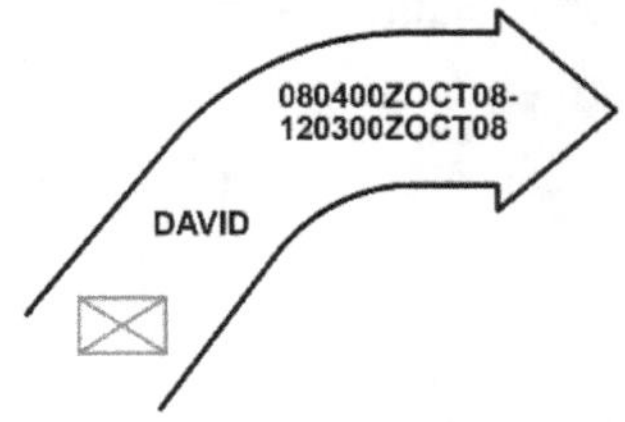Infantry brigade moving on supporting axis of advance DAVID from 2300 Zulu hour 31 October 2012</td></tr>
<tr><td rowspan="2">main axis of advance—The principal attack or effort into which the commander throws the full weight of the offensive power at his disposal.</td><td rowspan="2">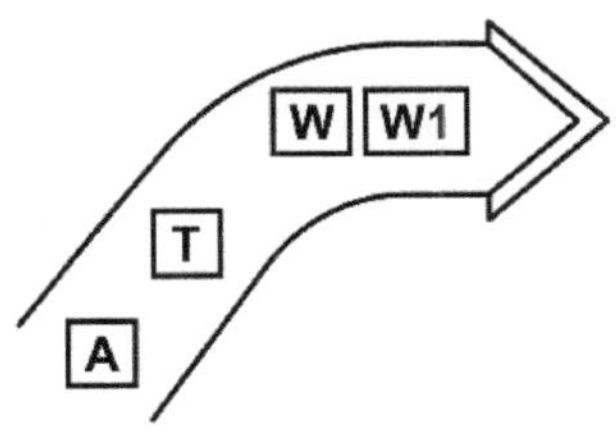</td><td>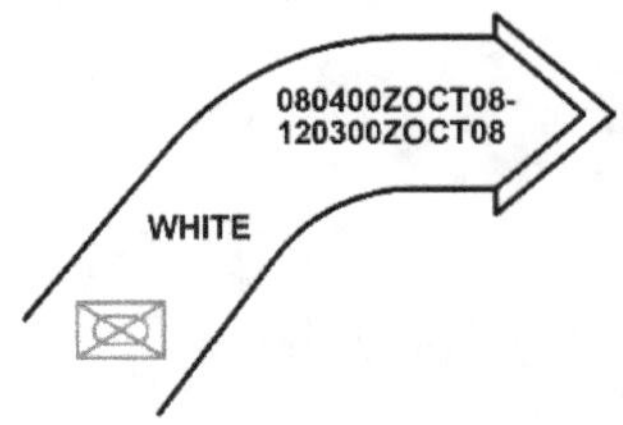Mechanized infantry unit moving on main axis of advance WHITE from 0400 Zulu hour 8 October 2008 to 0300 Zulu hour 12 October 2008</td></tr>
<tr><td>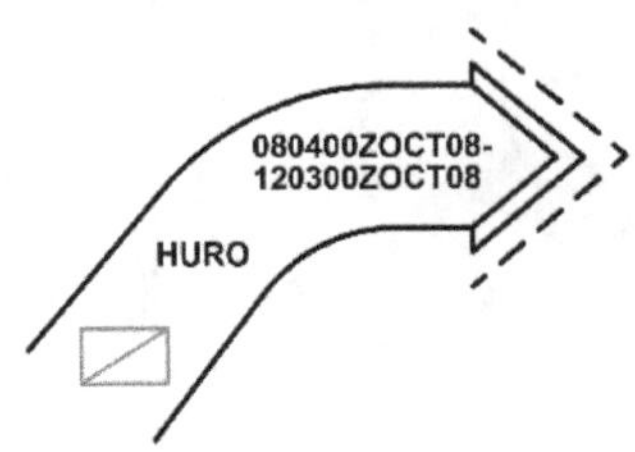Cavalry unit moving on feint main axis of advance HURO from 0400 Zulu hour 8 October 2008 to 0300 Zulu hour 12 October 2008</td></tr>
<tr><td colspan="3">Direction of Attack
A specific direction or assigned route a force uses and does not deviate from when attacking. (ADP 3-90)</td></tr>
<tr><td>aviation direction of attack</td><td>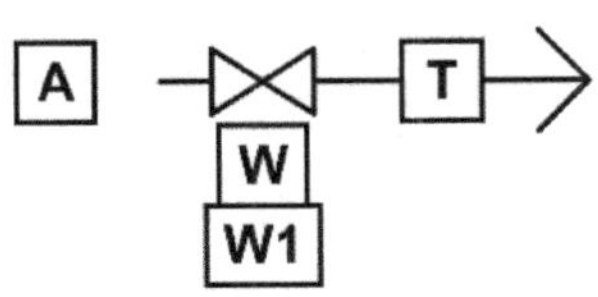</td><td>Aviation unit aviation direction of attack from 0400 Zulu hour 8 October 2008 to 0300 Zulu hour 12 October 2008</td></tr>
</table>

Table 5-9. Movement control measures and forms of maneuver symbols (*continued*)

Control Measure	***Template***	***Construct Example and Symbol Translation***
	Direction of Attack (*continued*)	
direction of main attack	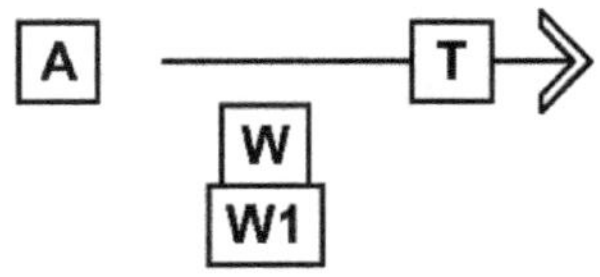	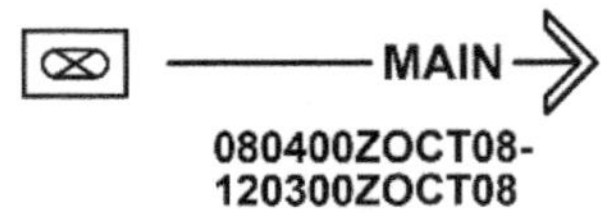 Combined arms unit direction of main attack from 0400 Zulu hour 8 October 2008 to 0300 Zulu hour 12 October 2008
		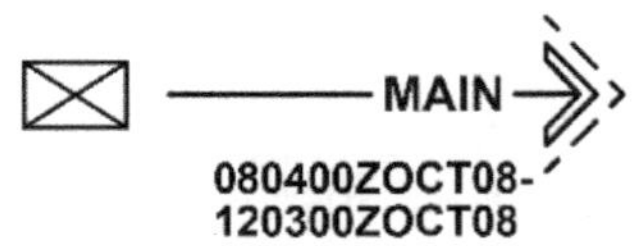 Infantry unit feint direction of main attack from 0400 Zulu hour 8 October 2008 to 0300 Zulu hour 12 October 2008
direction of supporting attack	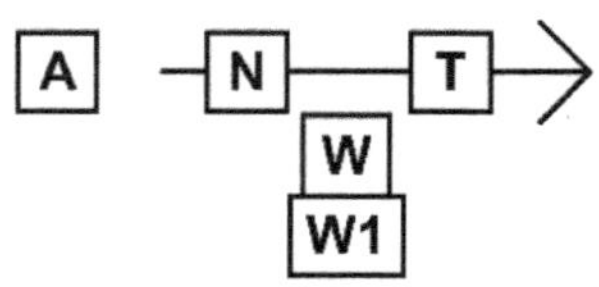	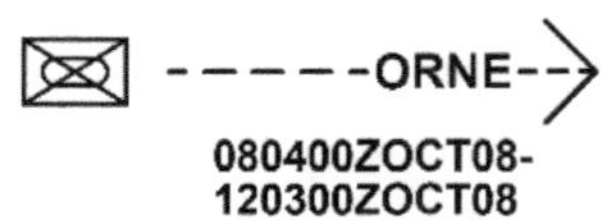 Mechanized armored unit planned direction of supporting attack ORNE from 0400 Zulu hour 8 October 2008 to 0300 Zulu hour 12 October 2008
		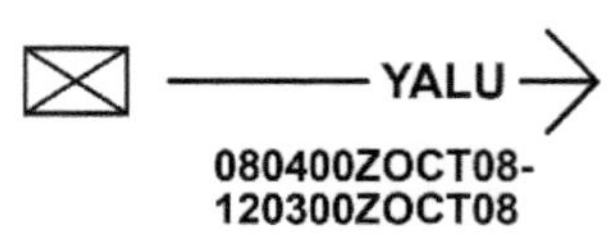 Infantry unit feint direction of supporting attack YALU from 0400 Zulu hour 8 October 2008 to 0300 Zulu hour 12 October 2008
		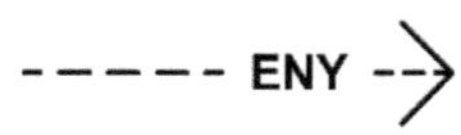 Enemy or suspected direction of supporting attack

BODY-200

Table 5-9. Movement control measures and forms of maneuver symbols (*continued*)

Control Measure	*Template*	*Construct Example and Symbol Translation*
	Direction of Attack (*continued*)	
infiltration lane—A control measure that coordinates forward and lateral movement of infiltrating units and fixes fire planning responsibilities. (FM 3-90)		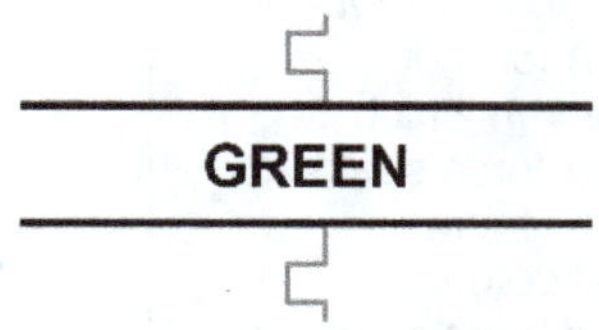 Infiltration lane GREEN through fortified line
	Forms of Maneuver	
envelopment—A form of maneuver in which an attacking force avoids an enemy's principal defense by attacking along an assailable flank. (FM 3-90)	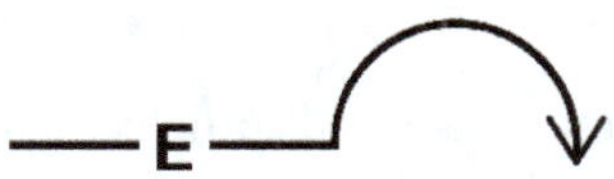	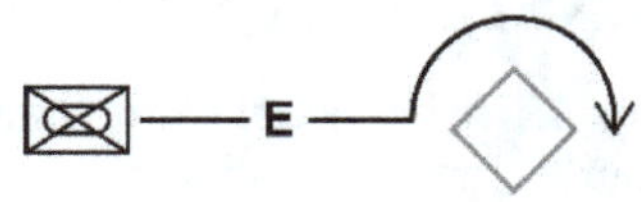
infiltration—(Army) A form of maneuver in which an attacking force conducts undetected movement through or into an area occupied by enemy forces. (FM 3-90)		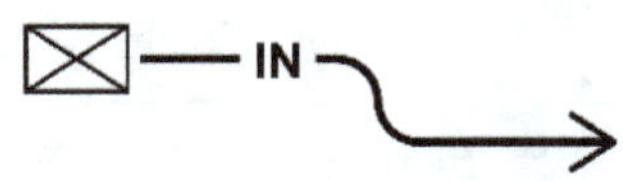
frontal attack—A form of maneuver in which the attacking force seeks to destroy a weaker enemy force or fix a larger enemy force in place over a broad front. (FM 3-90)	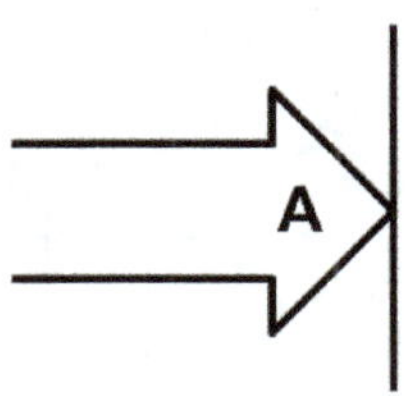	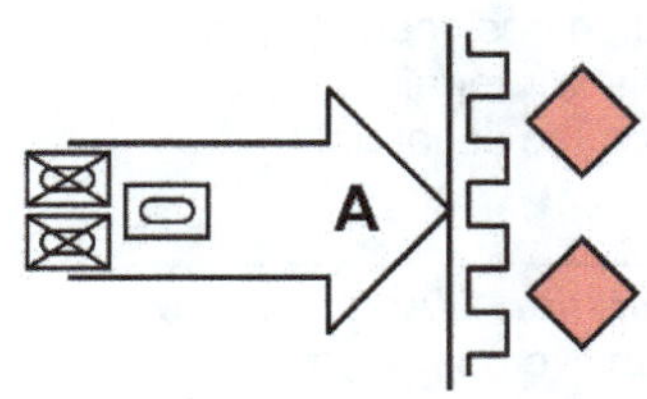
penetration—A form of maneuver in which a force attacks on a narrow front. (FM 3-90)	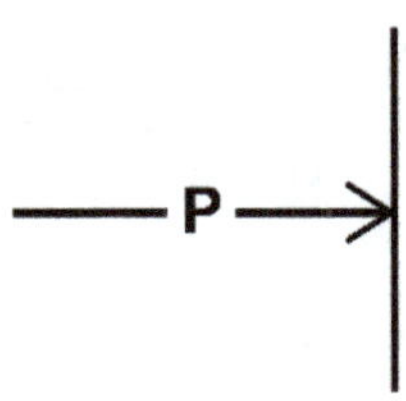	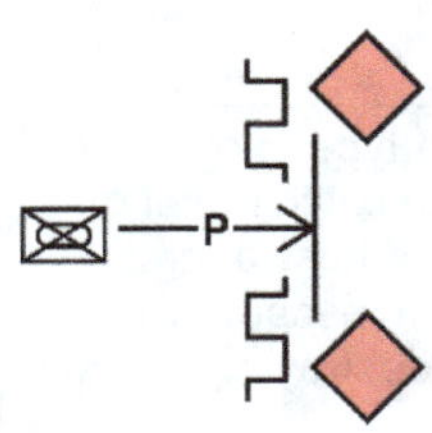

Table 5-9. Movement control measures and forms of maneuver symbols (*continued*)

Control Measure	*Template*	*Construct Example and Symbol Translation*
	Forms of Maneuver (*continued*)	
turning movement—(Army) A form of maneuver in which the attacking force seeks to avoid the enemy's principal defensive positions by attacking to the rear of their current positions forcing them to move or divert forces to meet the threat. (FM 3-90)	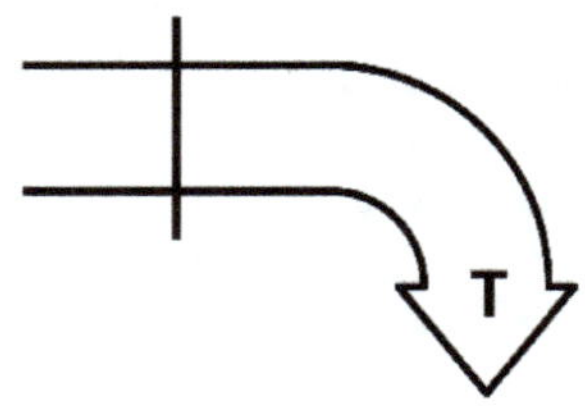	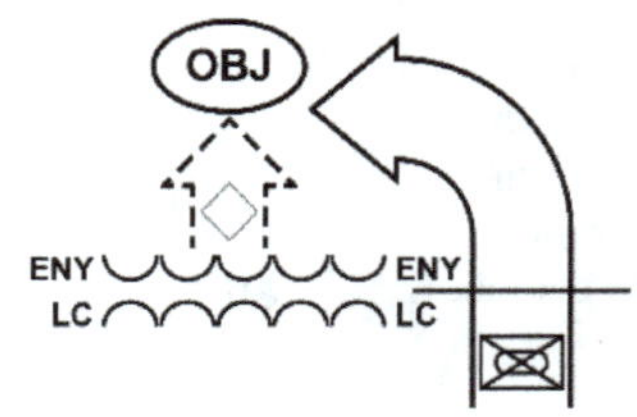

OFFENSE OPERATIONS PLANNING SYMBOLS

5-26. Offensive actions are combat operations conducted to defeat and destroy forces and seize terrain, resources, and population centers. (See FM 3-90 for more information on offensive actions).

5-27. **Offensive operation symbols.** Offensive operations symbols depict the general sequence of the offense and recommended formations. Table 5-10 includes offensive operation planning symbols.

Table 5-10. Offensive operations planning symbols

Control Measure	*Template*	*Construct Example and Symbol Translation*
movement to contact—(Army) A type of offensive operation designed to establish or regain contact to develop the situation. (FM 3-90)	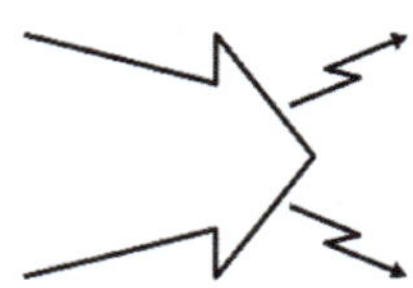	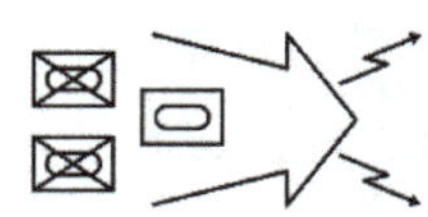
exploitation—(Army) A type of offensive operation following a successful attack to disorganize the enemy in depth. (FM 3-90)	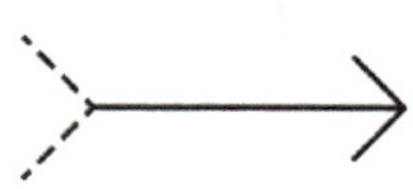	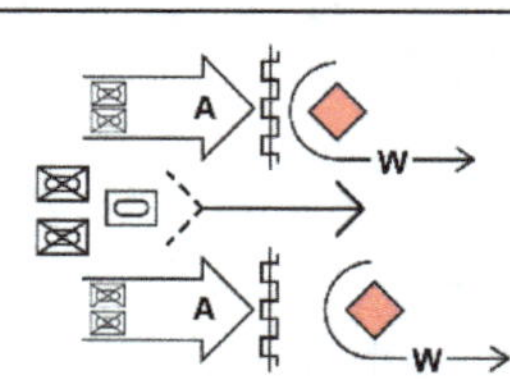
pursuit—A type of offensive operation to catch or cut off a disorganized hostile force attempting to escape, with the aim of destroying it. (FM 3-90)	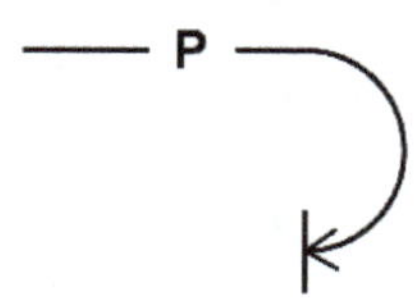	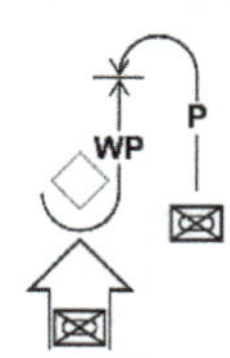

Table 5-10. Offensive operations planning symbols (*continued*)

Control Measure	*Template*	*Construct Example and Symbol Translation*
	Variation of movement to contact symbol	
cordon and search—A variation of movement to contact where a friendly force isolates and searches a target area. (FM 3-90)	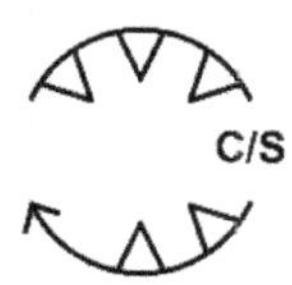	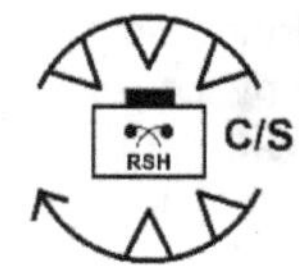 Cordon and search chemical, biological, radiological, and nuclear (CBRN) research facility

5-28. Table 5-11 includes variation of attack offensive planning symbols. An *attack* is a type of offensive operation that defeats enemy forces, seizes terrain, or secures terrain (FM 3-90).

Table 5-11. Attack operations planning symbols

Control Measure	*Template*	*Construct Example and Symbol Translation*
ambush—A variation of attack from concealed positions against a moving or temporarily halted enemy. (FM 3-90)	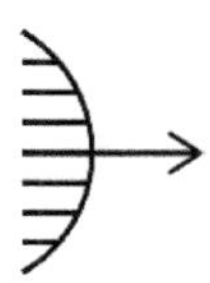	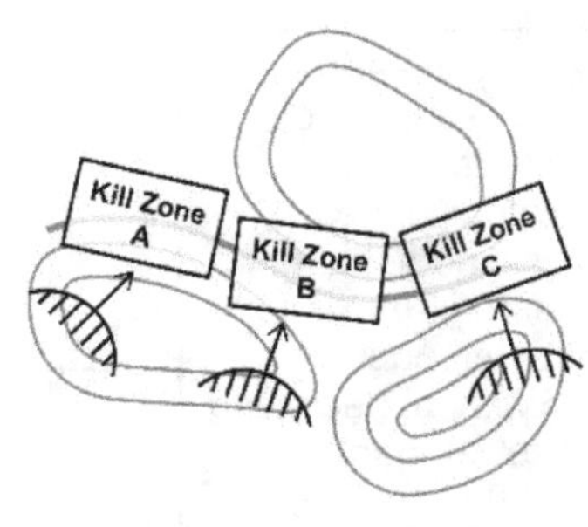
counterattack—A variation of attack by a defending force against an attacking enemy force. (FM 3-90) ***Note***. The dashed lines in this graphic shall be displayed in present and anticipated status.		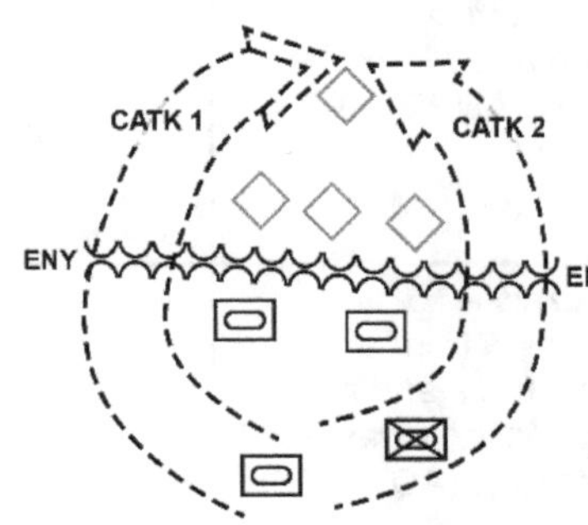

DEFENSE OPERATIONS PLANNING SYMBOLS

5-29. Table 5-12 includes types of defensive operations planning symbols and types of retrograde task symbols. Defensive operations are described in FM 3-90.

Table 5-12. Defensive operations planning symbols

Control Measure	*Template*	*Construct Example and Symbol Translation*
area defense—A type of defensive operation that concentrates on denying enemy forces access to designated terrain for a specific time rather than destroying the enemy outright. (ADP 3-90)	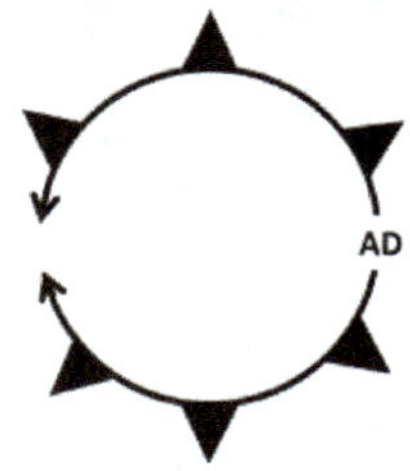	
mobile defense—A type of defensive operation that concentrates on the destruction or defeat of the enemy through a decisive attack by a striking force. (ADP 3-90)		
Retrograde task symbols		
delay—When a force under pressure trades space for time by slowing down the enemy's momentum and inflicting maximum damage on enemy forces without becoming decisively engaged. (ADP 3-90)	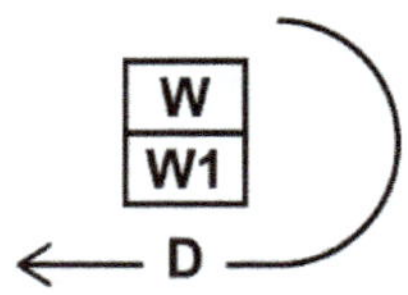	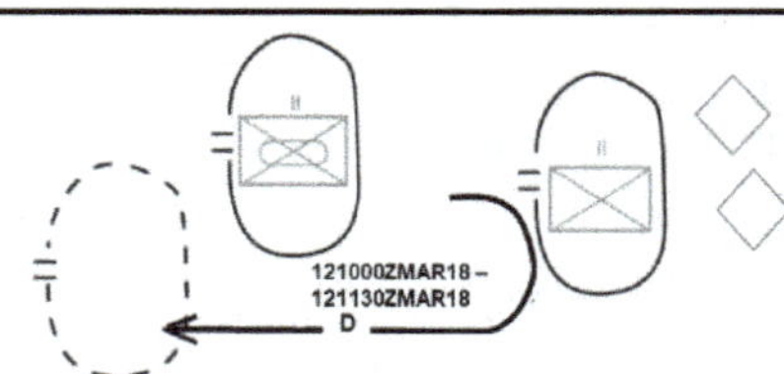
	Note. "W" and "W1" are optional amplifiers that identify the start date-time group and can be displayed alone or in conjunction with "W1" to identify the projected date-time group end date.	
retirement—When a force out of contact moves away from the enemy. (ADP 3-90)	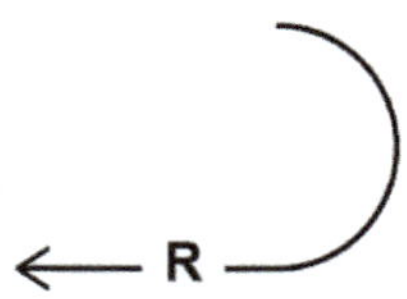	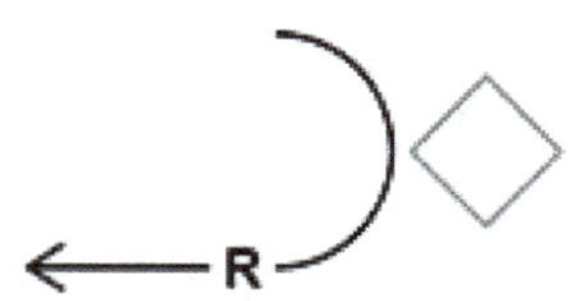
withdraw—To disengage from an enemy force and move in a direction away from the enemy. (ADP 3-90)	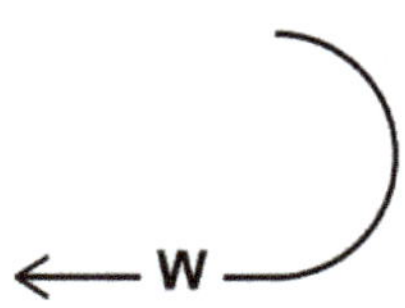	
withdraw under pressure	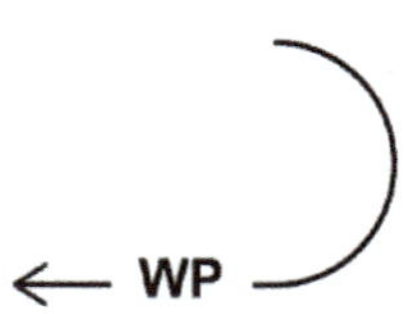	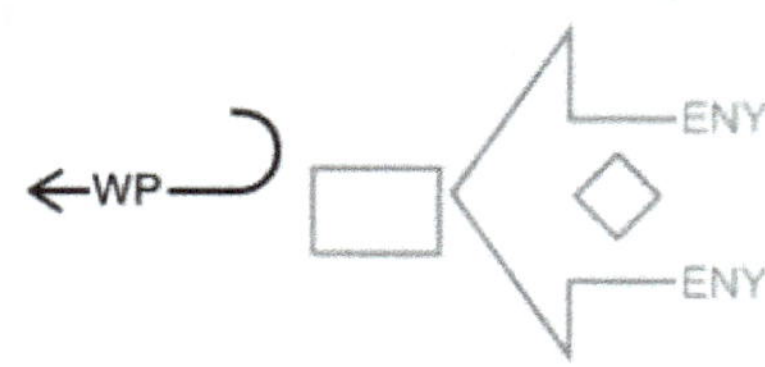

ENABLING OPERATIONS PLANNING SYMBOLS

5-30. Table 5-13 includes enabling operation planning symbols. *Enabling operation* is an operation that sets the friendly conditions required for mission accomplishment (FM 3-90). Enabling operations include security, reconnaissance, relief in place, and passage of lines.

Table 5-13. Enabling operations planning symbols

Security Operations Those operations performed by commanders to provide early and accurate warning of enemy operations, to provide the forces being protected with time and maneuver space within which to react to the enemy, and to develop the situation to allow commanders to effectively use their protected forces. (ADP 3-90)	
Draw rules: 1. Front of symbol faces enemy in a direction outward from friendly forces. 2. Letters (C,S, and G) have the option to rotate independently and can adjust based on the direction of symbol. 3. Unit symbol remains oriented for left-to-right legibility. 4. Arrows can originate from any point on the unit symbol and orient outward in any direction necessary to depict the task. 5. Arrows may be elongated to cover desired area.	
Horizontal	**Vertical**
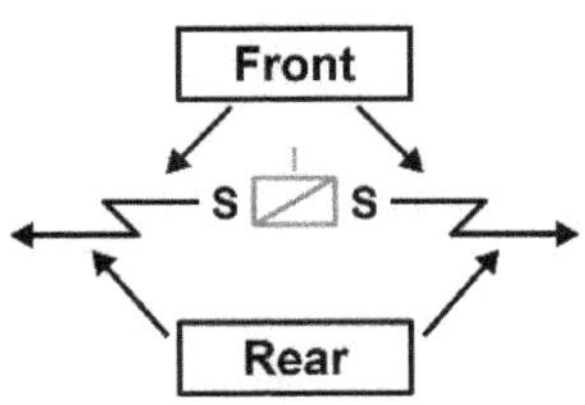	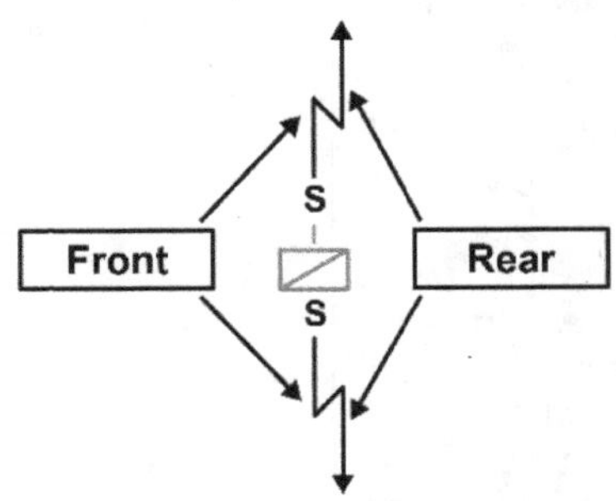

Control Measure	***Template***	***Construct Example and Symbol Translation*** ***Note.*** The friendly or hostile frame (gray) is not part of the symbol; it is for orientation only.
cover—A type of security operation done independent of the main body to protect them by fighting to gain time while preventing enemy ground observation of and direct fire against the main body. (ADP 3-90)	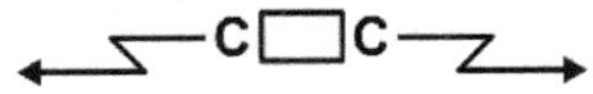	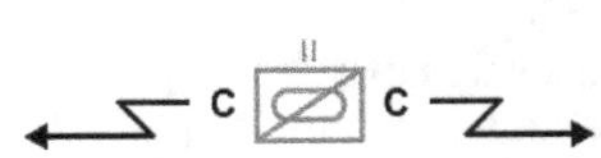

Table 5-13. Enabling operations planning symbols (*continued*)

Control Measure	*Template*	*Construct Example and Symbol Translation*
guard—A type of security operation done to protect the main body by fighting to gain time while preventing enemy ground observation of and direct fire against the main body. (ADP 3-90)		
screen—A type of security operation that primarily provides early warning to the protected force. (ADP 3-90)		
Passage of Lines An operation in which a force moves forward or rearward through another force's combat positions with the intention of moving into or out of contact with the enemy. (JP 3-18)		
forward passage of lines—Occurs when a unit passes through another unit's positions while moving toward the enemy. (ADP 3-90)	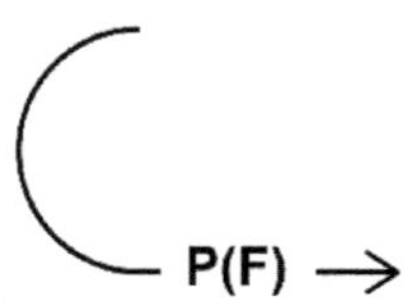	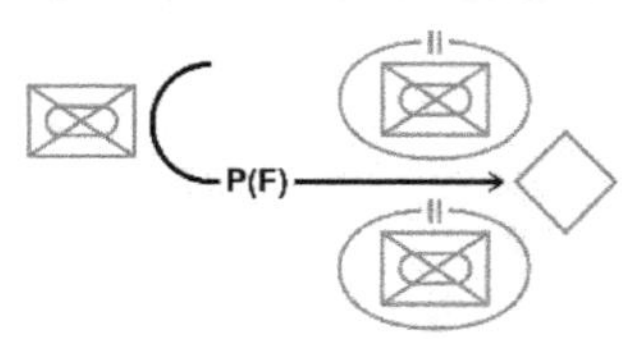
rearward passage of lines—Occurs when a unit passes through another unit's positions while moving away from the enemy. (ADP 3-90)	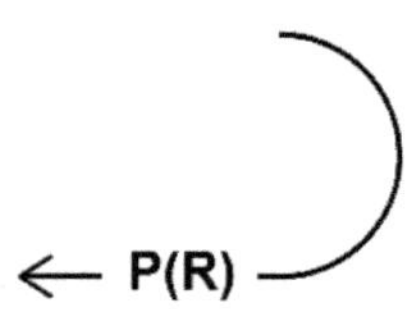	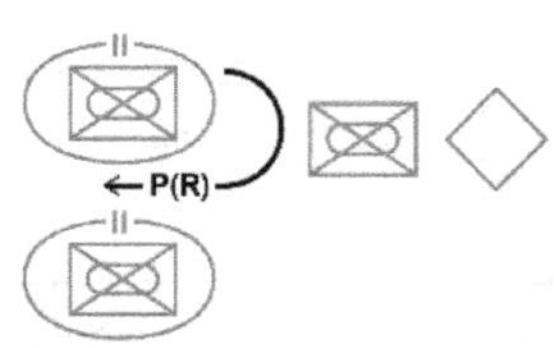
relief in place—An operation in which, by direction of higher authority, all or part of a unit is replaced in an area by the incoming unit and the responsibilities of the replaced elements for the mission and the assigned zone of operations are transferred to the incoming unit. (JP 3-07.3)	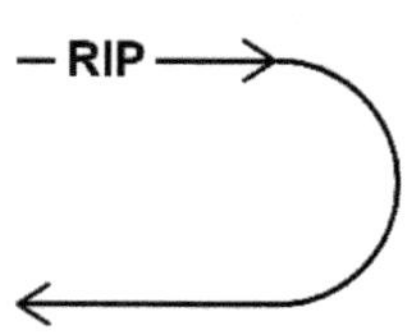	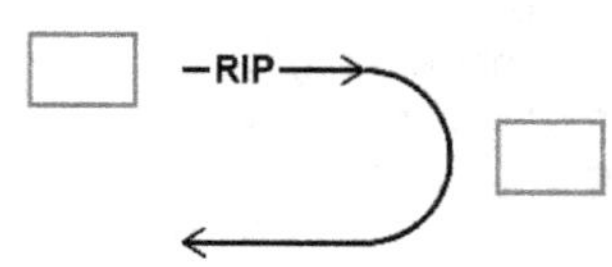

OBSERVATION POST CONTROL MEASURES

5-31. An observation post is a position from which military observations are made, or fire directed and adjusted, and which possesses appropriate communications. Table 5-14 on page 195 lists the observation post control measures.

Table 5-14. Observation post control measure symbols

Function	***Construct*** ***Note***. Symbols without amplifier fields cannot be modified and must be used as depicted.
observation post—A position from which observations are made, or fires directed and adjusted. (FM 3-90)	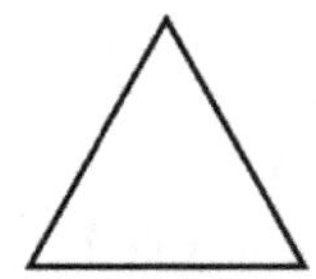
reconnaissance observation post—An observation post to conduct preliminary survey to gain information.	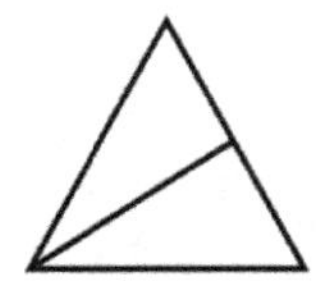
forward observer—An individual operating with front line troops trained to adjust ground or naval gunfire and pass back battlefield information. (JP 3-09)	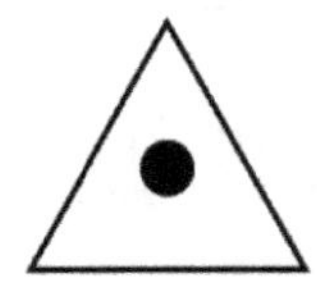
sensor observation post or listening post—An observation post with devices that responds to a physical stimulus and transmits a resulting impulse, or a location for monitoring electronic communications with the capability to observe or receive a signal or observable from a person or object. (See FM 2-0 for more information on sensors.)	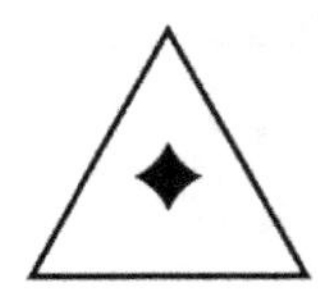
combat outpost—A reinforced observation post capable of conducting limited combat operations. (FM 3-90)	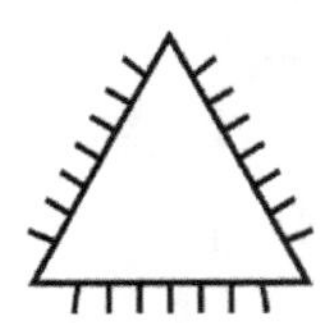

MILITARY DECEPTION SYMBOLS

5-32. *Military deception* is actions executed to deliberately mislead adversary military, paramilitary, or violent extremist organization decision makers, thereby causing the adversary to take specific actions (or inactions) that will contribute to the accomplishment of the friendly mission (JP 3-13.4). Military deception symbols are designed to identify actions executed to mislead the enemy by manipulation, distortion, or falsification of evidence to induce the enemy to react in a manner prejudicial to the enemy's interests. Table 5-15 on pages 196 and 197 demonstrates the military deception amplifier symbol, usage examples, and variations of tactical deception.

Table 5-15. Deception symbols

Feint, Decoy, or Dummy Indicator Line	
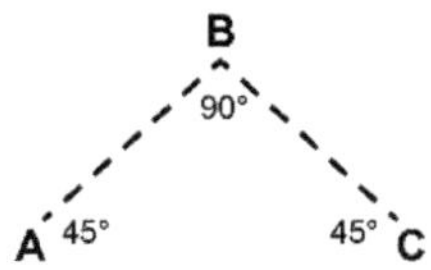	
Note. The decoy, dummy, feint, or phony symbol is added to another symbol, such as equipment, installations, or control measures to form the decoy, dummy, feint, or phony symbol.	
Examples	
point control measure	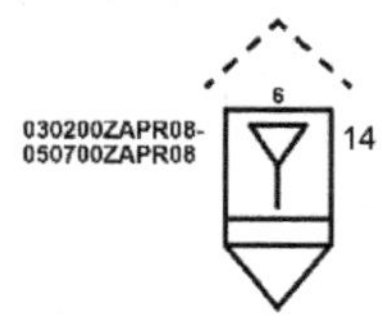 Dummy Class 3 supply point
equipment	 Dummy armored high mobility wheeled vehicle
installation	 Dummy electrical installation
minefield	 Dummy anti-tank minefield
minefield, dynamic or mined area	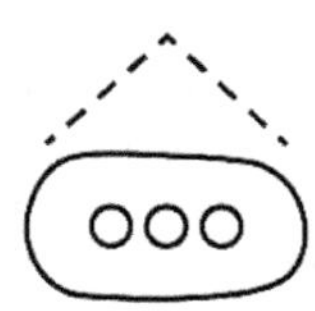 Dummy dynamic unspecified minefield

Table 5-15. Deception symbols (*continued*)

mined area, fenced	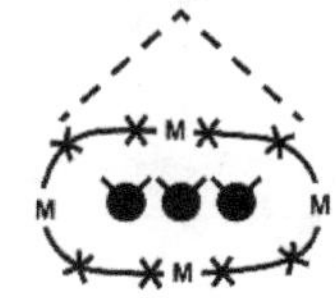 Antipersonnel fenced mined area
Variations of Tactical Deception	
feint—(Army) A variation of tactical deception that makes contact solely to deceive the adversary as to the location, time of attack or both. (FM 3-90)	 axis of advance feint
	 Main direction of attack feint
demonstration—(Army) A variation of tactical deception used as a show of force in an area where a unit does not seek a decision and attempts to mislead an adversary. (FM 3-90)	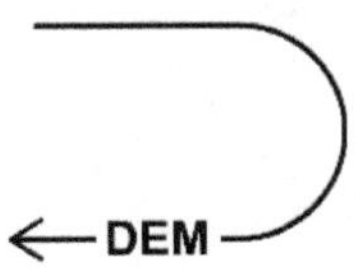

MOBILITY AND COUNTERMOBILITY CONTROL MEASURES

5-33. Mobility and countermobility are subordinate tasks within the movement and maneuver warfighting function. Mobility and countermobility are complementary opposites. Mobility is focused on friendly force movement and maneuver, and countermobility is focused on affecting or thwarting enemy mobility. (See ADP 3-90 and FM 3-90 for additional information on mobility and countermobility control measures.)

MOBILITY

5-34. *Mobility* is a quality or capability of military forces which permits them to move from place to place while retaining the ability to fulfill their primary mission (JP 3-36). Table 5-16 on pages 198 through 199 lists mobility related control measure symbols. (See ATP 3-90.4 for additional information on mobility and associated military symbol doctrine.)

Table 5-16. Mobility control measure symbols

Function	*Template*	*Construct Example* ***Note.*** Symbols colored gray are used to help explain how the control measure is used and are not a part of the control measure.
gap—An area free of obstacles that enables forces to maneuver in a tactical formation. (FM 3-90)	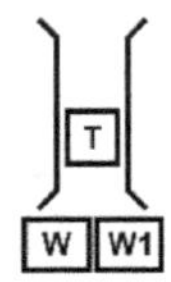	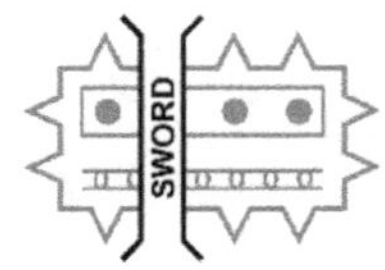
passage lane—A lane through an obstacle that provides safe passage for a passing force. (FM 3-90)	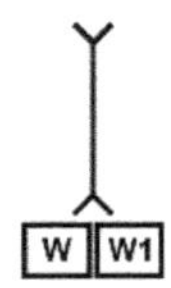	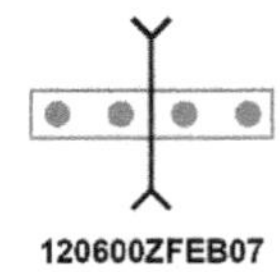 Passage lane through friendly anti-tank minefield at 0600 Zulu 12 FEB 2007
Water Crossing Site The location of a single bridge or rafting site, or in an initial assault a site for the crossing of assault boats or for the swimming or fording of vehicles on a broad front.		
assault crossing—A selected river crossing location where a short and violent well-ordered attack will take place against an objective.	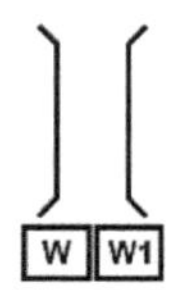	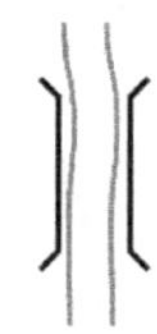
bridge—A structure carrying a pathway or roadway over a depression or obstacle.		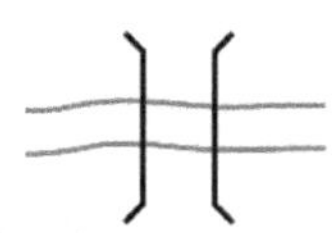
ford, easy—A shallow part of a body of water that may be easily crossed.		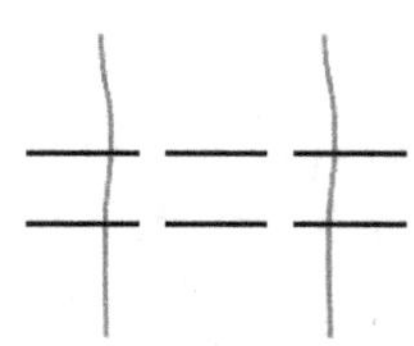
ford, difficult—A shallow part of a body of water that may be crossed with difficulty.	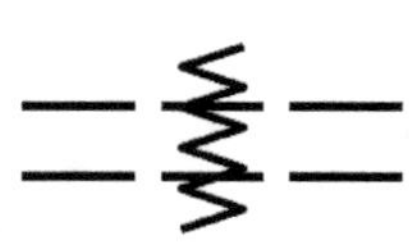	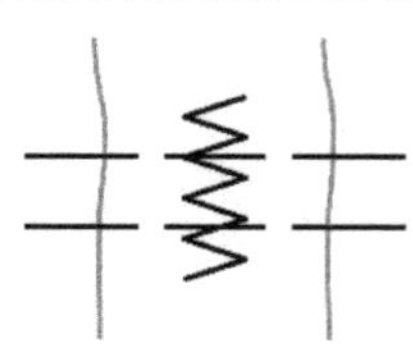

Table 5-16. Mobility control measure symbols (*continued*)

Function	*Template*	*Construct Example*
ferry crossing—Land locations where persons or things are routinely carried across a body of water.		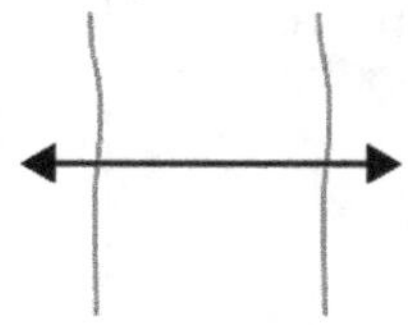

ROUTE CONTROL MEASURES

5-35. A *route* is the prescribed course to be traveled from a specific point of origin to a specific destination (FM 3-90). Routes may have different designated functions to effectively support freedom of movement. The commander may designate specific functions, restrictions, names, numbers, or alphanumeric designations to area of operations routes. Table 5-17 lists generic routes.

Table 5-17. Route control measures

Control Measure	*Template*	*Construct Example*
Routes		
route—The prescribed course to be traveled from a specific point of origin to a specific destination. (FM 3-90)	ROUTE T	**ROUTE FELA**
one-way traffic route ***Note***. The directional arrow may be turned to depict actual traffic direction.	ROUTE T →	**ROUTE CROW** →
alternating traffic route	ROUTE T ← ALT →	**ROUTE CABAN** ← ALT →
Sustainment Routes		
main supply route—The route or routes designated within an operational area upon which the bulk of traffic flows in support of military operations. (FM 1-02.1)	MSR T	**MSR MENDEZ**
main supply route, one-way traffic ***Note***. The directional arrow may be turned to depict actual traffic direction.	MSR T →	**MSR CLETO** →

Table 5-17. Route control measures (*continued*)

Control Measure	*Template*	*Construct Example*
main supply route, two-way traffic	MSR T	MSR SOTO
main supply route, alternating traffic	MSR T ALT	MSR PEDRO ALT
alternate supply route—A route or routes designated within an area of operations to provide for the movement of traffic when main supply routes become disabled or congested. (FM 4-01)	ASR T	ASR BORICUA
alternate supply route, one-way traffic ***Note***. The directional arrow may be turned to depict actual traffic direction.	ASR T	ASR LIZ
alternate supply route, alternating traffic	ASR T ALT	PEREZ ALT
alternate supply route, two-way traffic	ASR T	ASR HIDALGO

Convoy Symbols

5-36. A *convoy* is a group of vehicles organized for the purpose of control and orderly movement with or without escort protection that moves over the same route at the same time and under one commander (JP 3-02). Table 5-18 on page 201 lists convoy symbols.

Table 5-18. Convoy symbols

Function	*Template*	*Construct Example* *Note*. Symbols colored gray are used to help explain how the control measure is used and are not a part of the control measure.
moving convoy *Note*. The arrow points in the direction the convoy is moving.	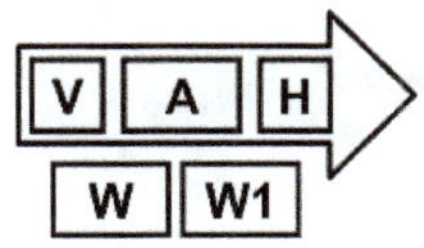	
halted convoy	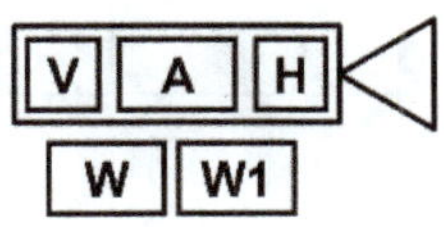	

Countermobility

5-37. Countermobility operations are those combined arms activities that use or enhance the effects of natural and man-made obstacles to deny enemy freedom of movement and maneuver, and they include proper obstacle integration with the maneuver plan, adherence to obstacle emplacement authority, and positive obstacle control. An *obstacle* is any natural or man-made obstruction designed or employed to disrupt, fix, turn, or block the movement of an opposing force, and to impose additional losses in personnel, time, and equipment on the opposing force (JP 3-15). Table 5-19 on pages 201 through 205 lists countermobility symbols. (See ATP 3-90.8 for additional information on countermobility and associated military symbol doctrine.)

Table 5-19. Countermobility symbols

Function	*Template*	*Example* **Note.** Symbols colored gray are used to help explain how the control measure is used and are not a part of the control measure.
Obstacle Effect Symbols Obstacle effect describes the effect the commander wants obstacles and fires to have on the enemy. (See ATP 3-90.8 for more information on obstacle effects.)		
block—An obstacle effect that integrates fire planning and obstacle effort to stop an attacker along a specific avenue of approach or prevent the attacking force from passing through an engagement area. (FM 3-90)		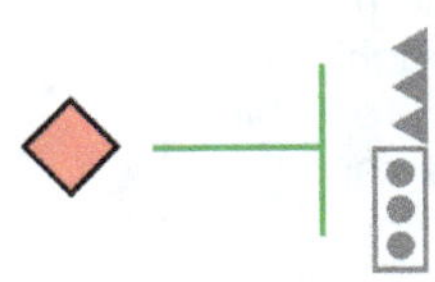

Table 5-19. Countermobility symbols (*continued*)

Function	*Template*	*Example*
disrupt—An obstacle effect that focuses fire planning and obstacle effort to cause the enemy to break up its formation and tempo, interrupt its timetable, commit breaching assets prematurely, and attack in a piecemeal effort. (FM 3-90)	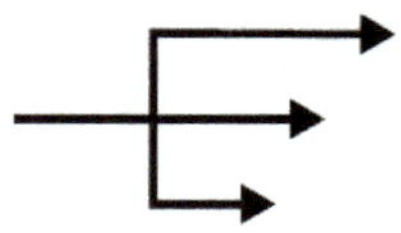	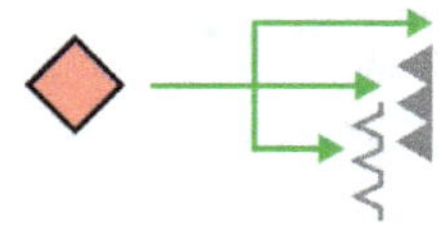
fix—An obstacle effect that focuses fire planning and obstacle effort to slow an attacker's movement within a specified area, normally an engagement area. (FM 3-90)		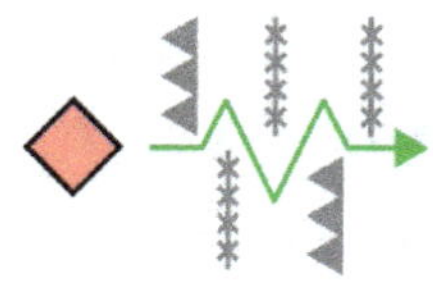
turn—A tactical obstacle effect that integrates fire planning and obstacle effort to divert an enemy formation from one avenue of approach to an adjacent avenue of approach or into an engagement area. (FM 3-90)		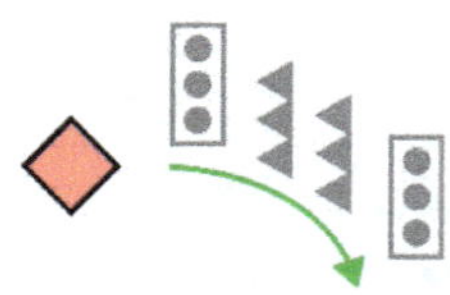
Obstacle Control Measure Symbols Specific measures that grant obstacle-emplacing authority. (FM 3-90)		
obstacle belt—A brigade-level command and control measure, normally depicted graphically, to show where within an obstacle zone the ground tactical commander plans to limit friendly obstacle employment and focus the defense. (JP 3-15)	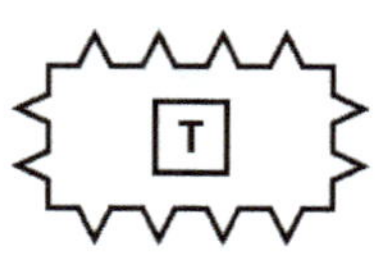	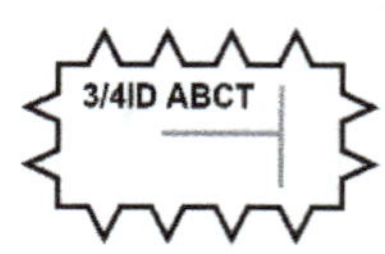
obstacle group—One or more individual obstacles that provide a specific obstacle effect. (FM 3-90)	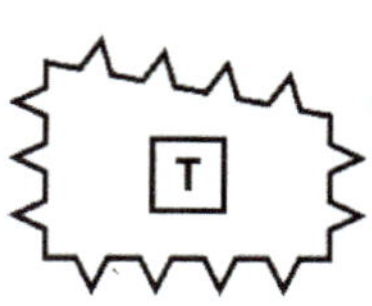	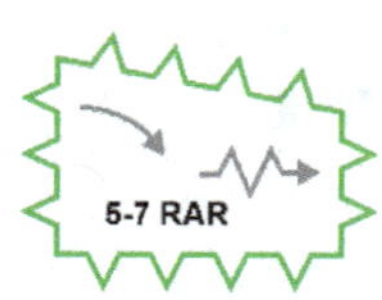

Table 5-19. Countermobility symbols (*continued*)

Function	*Template*	*Example*
obstacle zone—A division-level command and control measure to designate specific land areas where lower echelons are allowed to employ tactical obstacles. (JP 3-15)	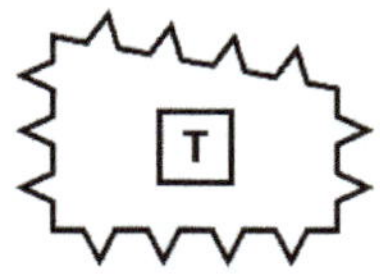	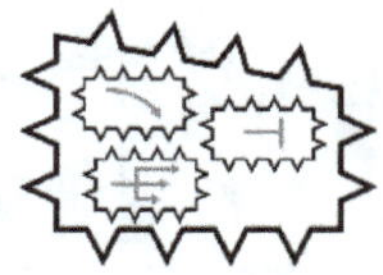
obstacle free area		
obstacle restricted area—A command and control measure used to limit the type or number of obstacles within an area. (JP 3-15)	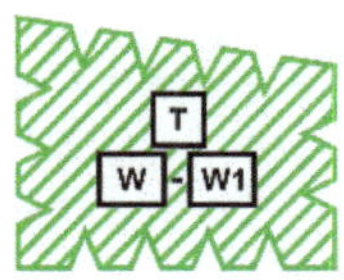	
obstacle line—A conceptual control measure used at battalion or brigade level to show placement intent without specifying a particular type of linear obstacle.		
Demolition Obstacle Symbol Obstacles created using explosives.		
planned explosive state of readiness		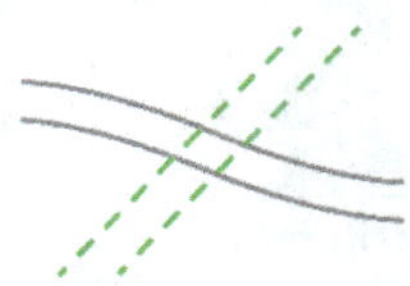
explosives, state of readiness 1 (safe)	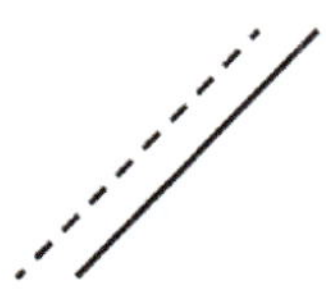	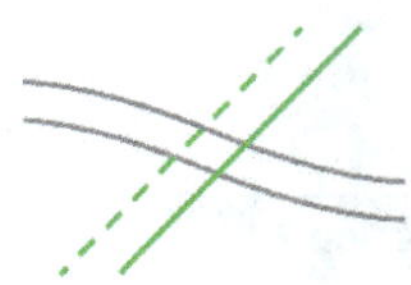
explosives, state of readiness 2 (armed but passable)	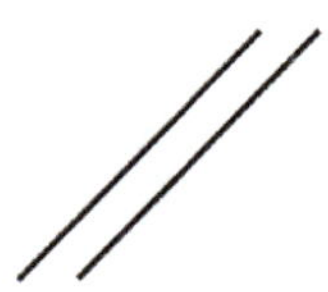	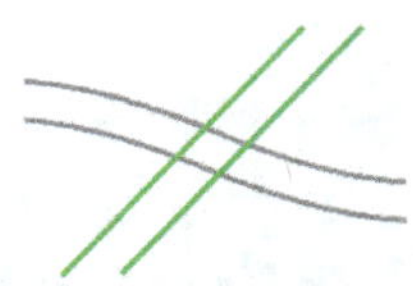

Table 5-19. Countermobility symbols (*continued*)

Function	***Template***	***Example***
roadblock complete (executed)	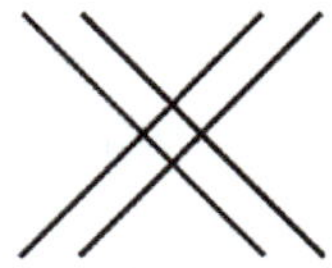	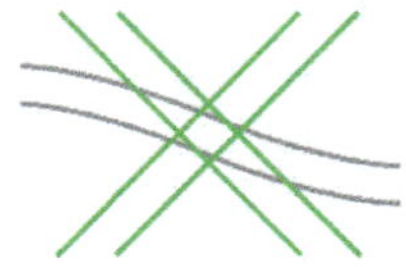
abatis—An obstacle constructed by the felling and interlacing of trees across a route.		
Constructed Obstacle Symbols Obstacles created with manual labor and or equipment. (See TM 3-34.85 for more information on constructed obstacle symbols.)		
Wire		
unspecified		
single fence		
double fence		
double apron fence	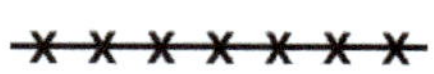	
low wire fence		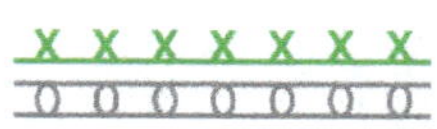

Table 5-19. Countermobility symbols (*continued*)

Function	*Template*	*Example*
high wire fence		
single concertina		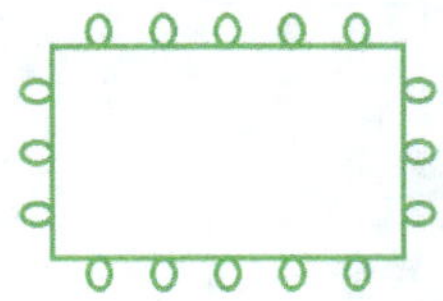
double strand concertina		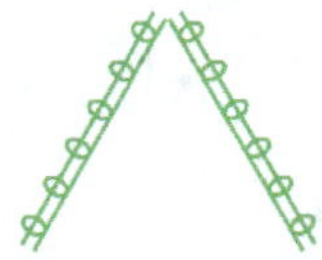
triple strand concertina		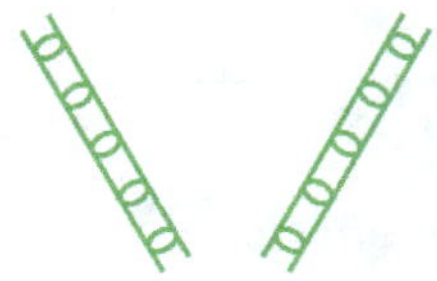
Anti-tank Ditch Triangular shaped or wide diches designed to stop tanks and armor fighting vehicles around a fortified position. (See TM 3-34.85 for more information on anti-tank ditches.)		
anti-tank ditch—under construction	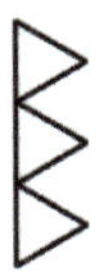	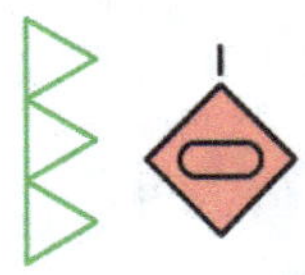
anti-tank ditch—completed		
anti-tank ditch reinforced, with anti-tank mines ***Note***. The teeth typically point toward enemy forces.		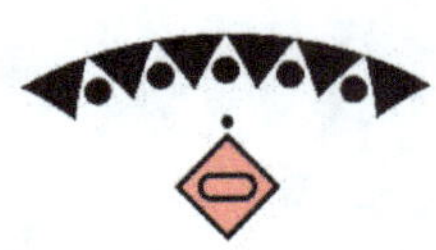

LAND MINE AND MINEFIELD CONTROL MEASURES

5-38. A land mine is a munition on or near the ground or other surface area that is designed to be exploded by the presence, proximity, or contact of a person or vehicle, and a minefield is an area which is dangerous because of the presence or suspected presence of land mines. (See JP 3-15 for more information on land mines.) Table 5-20 on pages 206 through 210 lists land mine symbols and minefield control measure symbols. The listed minefield control measures in table 5-21 (on page 211 through 214) include a modifier field to permit the use of mine modifiers found in table 5-25 (on page 229 through 232), and also have the option to use amplifier fields "H", "N", and "W" listed in table 5-2 (on page 147 through 149).

Table 5-20. Land mine and minefield control measure symbols

Function	***Template***	***Construct Example*** ***Note.*** Symbols colored gray are used to help explain how the control measure is used and are not a part of the control measure.
antipersonnel mine		 1 antipersonnel mine placed next to shelter
	Designed to be exploded by the presence, proximity, or contact of a person that will incapacitate, wound, or kill one or more persons.	
antipersonnel mine with directional effects	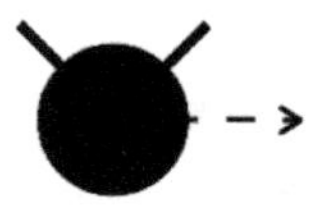	 1 antipersonnel mine with directional effects placed next to shelter
anti-tank mine		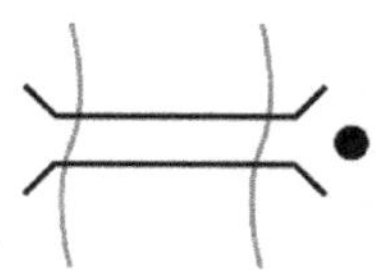 1 anti-tank mine at bridge entrance
	A mine designed to destroy a tank.	
anti-tank mine with anti-handling device		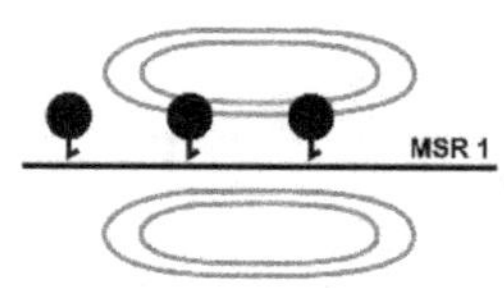 3 anti-tank mines with anti-handling device on main supply route (MSR) 1
	A device intended to protect a mine and which is part of, linked to, or attached to or placed under the mine and which activates when an attempt is made to tamper with or otherwise intentionally disturb the mine.	

Table 5-20. Land mine and minefield control measure symbols (*continued*)

Function	*Template*	*Construct Example*
Wide area anti-tank mine		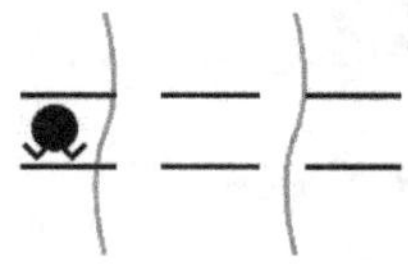 Wide area anti-tank mine at entrance of ford
	An anti-tank mine that detects and acquires targets then launches sub-ammunition that attacks the top of the targets.	
unspecified mine	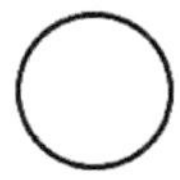	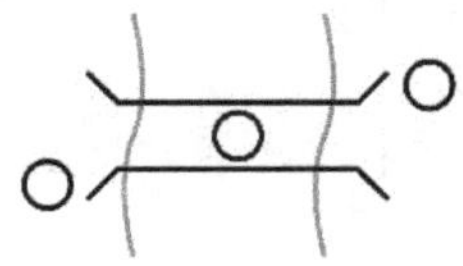 Bridge with 3 unspecified mines
booby trap—A device designed, constructed, or adapted to kill or injure, which functions when a person disturbs or approaches an apparently harmless object or performs an apparently safe act.		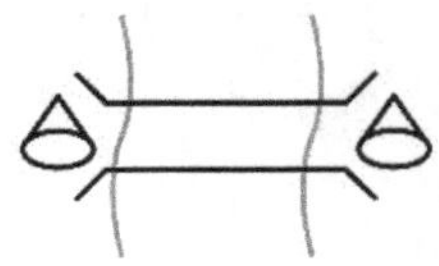 Bridge with 2 booby traps
tripwire—A passive triggering mechanism that activates an attached device by physical movement.	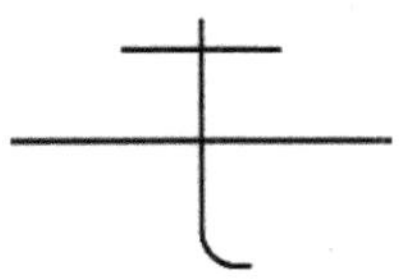	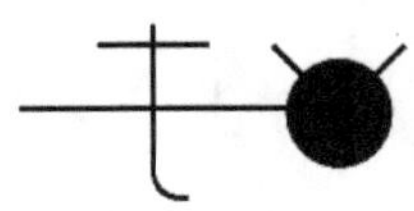 Trip wire attached to antipersonnel mine

Table 5-20. Land mine and minefield control measure symbols (*continued*)

Function	*Template*	*Construct Example*
Minefield In land warfare, an area of ground containing mines emplaced with or without a pattern. (See JP 3-15 for more information concerning minefields.) ***Note***. Minefield control measures use minefield sector 1 modifiers to identify type of minefield. (See table 5-21 on page 211 for depictions of minefield sector 1 modifiers.)		
static depiction minefield	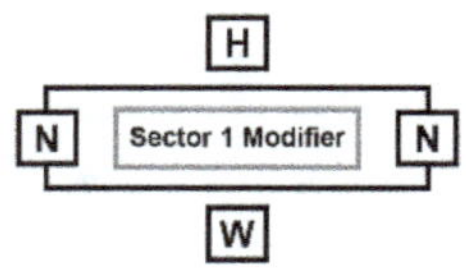	**Monochrome (friendly)** 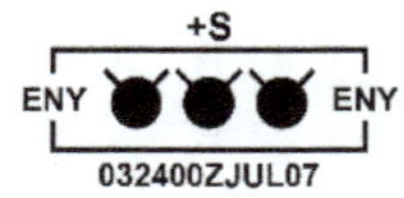 **Color (friendly)** 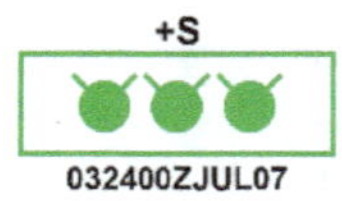
dynamic depiction minefield	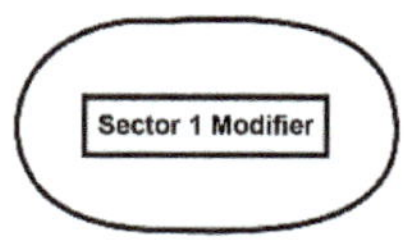	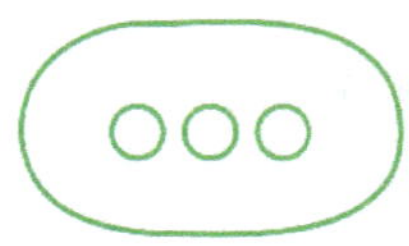 Unspecified dynamic depicted minefield
mined area	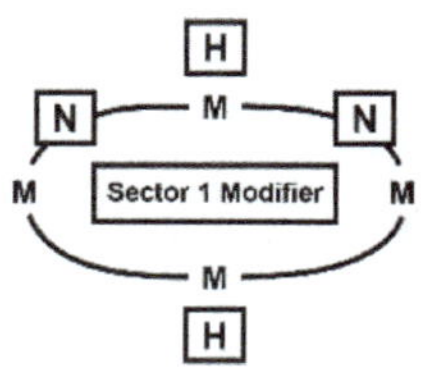	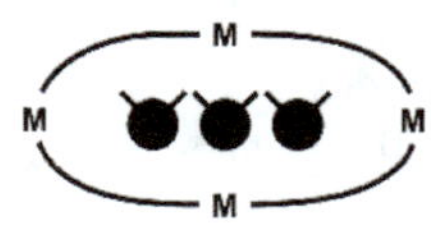 Antipersonnel mined area
mined area, fenced	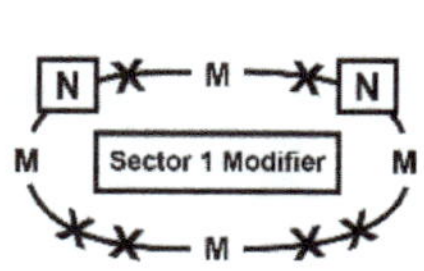	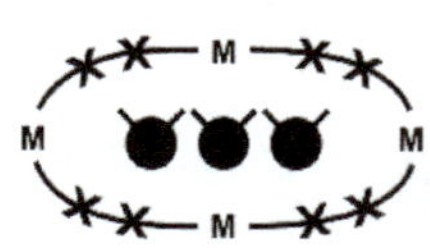 Fenced antipersonnel mined area

MINEFIELD SECTOR 1 MODIFIERS

5-39. Minefield sector 1 modifiers are used with minefield control measure symbols to identify type of minefield. Table 5-21, on pages 211 through 214, provides minefield sector 1 modifiers that can be used in minefield control measure symbols in table 5-24 on page 225 through 228.

Table 5-21. Minefield sector 1 modifiers

Description	*Modifier*
unspecified	
antipersonnel mine	
antipersonnel mine with directional effects	
anti-tank mine	
anti-tank mine with anti-handling device	
wide area anti-tank mine	
mine cluster	
antipersonnel mine and antipersonnel mine with directional effects	
antipersonnel mine and anti-tank mine	
antipersonnel mine and anti-tank mine with anti-handling device	
antipersonnel mine and wide area anti-tank mine	
antipersonnel mine and mine cluster	

Table 5-21. Minefield sector 1 modifiers (*continued*)

Description	*Modifier*
antipersonnel mine with directional effects and anti-tank mine	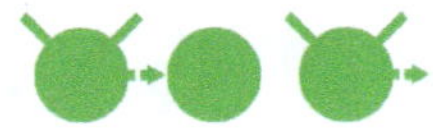
antipersonnel mine with directional effects and anti-tank mine with anti-handling device	
antipersonnel mine with directional effects and wide area anti-tank mine	
antipersonnel mine with directional effects and mine cluster	
anti-tank mine and anti-tank mine with anti-handling device	
anti-tank mine and wide area anti-tank mine	
anti-tank mine and mine cluster	
anti-tank mine with anti-handling device and wide area anti-tank mine	
anti-tank mine with anti-handling device and mine cluster	
wide area anti-tank mine and mine cluster	
antipersonnel mine, antipersonnel mine with directional effects, and anti-tank mine	
antipersonnel mine, antipersonnel mine with directional effects, and anti-tank mine with anti-handling device	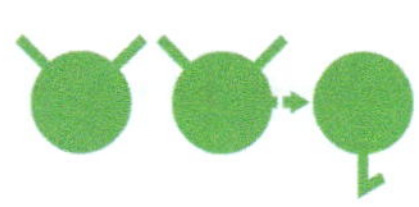

Table 5-21. Minefield sector 1 modifiers (*continued*)

Description	*Modifier*
antipersonnel mine, antipersonnel mine with directional effects, and wide area anti-tank mine	
antipersonnel mine, antipersonnel mine with directional effects, and mine cluster	
antipersonnel mine, anti-tank mine, and anti-tank mine with anti-handling device	
antipersonnel mine, anti-tank mine, and wide area anti-tank mine	
antipersonnel mine, anti-tank mine, and mine cluster	
antipersonnel mine, anti-tank mine with anti-handling device, and wide area anti-tank mine	
antipersonnel mine, anti-tank mine with anti-handling device, and mine cluster	
antipersonnel mine, wide area anti-tank mine, and mine cluster	
antipersonnel mine with directional effects, anti-tank mine, and anti-tank mine with anti-handling device	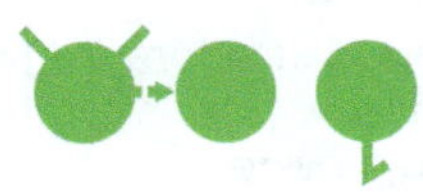
antipersonnel mine with directional effects, anti-tank mine, and wide area anti-tank mine	
antipersonnel mine with directional effects, anti-tank mine, and mine cluster	
antipersonnel mine with directional effects, anti-tank mine with anti-handling device, and wide area anti-tank mine	

Table 5-21. Minefield sector 1 modifiers (*continued*)

Description	*Modifier*
antipersonnel mine with directional effects, anti-tank mine with anti-handling device, and mine cluster	
antipersonnel mine with directional effects, wide area anti-tank mine, and mine cluster	

FIELD FORTIFICATION SYMBOLS

5-40. A field fortification is an emplacement or shelter of a temporary nature, constructed with reasonable facility by units requiring no more than minor engineer supervisory and equipment participation. Table 5-22 lists field fortification obstacle symbols.

Table 5-22. Field fortification symbols

Function	*Template*	*Construct Example and Symbol Translation*
shelter—Something that covers or affords protection.		 Secured shelter
above ground facility—Any visible permanent built-up structure used for housing, storage, business, maintenance, manufacturing, or power generating.		 Cordon and search of an above ground facility
underground facility—A sophisticated complex structure designed and built to be unobserved and to provide maximum protection. (See ATP 3-21.51 for more information on underground facilities.)		 Occupied underground facility
fort—A strong or fortified location.	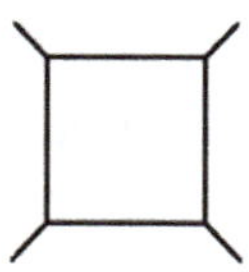	 Enemy guerrilla infantry unit in a fort

Table 5-22. Field fortification symbols (*continued*)

Function	*Template*	*Construct Example and Symbol Translation*
fortified/trench line—A built up or dug in line of defending troops with supporting armament organized to protect a location. ***Note***. The ramparts typically point toward enemy forces.	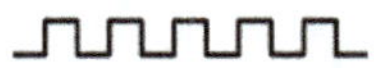	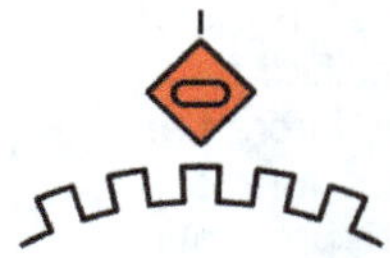
fighting position—A location on the ground from which fire is delivered by an individual, a fire unit (squad or fire team), or a crew served weapon.		

AIRSPACE COORDINATING MEASURES

5-41. Airspace coordinating measures employed to facilitate the efficient use of airspace to accomplish missions and simultaneously provide safeguards for friendly forces. (See JP 3-52 for more information on airspace coordinating measures.) Airspace coordinating measures are used to segregate, control, and reserve airspace for operations. Airspace coordinating measures are also used to—

- Enhance the effectiveness of accomplishing the commander's objectives.
- Prevent mutual interference.
- Facilitate air defense identification.
- Prevent fratricide.
- Help in safely accommodating the flow of all air traffic in the area of operations.

Table 5-23 on pages 213 through 221 lists airspace coordinating measure symbols. (See ATP 3-52.1 for additional information on airspace coordinating and associated military symbol doctrine.

Table 5-23. Airspace coordinating measure symbols

Air Corridors
Template
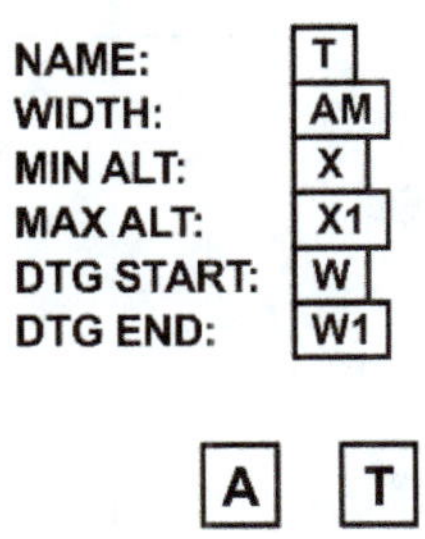
A—Main function symbol identifies function. AM—A numeric amplifier that displays a minimum, maximum, or specific distance (including range, radius, width, or length) in meters or feet. T—A free text amplifier that uniquely identifies the corridor. X, X1—Identifies the minimum, maximum, or specific altitude (in feet or meters in relation to a reference datum), flight level, or depth. W, W1—Identifies the date-time group associated with corridor.

Table 5-23. Airspace coordinating measure symbols (*continued*)

Function	*Main function symbol (Field A)*	*Construct Example and Symbol Translation*
air corridor—A restricted air route of travel specified for use by friendly aircraft and established for the purpose of preventing friendly aircraft from being fired on by friendly forces. (JP 3-52)	**AC**	**NAME: GOLD** **WIDTH: 1200FT** **MIN ALT: 1500FT AGL** **MAX ALT: 20000FT AGL** **DTG START: 270600ZMAY08** **DTG END: 271845ZMAY08** 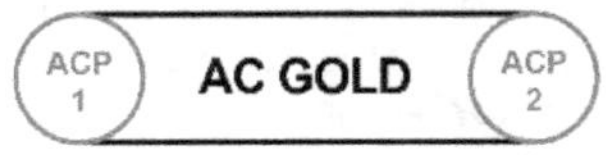 Air corridor GOLD including air control points
		NAME: GOLD **WIDTH: 1200FT** **MIN ALT: 1500FT AGL** **MAX ALT: 16000FT AGL** **DTG START: 240700ZSEP08** **DTG END: 280700ZSEP08** 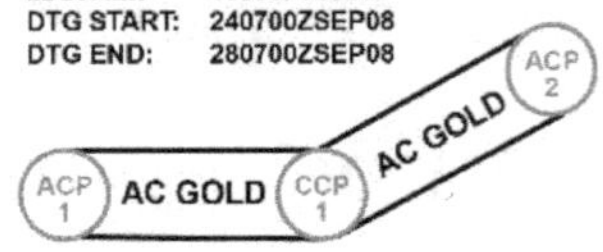 Air corridor GOLD with multiple segments, including air control points and a communication control point
low-level transit route—A temporary air corridor of defined dimensions established to minimize the risk to friendly aircraft from friendly air defenses. (JP 3-52)	**LLTR**	**NAME: COBRA** **WIDTH: 300FT** **MIN ALT: 150FT AGL** **MAX ALT: 3000FT AGL** **DTG START: 240500ZOCT08** **DTG END: 241845ZOCT08**  Low-level transit route air corridor Red, width 300 feet, minimum altitude 150 feet, maximum altitude 3000 feet, operational from 0500 Zulu 24 OCT 2008 to 1845 Zulu 24 OCT 2008 (between air control points 1 and 2)
minimum-risk route—A temporary air corridor of defined dimensions used by aircraft that presents the minimum known hazards to low-flying aircraft transiting the combat zone. Also called MRR. (JP 3-52)	**MRR**	**NAME: RED** **WIDTH: 1500FT** **MIN ALT: 3000FT AGL** **MAX ALT: 21000FT AGL** **DTG START: 110200ZSEP08** **DTG END: 140300ZSEP08**  Minimum-risk route air corridor RED, width 1500 feet, minimum altitude 3000 feet, maximum altitude 21000 feet, operational from 0200 Zulu 11 SEP 2008 to 0300 Zulu 14 SEP 2008 (between air control points 1 and 2)

Table 5-23. Airspace coordinating measure symbols (*continued*)

Function	*Main function symbol (Field A)*	*Construct Example and Symbol Translation*
safe lane—A bi-directional lane connecting an airbase, landing site, or base defense zone to adjacent routes or corridors. Safe lanes may also be used to connect adjacent activated routes or corridors. (See ATP 3-52.1 for more information on safe lanes.)	**SL**	NAME: RED WIDTH: 1500FT MIN ALT: 3000FT AGL MAX ALT: 21000FT AGL DTG START: 110200ZSEP08 DTG END: 140300ZSEP08 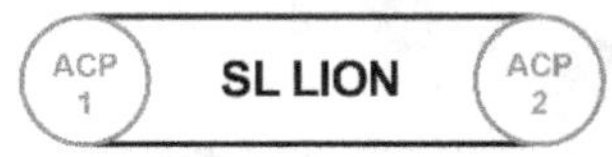 Safe lane air corridor LION that includes air control points
special corridor—A corridor used to accommodate the special routing requirements of specific missions and are used for special operations aircraft. (See ATP 3-52.1 for more information on special corridors.)	**SC**	NAME: OWL WIDTH: 500M MIN ALT: 100M MAX ALT: 1200M DTG START: 220700ZJUN20 DTG END: 300700ZJUN20  Special corridor OWL including air control points
standard use Army aircraft flight route—Route established below the coordination level to facilitate the movement of Army aviation assets; it is normally located in the corps through brigade rear areas of operation and does not require approval by the airspace control authority. (JP 3-52)	**SAAFR**	NAME: BLUE WIDTH: 600FT MIN ALT: 150FT AGL MAX ALT: 3000FT AGL DTG START: 260930ZMAY08 DTG END: 280700ZMAY08 
transit corridor—Bi-directional in the rear area and established to route aircraft through air defenses. (See ATP 3-52.1 for more information on transit corridor.)	**TC**	NAME: KING WIDTH: 900FT MIN ALT: 2100FT AGL MAX ALT: 6000FT AGL DTG START: 260700ZMAR08 DTG END: 280700ZMAR08 

Table 5-23. Airspace coordinating measure symbols (*continued*)

Function	*Main function symbol (Field A)*	*Construct Example and Symbol Translation*
unmanned aircraft (UA) corridor—Airspace established by the airspace control authority, based on the UA mission. (See ATP 3-52.1 for more information on an UA corridor.)	UA	NAME: DRAGON WIDTH: 1200FT MIN ALT: 1500FT AGL MAX ALT: 12000FT AGL DTG START: 240700ZMAY08 DTG END: 210700ZMAY08 ACP 1 UA DRAGON ACP 2
Control Points		
air control point—A point easily defined and used for navigation, command and control, and communication. The point may be a terrain feature or an electronic navigational air. (See JP 3-52, ATP 3-52.1, and ATP 3-04.1 for more information on air control points.)	ACP T	NAME: GOLD WIDTH: 700M MIN ALT: 500M MAX ALT: 400M DTG START: 240700ZMAY12 DTG END: 290700ZMAY12 ACP 1 AC GOLD ACP 2 Air corridor GOLD with air control points 1 and 2
Identification Lines ***Note***. Symbols without amplifier fields cannot be modified and must be used as depicted.		
identification, friend-or-foe switch off-line—Line marks where friendly aircraft stop emitting an identification, friend-or-foe signal. (See ATP 3-52.1 for more information on identification, friend-or-foe switch-off lines.)	IFF OFF IFF OFF	
identification, friend-or-foe switch on-line—Line marks where friendly aircraft start emitting an identification, friend-or-foe signal. (See ATP 3-52.1 for more information on identification, friend-or-foe switch-on lines.)	IFF ON IFF ON	

Table 5-23. Airspace coordinating measure symbols (*continued*)

Function	***Main function symbol (Field A)***	***Construct Example and Symbol Translation***
Engagement Zone Symbols with Amplifier Fields		
Template		
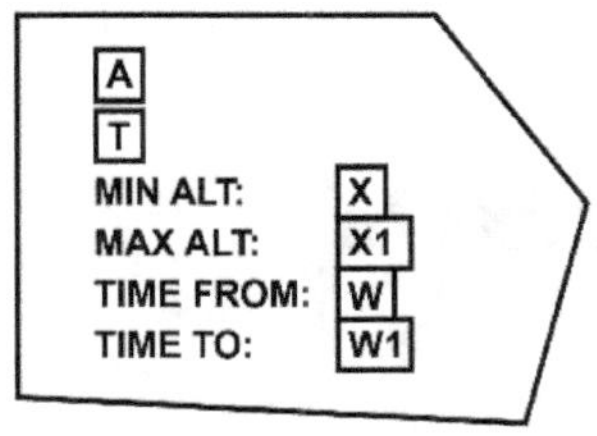 A—Main function symbol that identifies the function. T—Identifies the unique unit designation servicing the zone. X, X1—Identifies the minimum, maximum, and/or specific altitude (in feet or meters in relation to a reference datum), flight level, or depth. W, W1—Identifies the date-time group associated with zone.		
high-density airspace control zone—Airspace designated in an airspace control plan or airspace control order in which there is a concentrated employment of numerous and varied weapons and airspace users. (JP 3-52)	**HIDACZ**	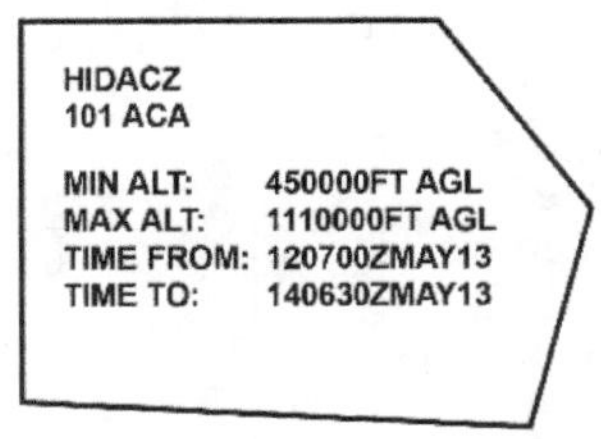 High-density airspace control zone, established by 101st airborne control authority for a minimum altitude of 450000 feet above ground level and a maximum altitude of 1110000 feet above ground level, operational from 0700 Zulu hour 14 MAY 2013
Restricted Operations Zones		
restricted operations zone—Airspace reserved for specific activities in which the operations of one or more airspace users is restricted. (JP 3-52)	**ROZ**	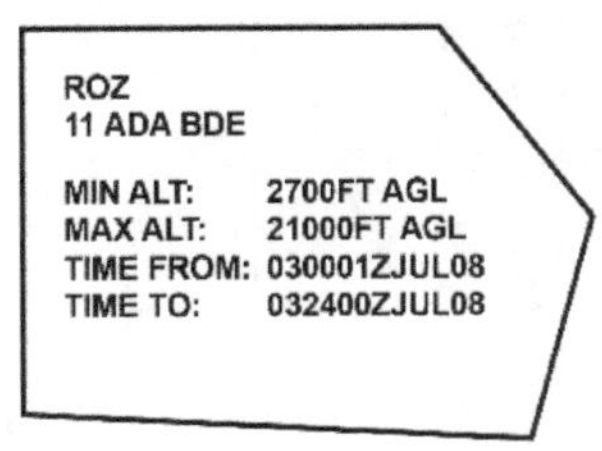 Restricted operations zone, designated by 11th Air Defense Artillery Brigade for a minimum altitude of 2700 feet above ground level and a maximum altitude of 21000 feet above ground level, operational from 0001 Zulu hour 03 JULY 2008 to 2400 Zulu hour 03 JULY 2008

Table 5-23. Airspace coordinating measure symbols (*continued*)

<table>
<tr><th>Function</th><th>Main function symbol (Field A)</th><th>Construct Example and Symbol Translation</th></tr>
<tr><td>air-to-air refueling restricted operations zone—Airspace of defined dimensions set aside for aerial refueling operations. (See ATP 3-52.1 for more information on air-to-air refueling restricted operations zones.)</td><td>AARROZ</td><td>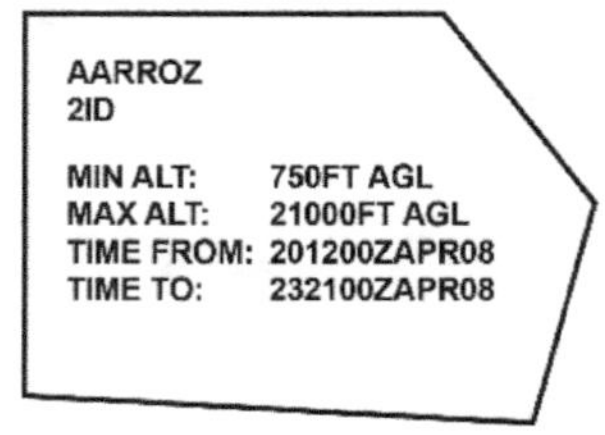

Air-to-air refueling restricted operations zone, designated by 2 Infantry Division for a minimum altitude of 750 feet above ground level, operational from 1200 Zulu hour 20 APR 2013 to 2100 Zulu hour 23 APR 2013</td></tr>
<tr><td>unmanned aircraft restricted operations zone—Airspace of defined dimensions created for unmanned aircraft system operations. (See ATP 3-52.1 for more information on unmanned aircraft restricted operations zones.)</td><td>UAROZ</td><td>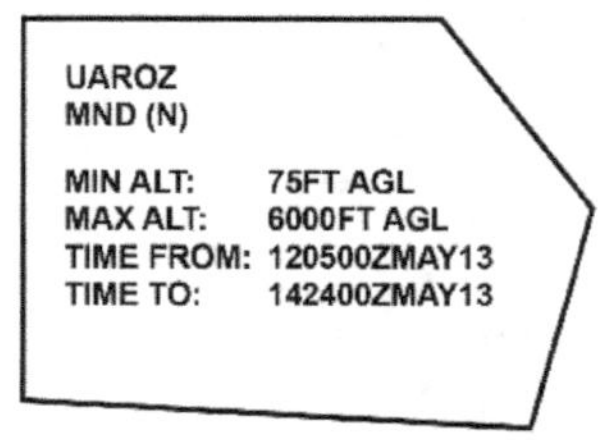

Unmanned aircraft restricted operations zone, designated by Multinational Division (North) for a minimum altitude of 75 feet above ground level and a maximum altitude of 6000 feet above ground level, operational from 0500 Zulu hour 12 MAY 2013 to 2400 Zulu hour 14 May 2013</td></tr>
<tr><td colspan="3">Engagement Zones</td></tr>
<tr><td>weapon engagement zone—A defined airspace in which a particular weapon system engages air and missile threats. (JP 3-01)</td><td>WEZ</td><td>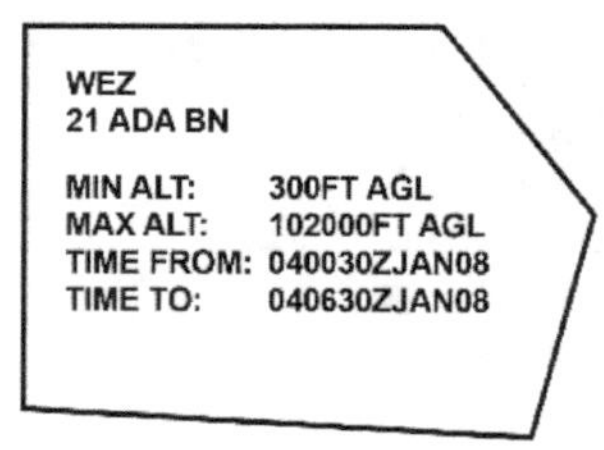

Weapon engagement zone, designated by 21st Air Defense Artillery for a minimum altitude of 300 feet above ground level and a maximum altitude of 102000 feet above ground level, operational from 0030 Zulu hour 4 JAN 2008 to 0630 Zulu hour 14 JAN 2008</td></tr>
</table>

Table 5-23. Airspace coordinating measure symbols (*continued*)

Function	*Main function symbol (Field A)*	*Construct Example and Symbol Translation*
joint engagement zone—A defined airspace in which surface-to-air missiles and aircraft are simultaneously employed to engage air and missile threats. (JP 3-01)	**JEZ**	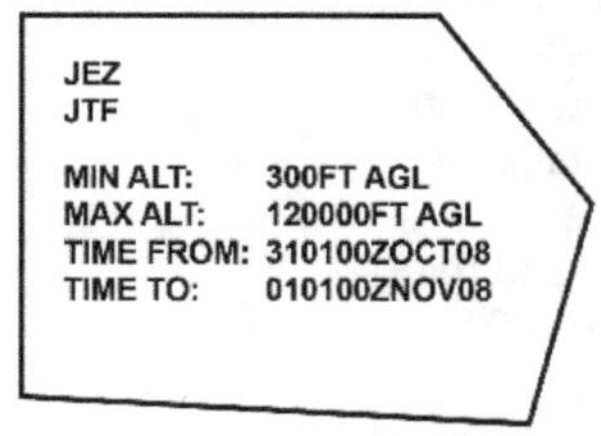 Joint engagement zone, designated by joint task force for a minimum altitude of 300 feet above ground level and a maximum altitude of 120000 feet above ground level, operational from 0100 Zulu hour 31 OCT 2008 to 0100 Zulu hour 1 NOV 2008
missile engagement zone—In air and missile defense, that airspace of defined dimensions within which the responsibility for engagement of air and missile threats normally rests with surface-to-air missile systems. (JP 3-01)	**MEZ**	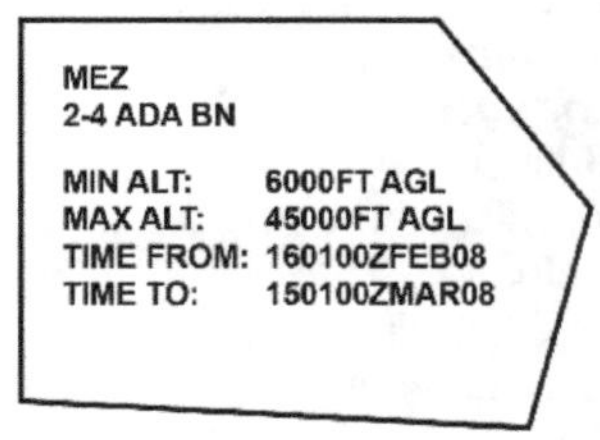 Missile engagement zone, designated by 2nd Company, 4th Air Defense Artillery Battalion for a minimum altitude of 6000 feet above ground level and a maximum altitude of 45000 feet above ground level, operational from 0100 Zulu hour 16 FEB 2008 to 0100 Zulu hour 15 MAR 2008
low-altitude missile engagement zone—A defined airspace in which low- to medium altitude surface-to-air missiles engage air and missile threats. (JP 3-01)	**LOMEZ**	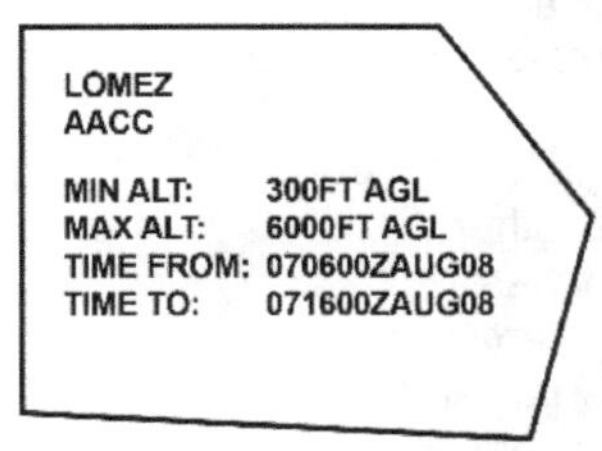 Low (altitude) missile engagement zone, designated by anti-air coordination center, for a minimum altitude of 300 feet above ground level and a maximum altitude of 6000 feet above ground level, operational from 0600 Zulu hour 7 AUG 2008 to 1600 Zulu hour 7 AUG 2008

Table 5-23. Airspace coordinating measure symbols (*continued*)

<table>
<tr><th>Function</th><th>Main function symbol (Field A)</th><th>Construct Example and Symbol Translation</th></tr>
<tr><td>high-altitude missile engagement zone—A defined airspace in which high-altitude surface-to-air missiles engage air and missile threats. (JP 3-01)</td><td>HIMEZ</td><td>
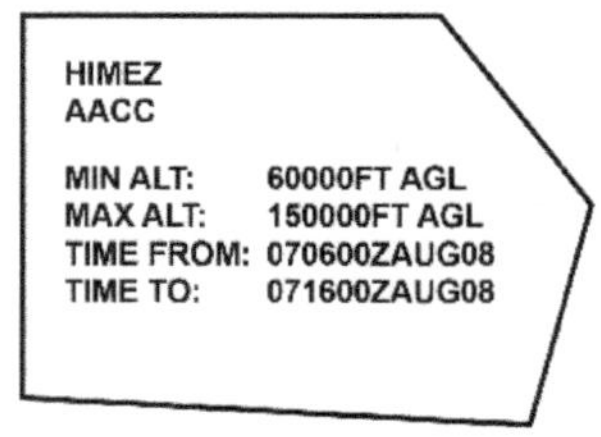

High (altitude) missile engagement zone, designated by anti-air coordination center, for a minimum altitude of 6000 feet above ground level and a maximum altitude of 150000 feet above ground level, operational from 0600 Zulu hour 7 AUG 2008 to 1600 Zulu hour 7 AUG 2008</td></tr>
<tr><td>short-range air defense engagement zone—A defined airspace in which short-range air defense weapons engage air threats and which may be established within a low- or high-altitude missile engagement zone. (JP 3-01)</td><td>SHORAD
EZ</td><td>
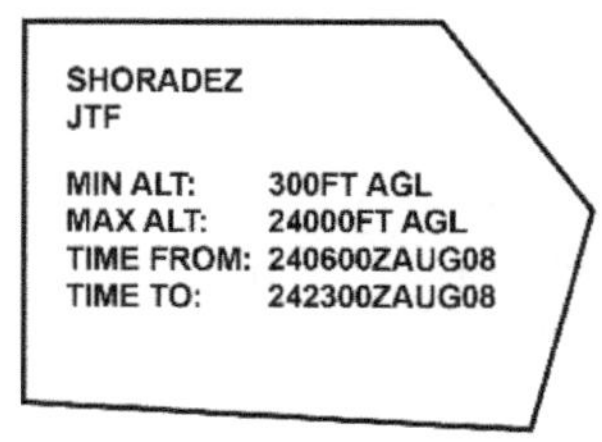

Short-range air defense engagement zone, designated by joint task force, for a minimum altitude of 300 feet above ground level and a maximum altitude of 24000 feet above ground level, operational from 0600 Zulu hour 24 AUG 2008 to 2300 Zulu hour 24 AUG 2008</td></tr>
<tr><td>weapons free zone—An air defense zone established for the protection of key assets or facilities, other than air bases, where weapon systems may be fired at any target not positively recognized as friendly. (JP 3-01)</td><td>WFZ</td><td>
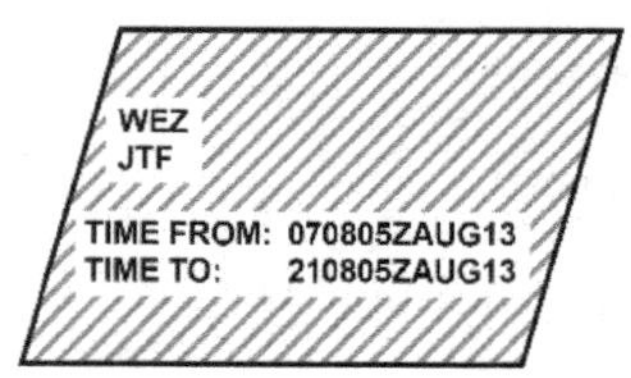

Weapons free zone, designated by joint task force, operational from 0805 Zulu 7 AUG 2013 to 0805 Zulu hour 21 AUG 2013</td></tr>
</table>

Table 5-23. Airspace coordinating measure symbols (*continued*)

Function	*Template*	*Usage Example*
Airspace coordination area A three-dimensional block of airspace in a target area, established by the appropriate commander, in which friendly aircraft are reasonably safe from friendly surface fires. (JP 3-09.3)		
airspace coordination area, irregular	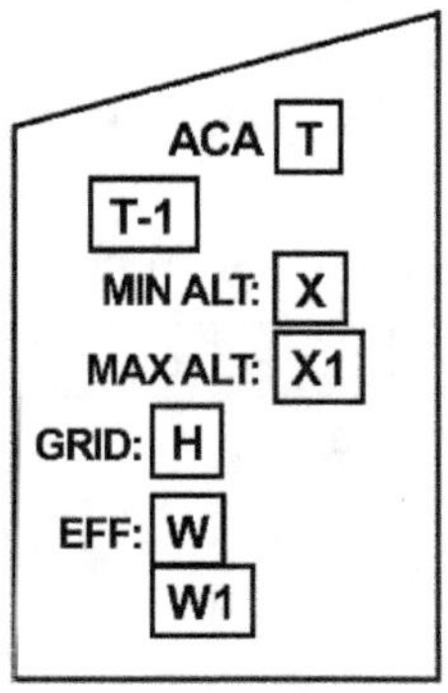	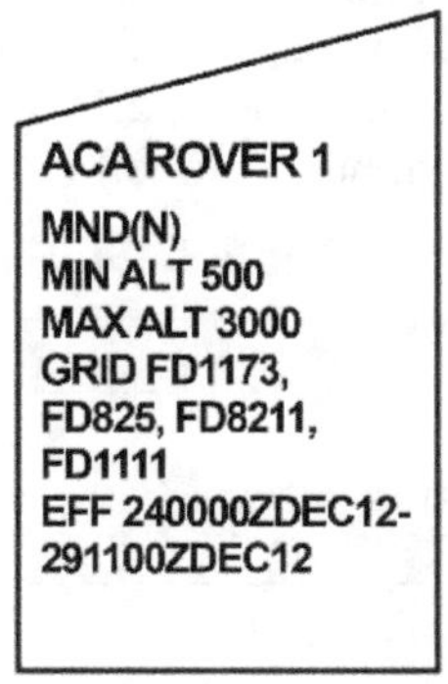
airspace coordination area, rectangular	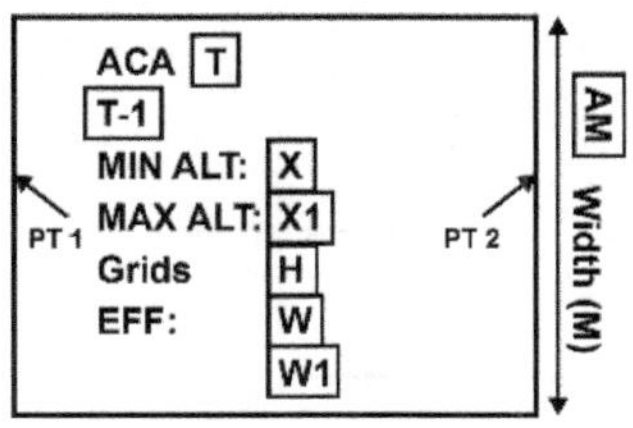	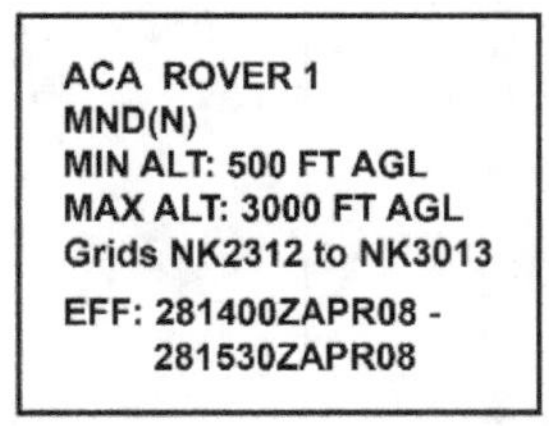
airspace coordination area, circular	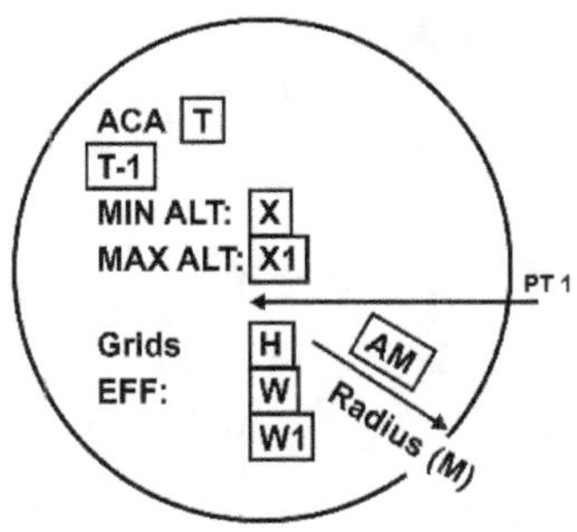	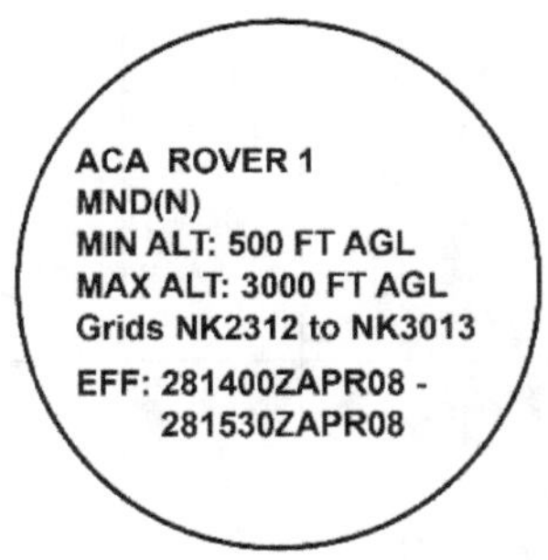
Zones Without Amplifier Fields		
base defense zone—An air defense zone established around an air base and limited to the engagement envelope of short-range air defense weapons systems defending that base. (JP 3-52)		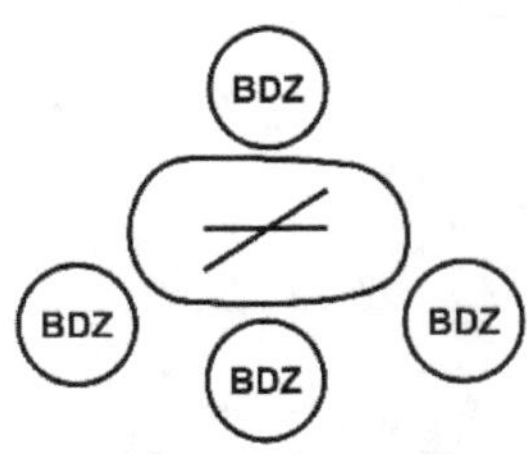

FIRE SUPPORT COORDINATION CONTROL MEASURES

5-42. Fire support coordination measures are employed by land or amphibious commanders to facilitate the rapid engagement of targets and simultaneously provide safeguards for friendly forces. Fire support coordination control measures should be labeled with the abbreviation of the control measures, the controlling headquarters (Field T), and the effective times (Fields W and W1). Lines should have this labeling on both ends of the line and repeated as often as necessary for clarity along any line that passes through boundaries. Table 5-24 on pages 222 through 225 lists fire support coordination control measure symbols. (See JP 3-0, FM 3-09, and ATP 3-52.1 for additional information on fire support coordination measures and associated military symbol doctrine.)

Table 5-24. Fire support coordination control measure symbols

Function	*Template*	*Construct Example*
Free-fire area A specific area into which any weapon system may fire without additional coordination with the establishing headquarters. (JP 3-09)		
free-fire area, irregular	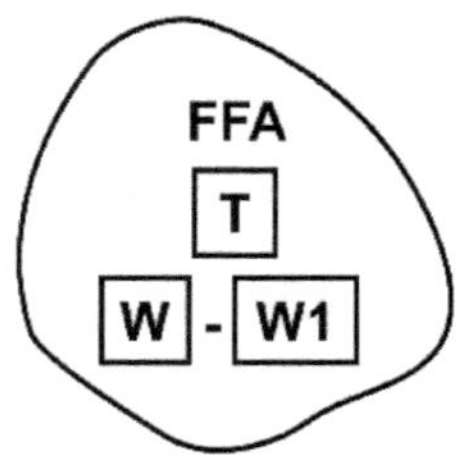	
free-fire area, rectangular	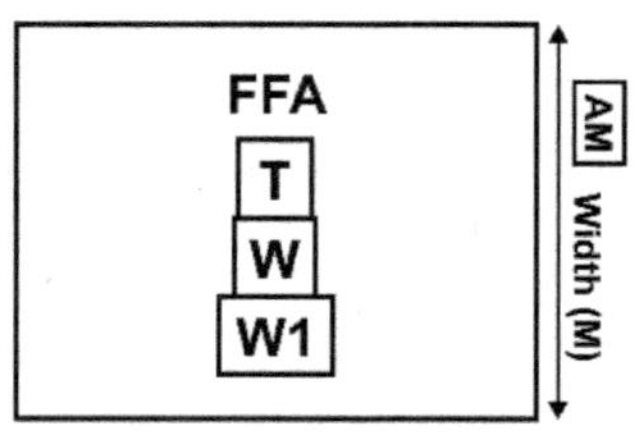	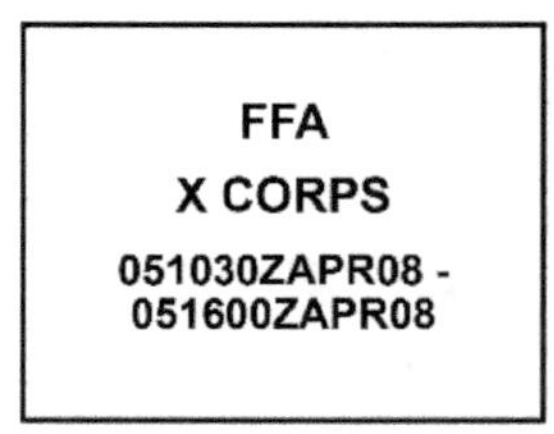
free-fire area, circular	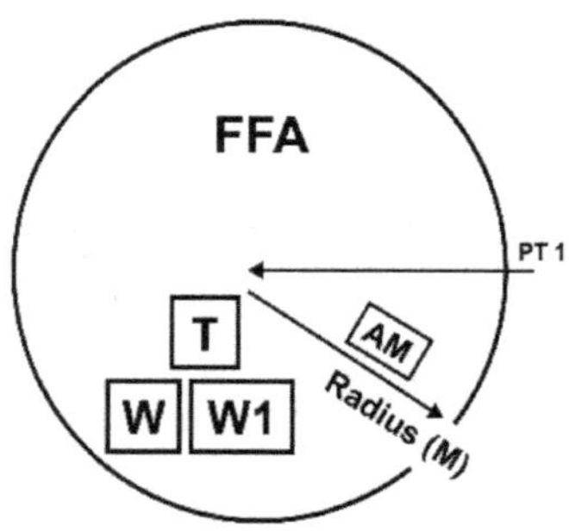	
No-fire area An area designated by the appropriate commander into which fires or their effects are prohibited. (JP 3-09.3)		

BODY-234

Table 5-24. Fire support coordination control measure symbols (*continued*)

Function	*Template*	*Construct Example*
no-fire area, irregular	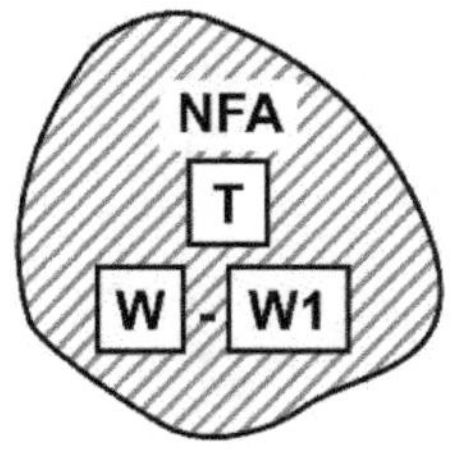	
no-fire area, rectangular	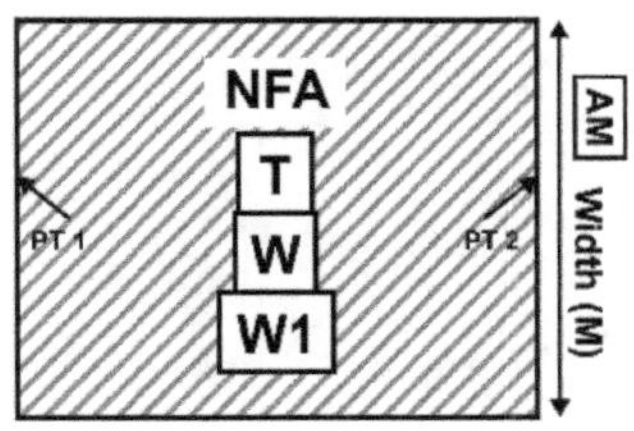	
no-fire area, circular	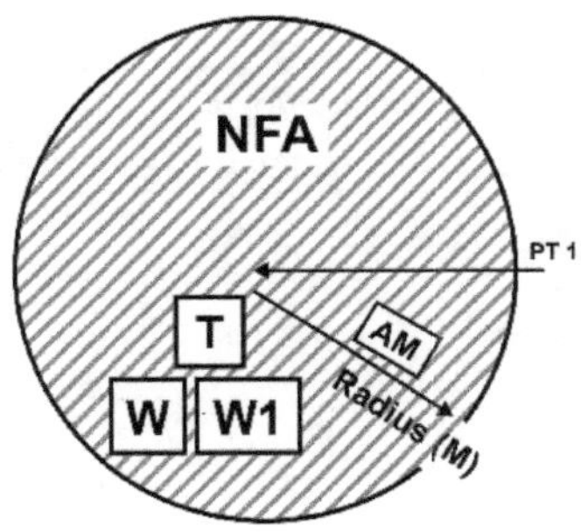	
Restrictive fire area A location in which specific restrictions are imposed and into which fires that exceed those restrictions will not be delivered without coordination with the establishing headquarters. (JP 3-09)		
restrictive fire area, irregular	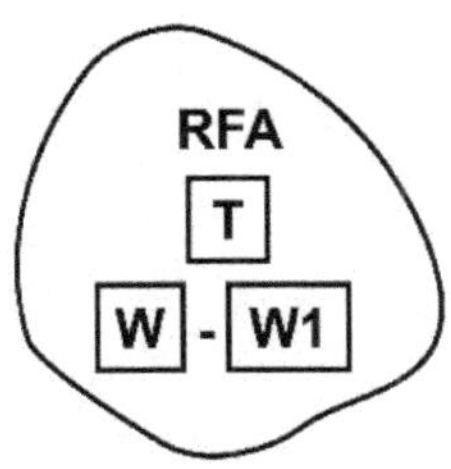	

Table 5-24. Fire support coordination control measure symbols (*continued*)

Function	*Template*	*Construct Example*
restrictive fire area, rectangular	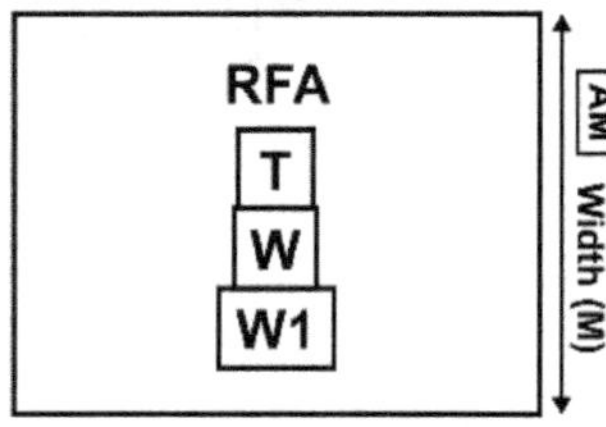	
restrictive fire area, circular	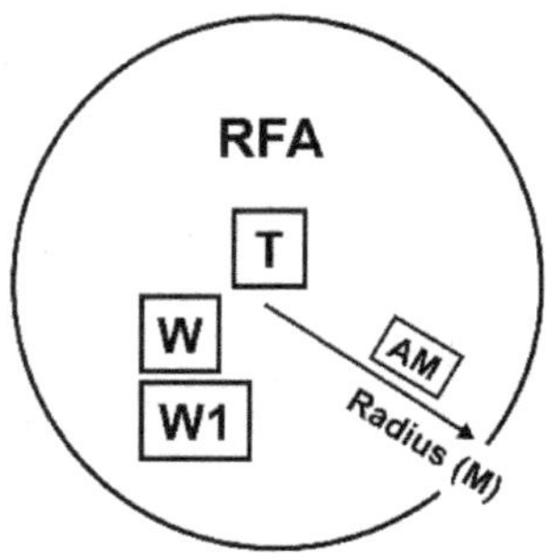	
Position area for artillery An area assigned to an artillery unit to deliver surface to surface fires. (FM 3-90)		
position area for artillery, irregular	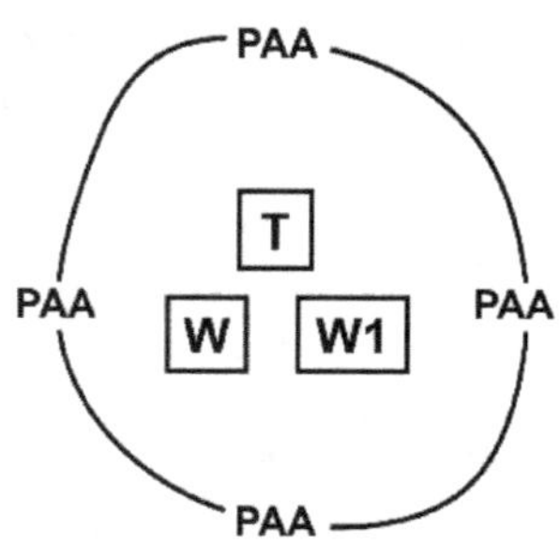	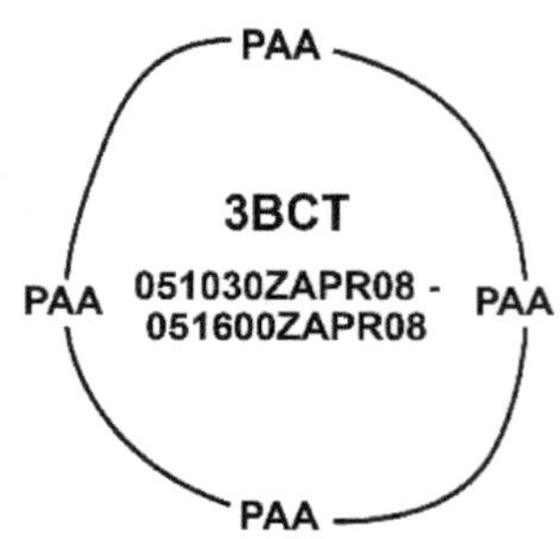
position area for artillery, rectangular	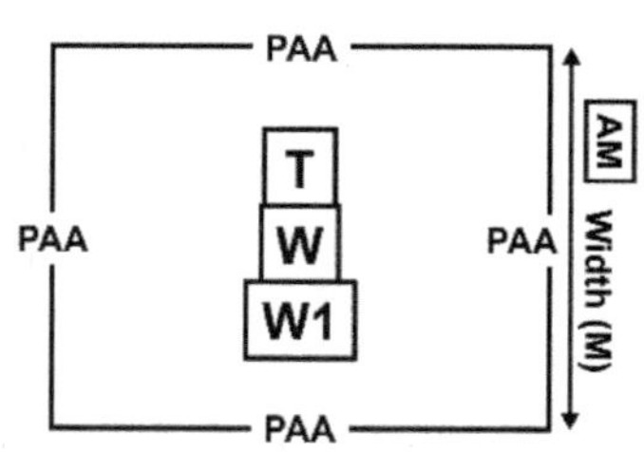	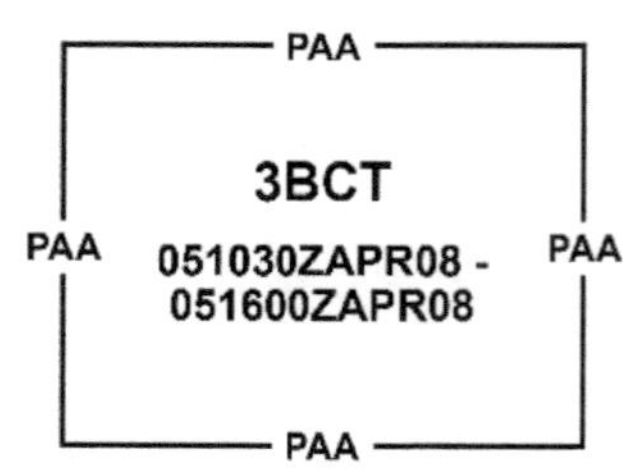

Table 5-24. Fire support coordination control measure symbols (*continued*)

Function	*Template*	*Construct Example*
position area for artillery, circular	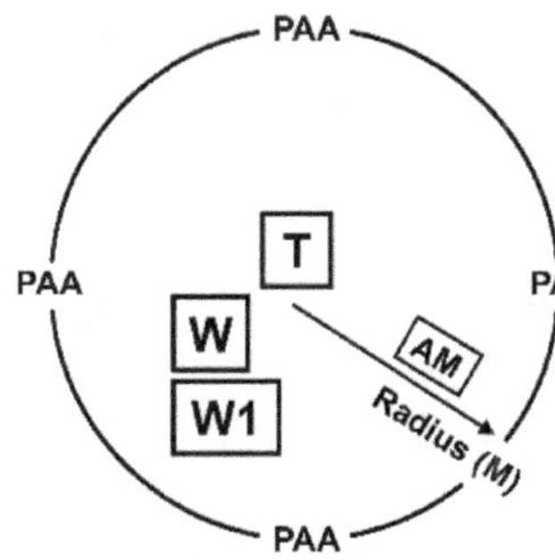	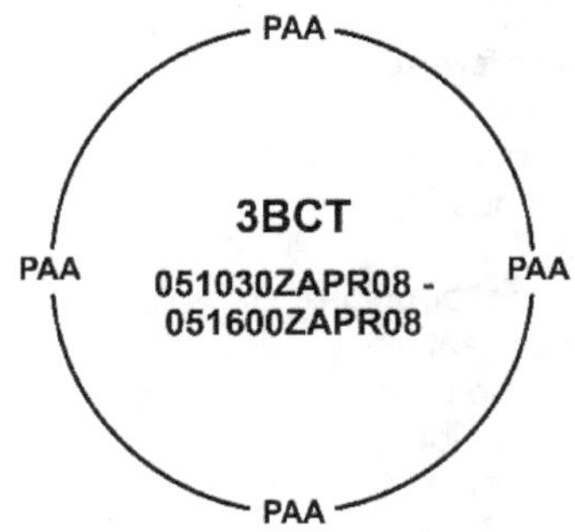
Direction of fire The direction on which a fire unit is laid to the most significant threat in the target area, to the chart direction to the center of the zone of fire, or to the target. (ATP 3-09.50)		
fields of fire/sector of fire—The area into which you must observe and fire. (See TC 3-21.75 for more information about fields of fire/sector of fire.)	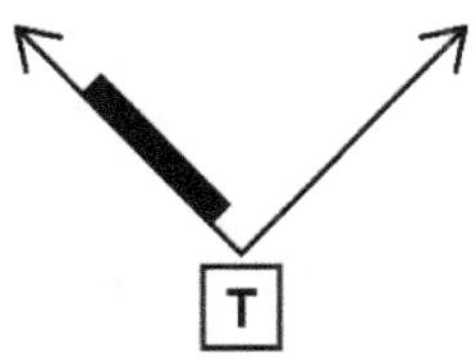	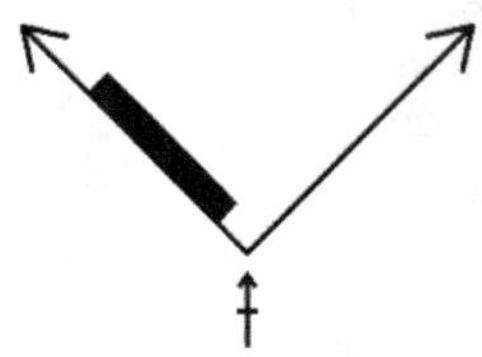
munition flight path (MFP)—Used to display the trajectory of inflight munition from point of origin to target area, (See ATP 3-09.30 or ATP 3-09.23 for more information about munition flight paths.)	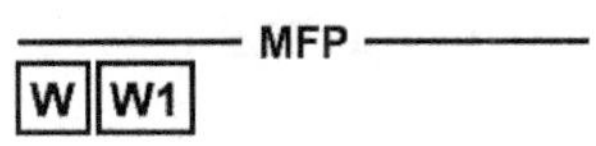	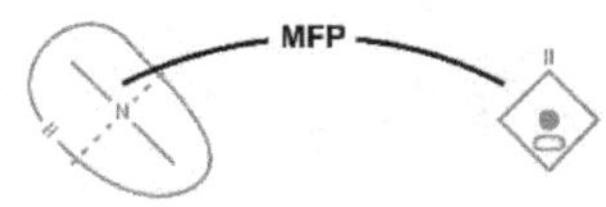 A munition flight path of an enemy artillery battery neutralizing a battalion battle position
	Note 1. "MFP" displayed once at the approximate center of the overall length of the munition flight path. ***Note 2.*** The munition flight path begins at a weapons system or surface-to-surface fires unit and terminates at a target. ***Note 3.*** Date-time groups is optional. The effective date-time group of the munition flight path is the shot or launch time of the projectile, and the expiration date-time group is the splash or time of impact of the projectile.	

TARGET CONTROL MEASURES

5-43. A target is the object of a particular action, for example a geographic area, a complex, an installation, a force, equipment, an individual, a group, or a system, planned for capture, exploitation, neutralization, or destruction by military forces. Table 5-25 on pages 226 through 229 lists target control measures symbols. (See FM 3-09 for additional information on target control measures and associated military symbol doctrine.)

Table 5-25. Target control measure symbols

Function	*Template*	*Construct Example*
	Target Points	
target reference point—A predetermined point of reference, normally a permanent structure or terrain feature that can be used when describing a target location. (JP 3-09.3)	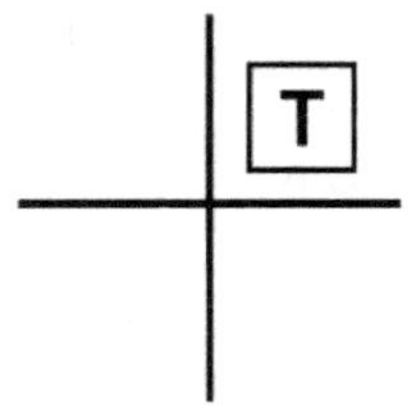 T	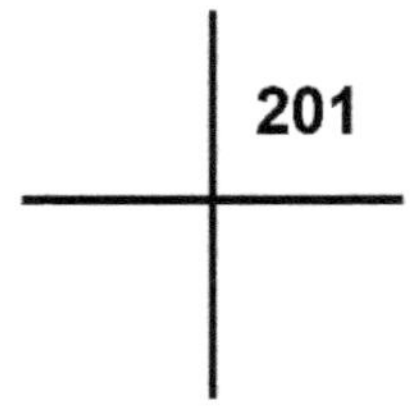 201 Target reference point 201
point target—A target that is less than or equal to 200 meters in width and length. (See ATP 3-09.30 for more information on point targets.)	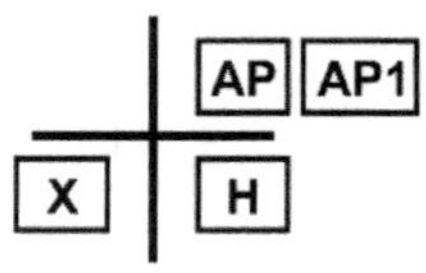 AP AP1 X H	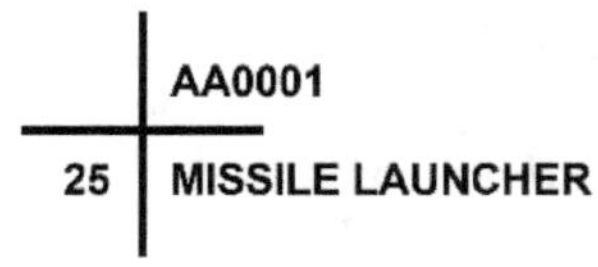 AA0001 25 MISSILE LAUNCHER 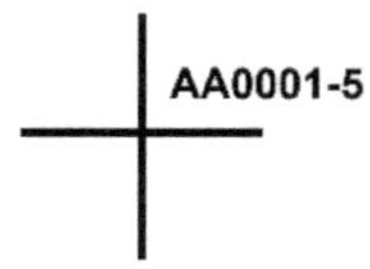 AA0001-5
	Linear Targets Targets that are greater than 200 meters in length and less than or equal to 200 meters in width. (See ATP 3-09.30 for more information on linear targets.)	
linear target	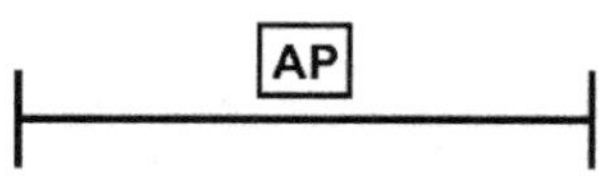 AP	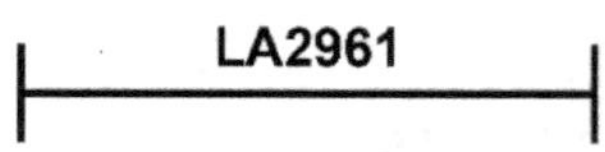 LA2961
linear smoke target	 AP SMOKE	 VB1910 SMOKE
final protective fire—An immediately available prearranged barrier of fire designed to impede enemy movement across defensive lines or areas. (JP 3-09.3)	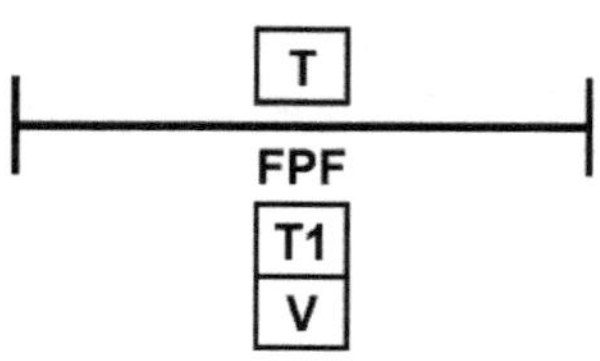 T FPF T1 V	 QC1968 FPF 12 IN BN MORTAR

Table 5-25. Target control measure symbols (*continued*)

Function	*Template*	*Construct Example*
	Area Targets	
target area, irregular—(See ATP 3-09.30 for more information on area targets.)	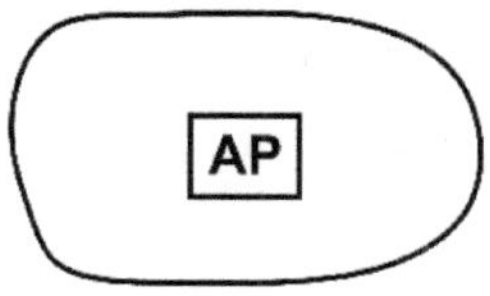	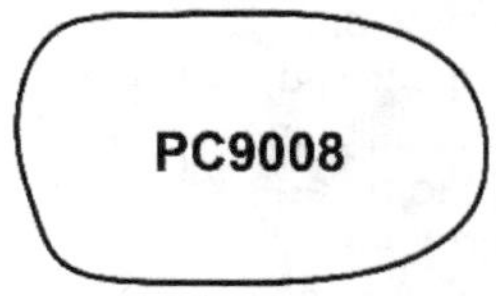
target area, rectangular—A target that is greater than 200 meters in length and width described by four grids or by a center grid, a length, width, and an altitude. (See ATP 3-09.30 for more information on rectangular targets.)	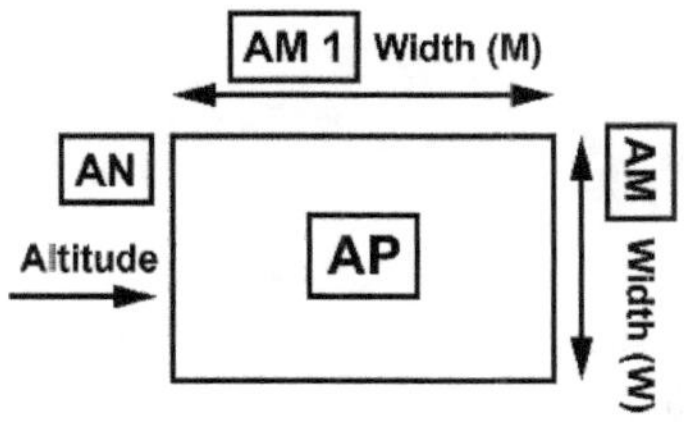	
target area, circular—A target that is in a circular pattern or is vague as to exact composition and has a radius greater than 100 meters. (See ATP 3-09.30 for more information on circular targets.)	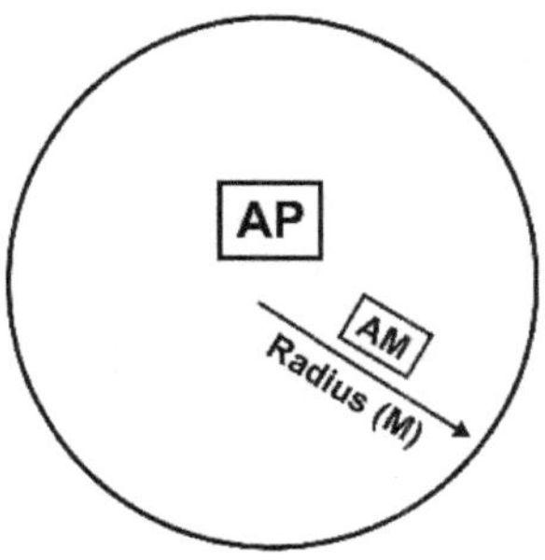	

Table 5-25. Target control measure symbols (*continued*)

Function	*Template*	*Construct Example*
group of targets—Consists of two or more targets on which fire desired simultaneously and is designed by a letter and number combination or a nickname.		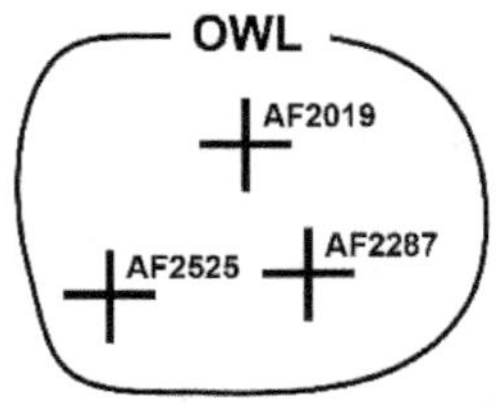
series of targets—A number of targets or group(s) of targets planned to be fired in a predetermined sequence to support a maneuver operation. (See ATP 3-09.30 for more information on series of targets.)	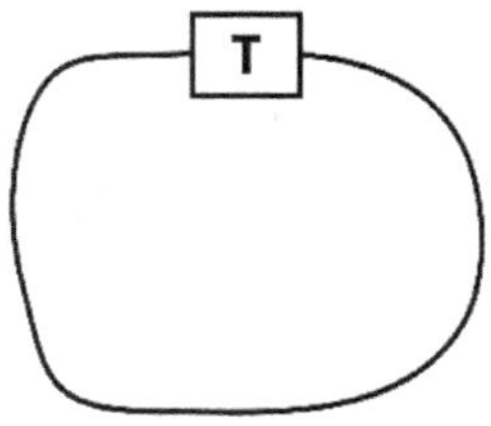	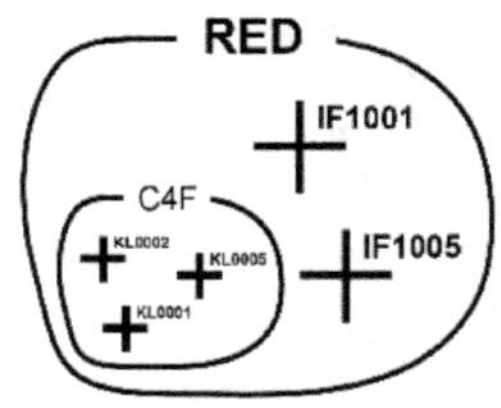
		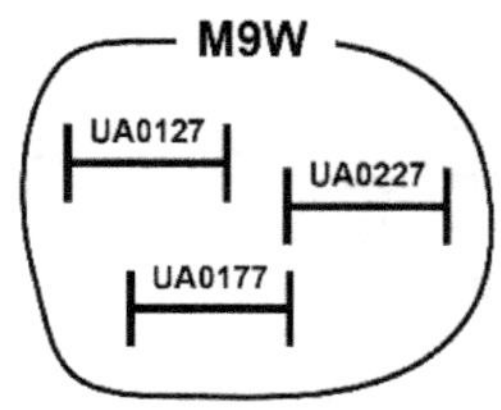 Linear targets
		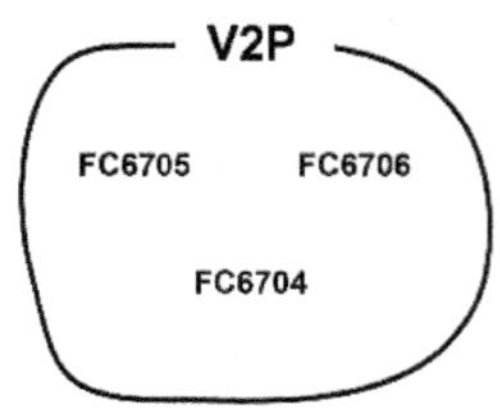 Area targets
smoke obscurant—Airborne particulates and gases generated from combustion or pyrolysis to mask the movement or location of military units.		
		 Planned

Table 5-25. Target control measure symbols (*continued*)

Function	*Template*	*Construct Example*
Fire Support Area An appropriate maneuver area assigned to fire support ships by the naval force commander from which they can deliver gunfire support to an amphibious operation. (JP 3-09)		
fire support area, irregular	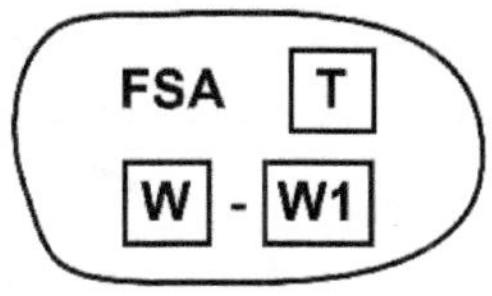	
fire support area, rectangular	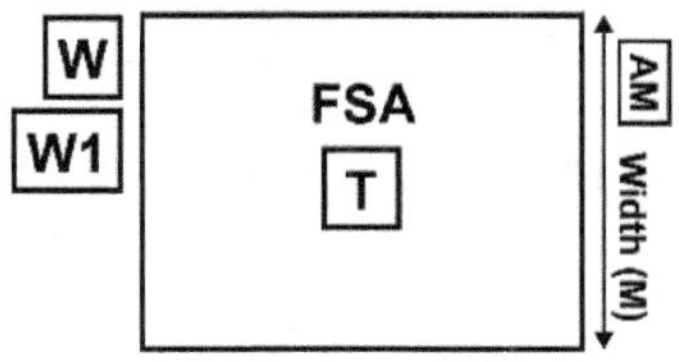	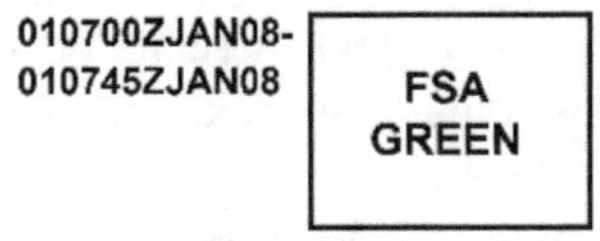
fire support area, circular	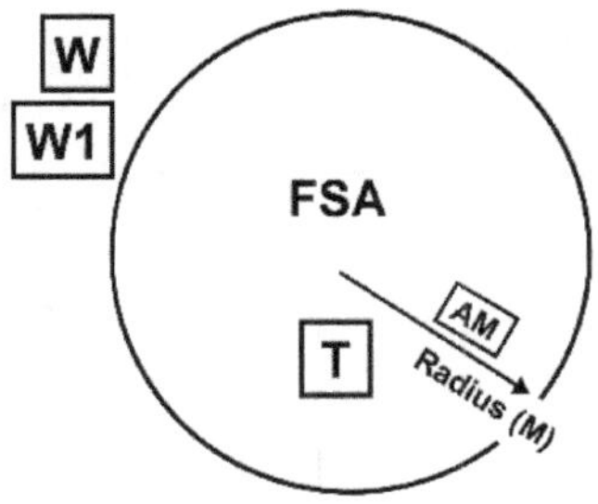	
Naval Gunfire		
fire support station—An exact location at sea within a fire support area from which a fire support ship delivers fire. (JP 3-02)		

TARGET ACQUISITION CONTROL MEASURES

5-44. *Target acquisition* is the detection, identification, and location of a target in sufficient detail to permit the effective employment of capabilities that create the required effects (JP 3-60). (See ATP 3-09.12 for additional information on field artillery target acquisition systems and associated military symbol doctrine.) Table 5-26 on pages 230 through 234 lists target acquisition control measure symbols.

Table 5-26. Target acquisition control measure symbols

Templates		
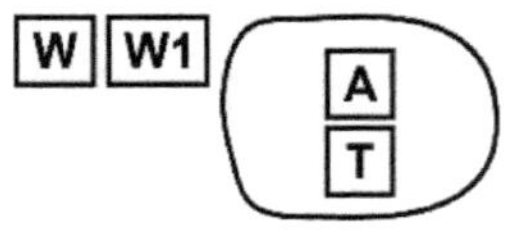		
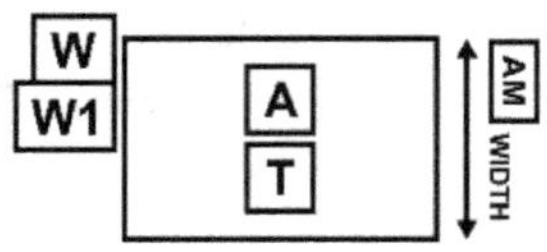	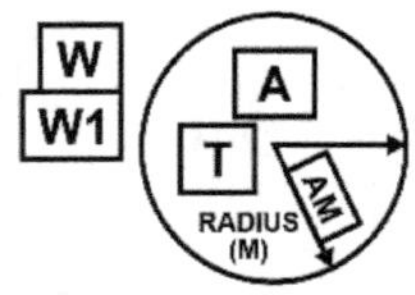	
A—Main function symbol that identifies the function. AM—A numeric amplifier that permits displaying minimum, maximum, or a specific distance (range, radius, width, or length). T—Text amplifier that uniquely identifies the zone. W, W1—Identifies the date-time group associated with zone.		
Function	***Main function symbol (Field A)***	***Construct Example***
Artillery Target Intelligence Zone A weapons locating radar search area in enemy territory that the commander monitors closely to detect and report any weapon ahead of all acquisitions other than those from critical friendly zones or call for fire zones. (FM 3-09)		
artillery target intelligence zone, irregular	**ATI ZONE**	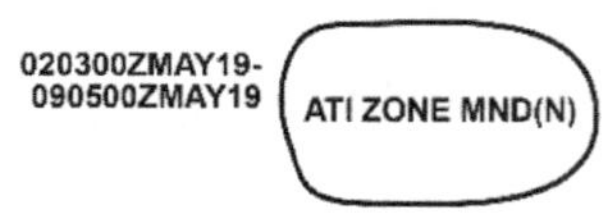
artillery target intelligence zone, rectangular		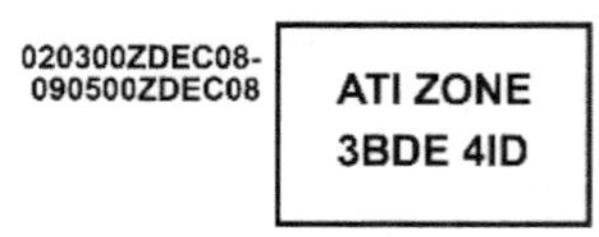
artillery target intelligence zone, circular		

Table 5-26. Target acquisition control measure symbols (*continued*)

Function	*Main function symbol (Field A)*	*Construct Example*
Call for Fire Zone A weapons locating radar search area from which the commander wants to attack hostile firing systems. (FM 3-09)		
call for fire zone, irregular	**CFF ZONE**	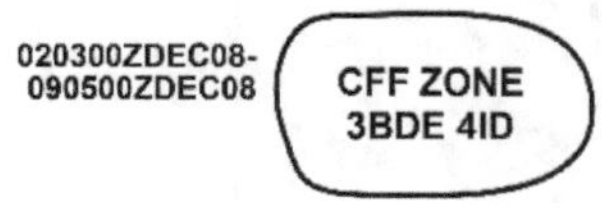
call for fire zone, rectangular		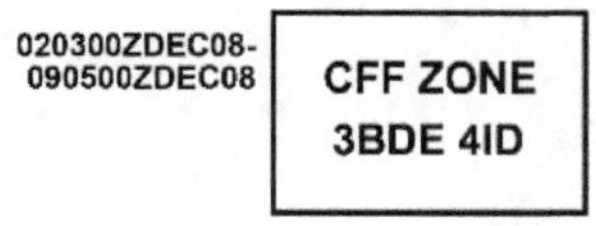
call for fire zone, circular		
Censor Zone An area from which the weapons locating radar is prohibited from reporting acquisitions. (FM 3-09)		
censor zone, irregular	**CENSOR ZONE**	
censor zone, rectangular		
censor zone, circular		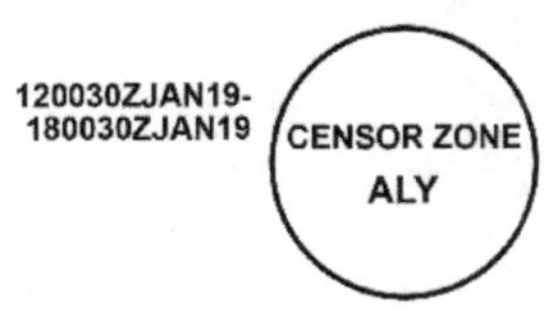

Table 5-26. Target acquisition control measure symbols (*continued*)

Function	*Main function symbol (Field A)*	*Construct Example*
Critical Friendly Zone A friendly area of coverage employed by weapons locating radar which the maneuver commander designates as critical to the protection of an asset whose loss would seriously jeopardize the mission. (FM 3-09)		
critical friendly zone, irregular	**CF ZONE**	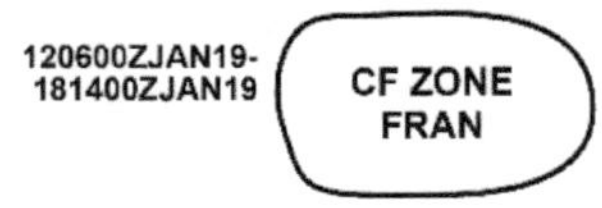
critical friendly zone, rectangular		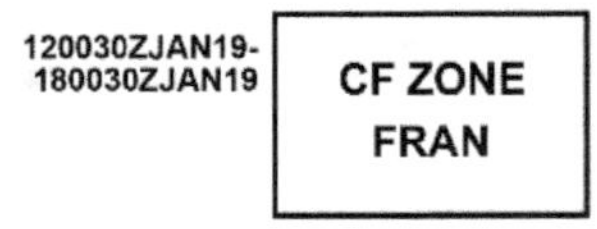
critical friendly zone, circular		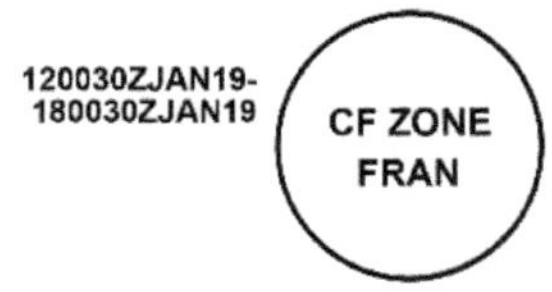
Dead Space Area An area where hostile weapons cannot be detected.		
dead space area, irregular	**DA**	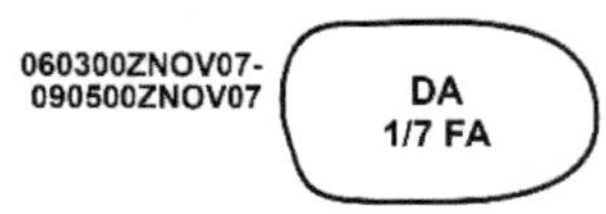
dead space area, rectangular		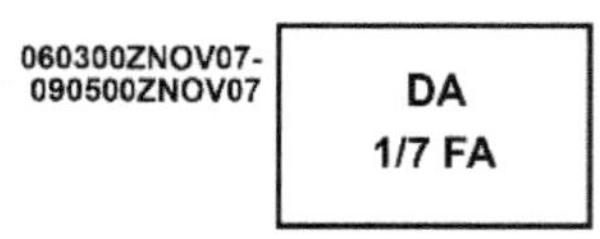
dead space area, circular		

Table 5-26. Target acquisition control measure symbols (*continued*)

Function	*Main function symbol (Field A)*	*Construct Example*
Blue Kill Box A fire support and airspace coordination measure that facilitates attacking surface targets with air-to-surface munitions without further coordination with the area of operations commanders' headquarters. (ATP 3-09.34)		
blue kill box, irregular	**BKB**	020300ZMAY08- 090500ZMAY08 BKB X CORPS
blue kill box, rectangular	**BKB**	020300ZMAY08- 090500ZMAY08 BKB X CORPS
blue kill box, circular	**BKB**	020300ZMAY08- 090500ZMAY08 BKB X CORPS
Purple Kill Box A fire support and airspace coordination measure that facilitates attacking surface targets with subsurface-to-surface, surface-to-surface, and air-to-surface munitions without further coordination with the area of operations commander's headquarters. (ATP 3-09.34)		
purple kill box, irregular	**PKB**	020300ZMAY08- 090500ZMAY08 PKB X CORPS
purple kill box, rectangular	**PKB**	020300ZMAY08- 090500ZMAY08 PKB X CORPS
purple kill box, circular	**PKB**	020300ZMAY08- 090500ZMAY08 PKB X CORPS

Table 5-276. Target acquisition control measure symbols (*continued*)

<table>
<tr><th>Function</th><th>Main function symbol (Field A)</th><th>Construct Example</th></tr>
<tr><td colspan="3">Range Fans
Identifies the effective distance, altitude, area, and or the selected sectors to engage a weapon or sensor.</td></tr>
<tr><td>weapon or sensor range fan, circular
Note. Symbols colored blue are used to help explain how the control measure is used and are not a part of the control measure.</td><td>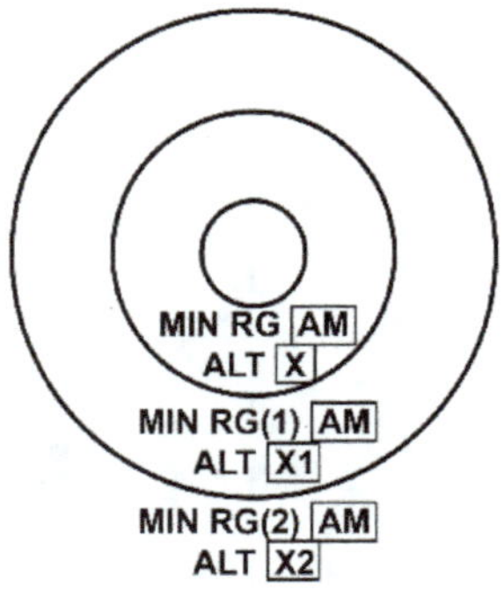
</td><td>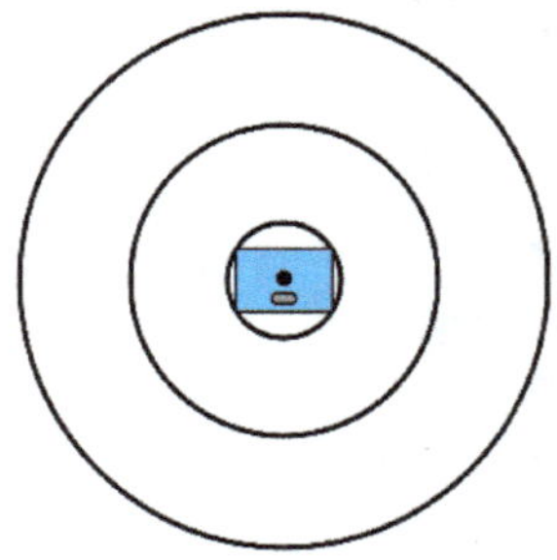</td></tr>
<tr><td colspan="3">Note. The coordinate which pinpoints the current physical location of a specific unit, weapon, or acquisition system may change with the movement of the object. The symbol for that object is located at the anchor point.</td></tr>
<tr><td>weapon or sensor range fan
Note. Symbols colored blue are used to help explain how the control measure is used and are not a part of the control measure.</td><td>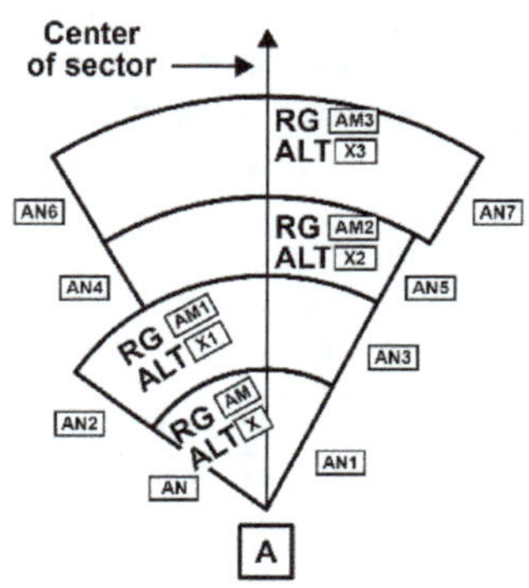
</td><td>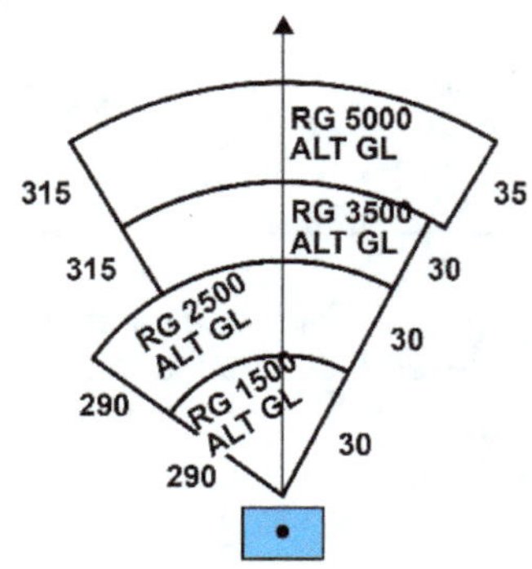

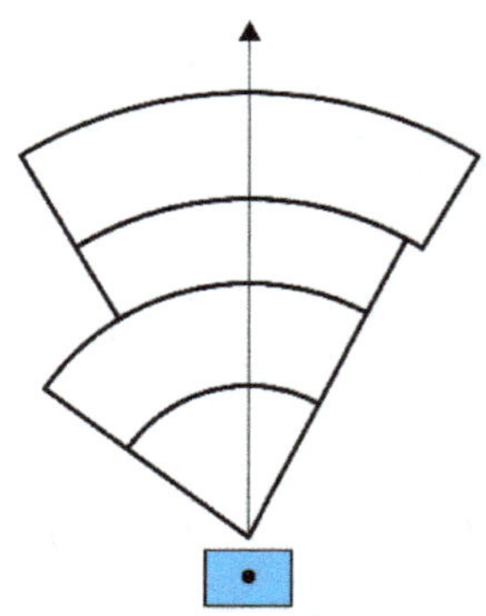</td></tr>
</table>

CBRN CONTAMINATION MARKERS

5-45. CBRN contamination markers symbols include events and contaminated area control measures used to depict those conditions found in an area resulting from immediate or persisting effects of CBRN attacks or events. Event symbols are used to identify CBRN detonation or spillage location, incident date, type, size, and direction of movement. A warning system formats, processes, and broadcasts CBRN event plumes symbols to provide common operational picture effects throughout the operational environment. Table 5-27 on 235 through 236 depicts CBRN events symbols. Contaminated area control measures are used to depict a stretch of surface that has been surveyed, located, marked, and reported to visually identify and warn of specific hazards that are dangerous to units and personnel. Table 5-28 on pages 236 through 237 depicts CBRN contaminated area control measures. (See ATP 3-11.36 and ATP 3-11.37 for additional information on CBRN planning, reconnaissance, and associated military symbol doctrine.)

Table 5-27. CBRN contaminated marker symbols (event)

CBRN Event Template		
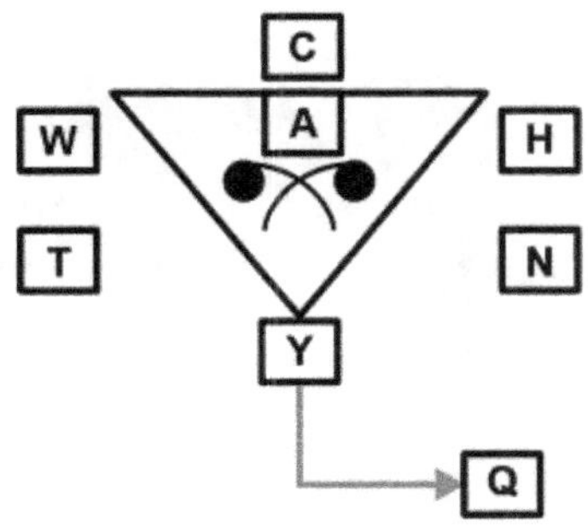		
A—Symbol that identifies the main function. C—Quantity, identifies the number of items present. H—Additional information. Content is implementation specific. N—Enemy: Letters "ENY" denote hostile. T—Identifies a unique designation. W—Identifies the date-time group associated with point. Q—Identifies the direction of movement or intended movement. Y—Displays a symbol's location in degrees, minutes, and decimal minutes, or other applicable display formats.		
Function	***Main function symbol***	***Construct Example and Symbol Translation***
chemical	C	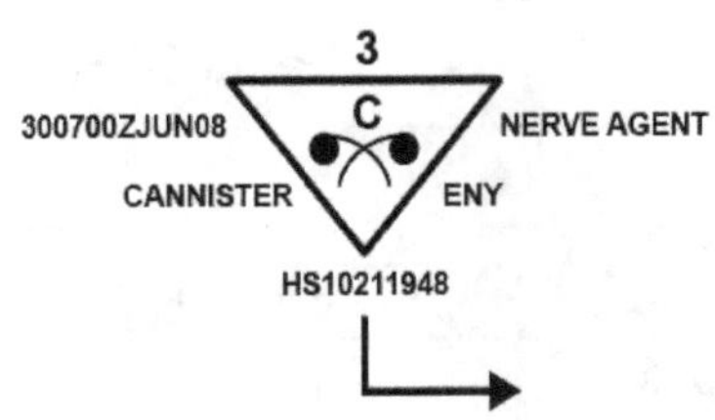 3 enemy chemical nerve agent event delivered by canister at 0700 Zulu hour, on 30 JUN 2008, location HS10211948

Table 5-27. CBRN contaminated marker symbols (event) (*continued*)

Function	*Main function symbol*	*Construct Example and Symbol Translation*
biological	**B**	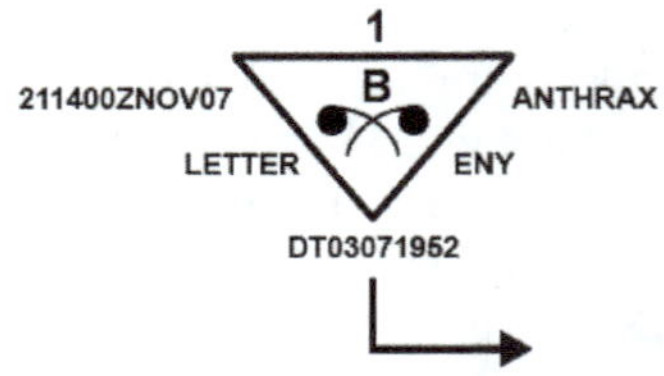 1 enemy biological anthrax agent event delivered by letter at 1400 Zulu hour, on 21 NOV 2007, location DT03071952
nuclear or nuclear fallout-producing	**N**	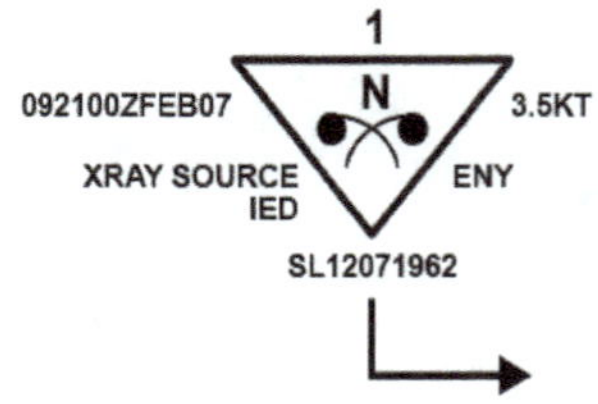 1 enemy nuclear 3.5 kiloton event delivered by improvised explosive device at 2100 Zulu hour, on 9 FEB 2007, location SL12071962
radiological	**R**	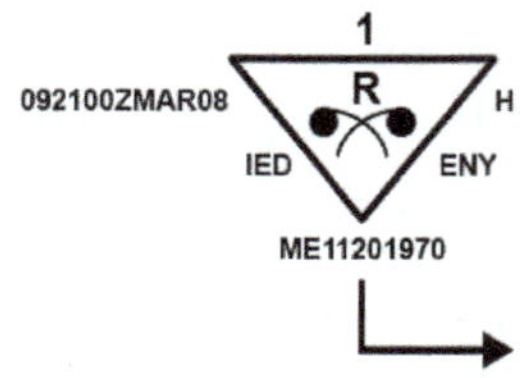 1 enemy radiological event delivered by improvised explosive device at 2100 Zulu hour, on 9 MAR 2008, location ME11201970

Table 5-28. CBRN contaminated marker symbols (area)

Contaminated area Sites or regions that are damaged, harmed, or made unfit for use by the introduction of unwanted substances.	
Function	*Symbol*
biological	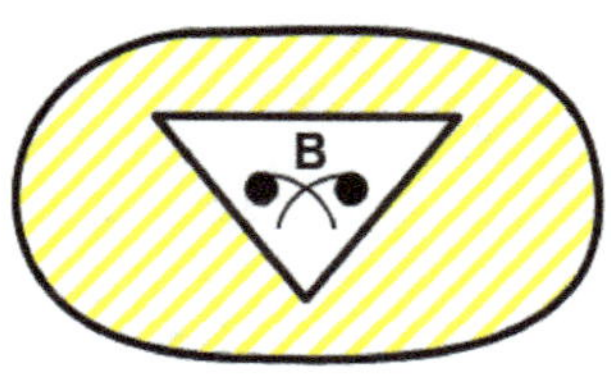

Table 5-28. CBRN contaminated marker symbols (area) (*continued*)

Function	*Symbol*	
chemical	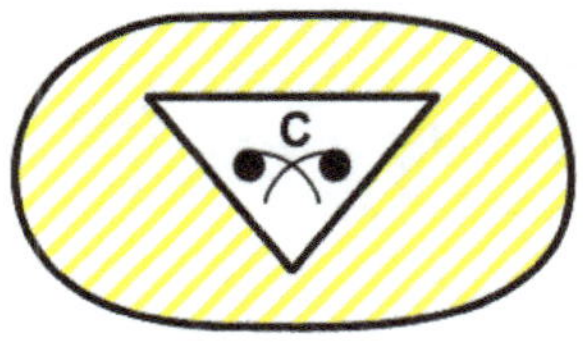	
nuclear		
radiological	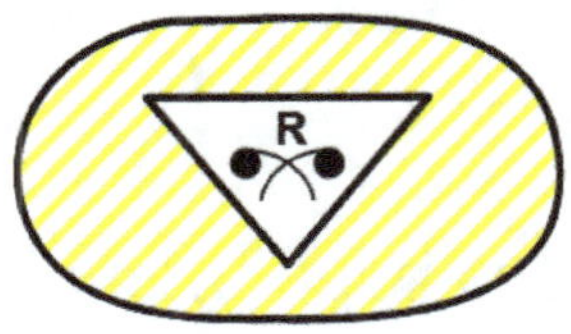	
CBRN Contour Lines		
Function	*Template Example*	
minimum safe distance zone—A zone designed to ensure that friendly individuals are far enough away from the effects of impact or contaminated area so the risk to them is negligible.	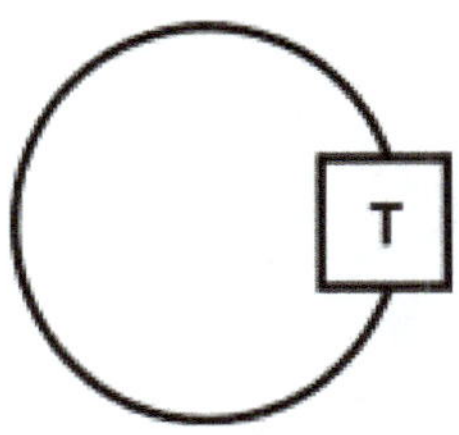	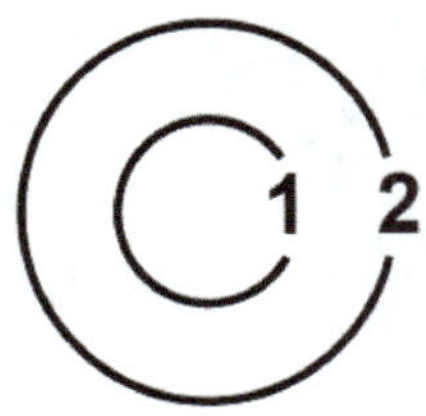
radiation dose rate contour line—Depicts contour lines for radiation dose rate caused by radiological contamination fallout at a given time. (See ATP 3-11.37 or TM 3-11.32 for more information on radiation dose contour lines.)	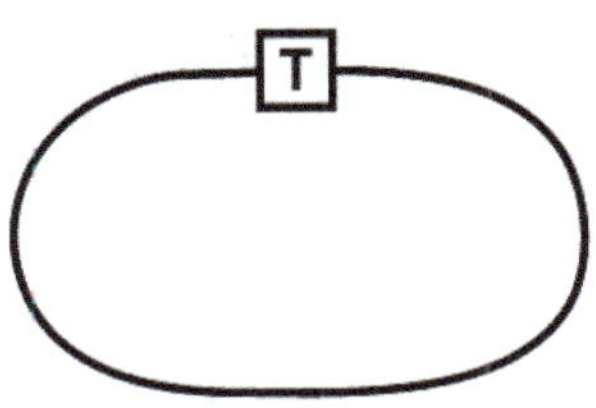	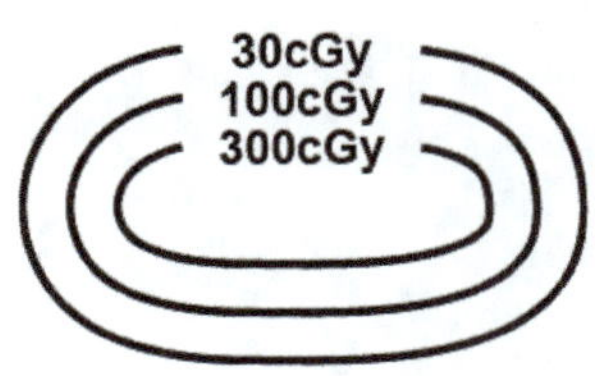

This page intentionally left blank.

Chapter 6
Tactical Mission Tasks

This chapter provides symbols for tactical mission tasks.

TACTICAL MISSION TASKS DEFINED

6-1. A task is a clearly defined and measurable activity accomplished by individuals or organizations. A *tactical mission task* is the specific activity a unit performs while executing a tactical operation or form of maneuver. (FM 3-90). The tactical mission tasks describe the results or effects commanders want to achieve.

SYMBOLS FOR TACTICAL MISSION TASKS

6-2. Table 6-1 on pages 239 through 243 shows the tactical mission tasks that have symbols. Not all tactical mission tasks have symbols. Tactical mission task symbols can be used with unit symbols, but they do not have modifiers. Tactical mission task symbols are used in course of action sketches, synchronization matrices, and maneuver sketches. They do not replace any part of the operation order. Tactical mission task symbols are sized to accommodate the scale of the display or map being used.

Table 6-1. Tactical mission task symbols

Task	***Symbol***	***Construct Usage Example*** ***Note***. The friendly or hostile frame (gray) is not part of the symbol; it is for orientation only.
attack by fire—A tactical mission task using direct and indirect fires to engage an enemy from a distance. (FM 3-90)	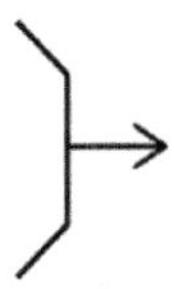	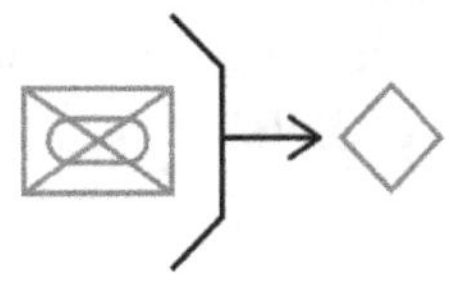 Mechanized infantry unit attacks by fire an enemy unit
block—A tactical mission task that denies the enemy access to an area or an avenue of approach. (FM 3-90)	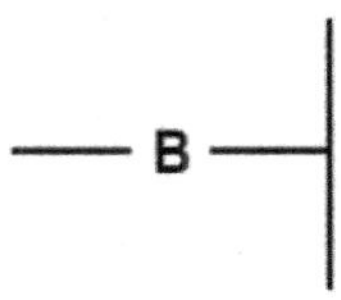	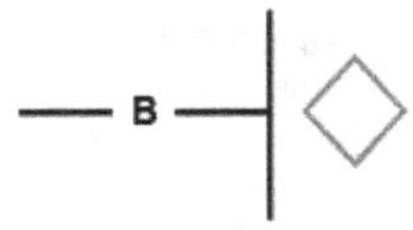 Block enemy unit
breach—A tactical mission task in which a unit breaks through or establishes a passage through an enemy obstacle. (FM 3-90)	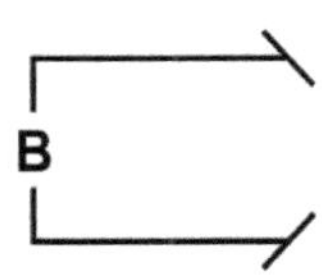	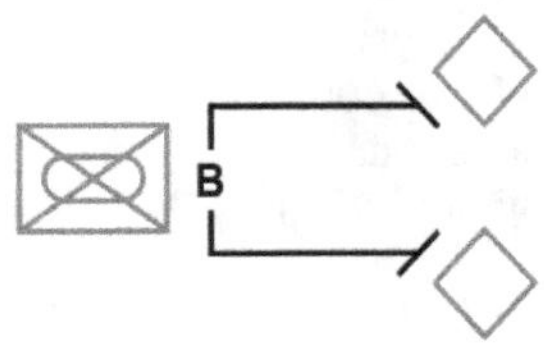 Mechanized infantry unit breaches enemy units

Table 6-1. Tactical mission task symbols (*continued*)

Task	*Symbol*	*Construct Usage Example*
bypass—A tactical mission task in which a unit deliberately avoids contact with an obstacle or enemy force. (FM 3-90)	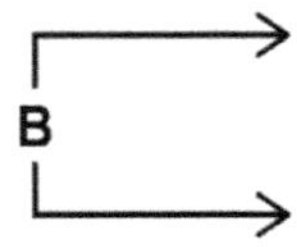	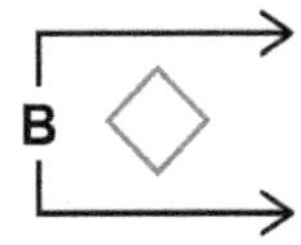 Bypassing enemy unit
canalize—A tactical mission task in which a unit restricts enemy movement to a narrow zone. (FM 3-90)	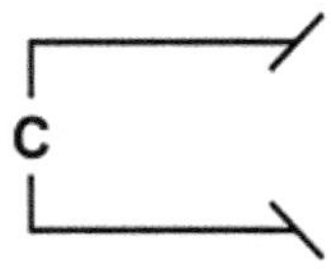	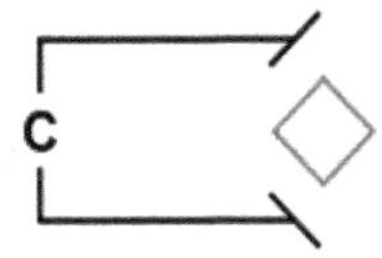 Canalize enemy unit
clear—A tactical mission task in which a unit eliminates all enemy forces within an assigned area. (FM 3-90)	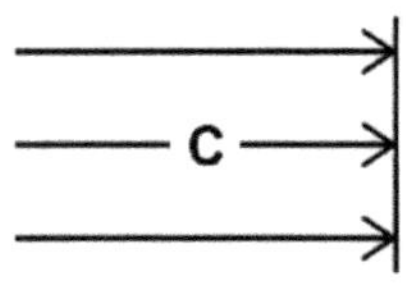	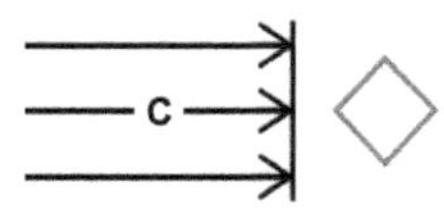 Clear enemy unit
contain—A tactical mission task in which a unit stops, holds, or surrounds an enemy force. (FM 3-90)		 Contain enemy unit
control—(Army) A tactical mission task in which a unit maintains physical influence over an assigned area. (FM 3-90)	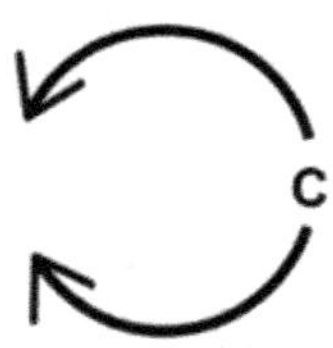	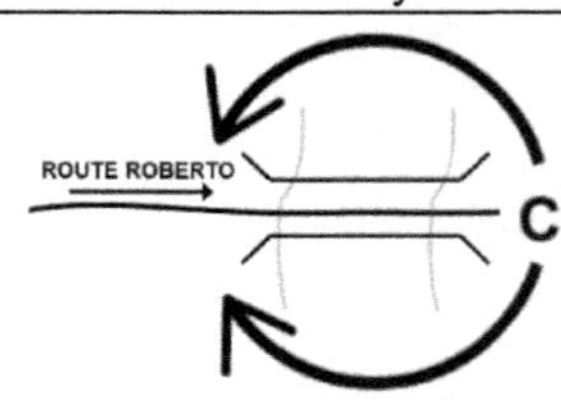 Control bridge on one-way traffic Route Roberto
destroy—A tactical mission task that physically renders an enemy force combat-ineffective until it is reconstituted. (FM 3-90)	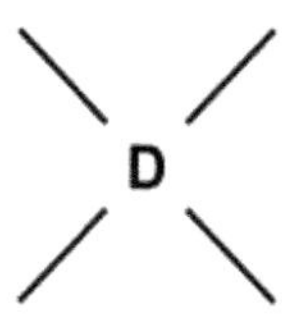	 Destroy enemy unit
disengage—A tactical mission task in which a unit breaks contact with an enemy to conduct another mission or to avoid becoming decisively engaged. (FM 3-90)	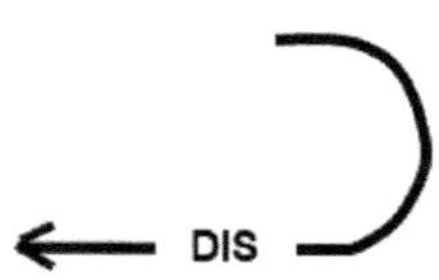	 Disengage with enemy unit

Table 6-1. Tactical mission task symbols (*continued*)

Task	*Symbol*	*Construct Usage Example*
disrupt—A tactical mission task in which a unit upsets an enemy's formation or tempo and causes the enemy force to attack prematurely or in a piecemeal fashion. (FM 3-90)	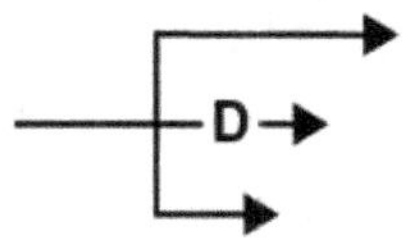	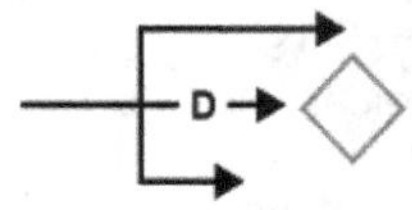 Disrupt enemy unit
exfiltrate—A tactical mission task in which a unit removes Soldiers or units from areas under enemy control by stealth, deception, surprise, or clandestine means. (FM 3-90)		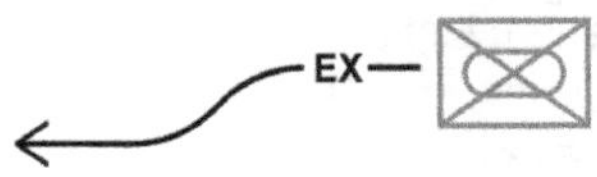 Mechanized infantry unit exfiltrates
fix—A tactical mission task in which a unit prevents the enemy from moving from a specific location for a specific period. (FM 3-90)		 Fix enemy unit
follow and assume—A tactical mission task in which a committed force follows a lead force conducting an offensive operation and continues the mission if the lead force cannot continue. (FM 3-90)	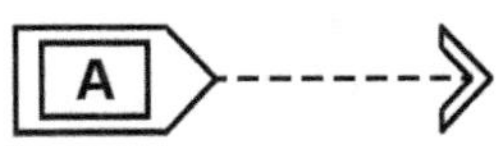	 Mechanized infantry unit follows and assumes a mechanized infantry unit
follow and support—A tactical mission task in which a committed force follows and supports a lead force conducting an offensive operation. (FM 3-90)	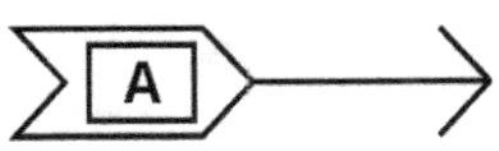	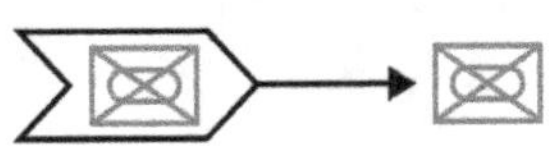 Mechanized infantry unit follows and supports a mechanized infantry unit
interdict—A tactical mission task in which a unit prevents, disrupts, or delays the enemy's use of an area or route in any domain. (FM 3-90)	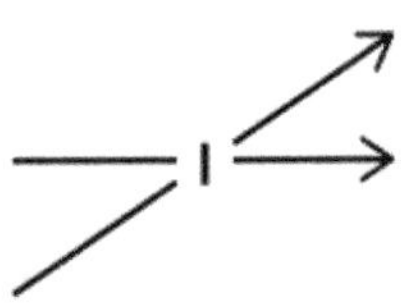	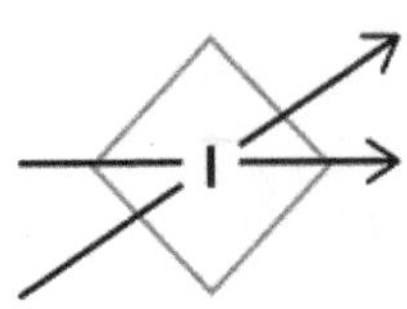 Interdict enemy unit

Table 6-1. Tactical mission task symbols (*continued*)

Task	*Symbol*	*Construct Usage Example*
isolate—A tactical mission task in which a unit seals off an enemy, physically and psychologically, from sources of support and denies it freedom of movement. (FM 3-90)		 Isolate enemy unit
neutralize—A tactical mission task in which a unit renders the enemy incapable of interfering with an operation. (FM 3-90)	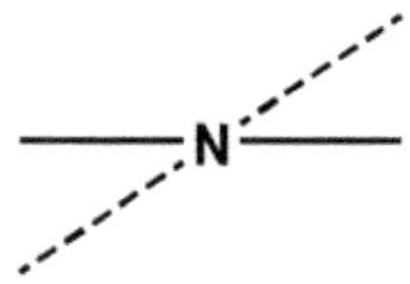	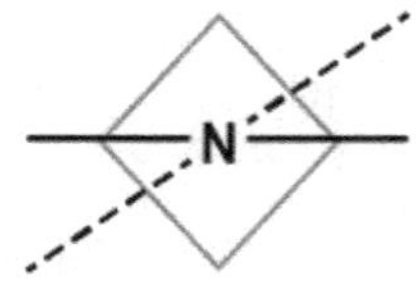 Neutralize enemy unit
occupy—A tactical mission task in which a unit moves into an area to control it without enemy opposition. (FM 3-90)		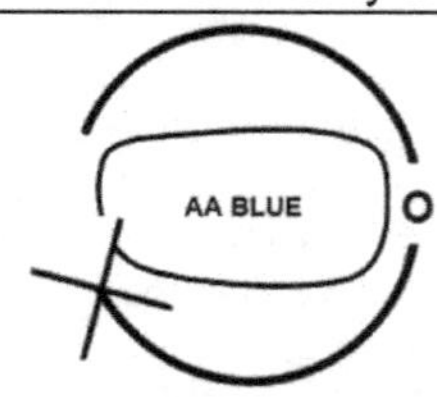 Occupy assembly area blue
retain—A tactical mission task in which a unit prevents enemy occupation or use of terrain. (FM 3-90)	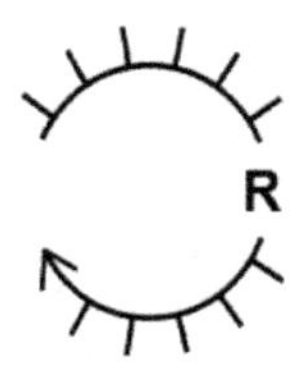	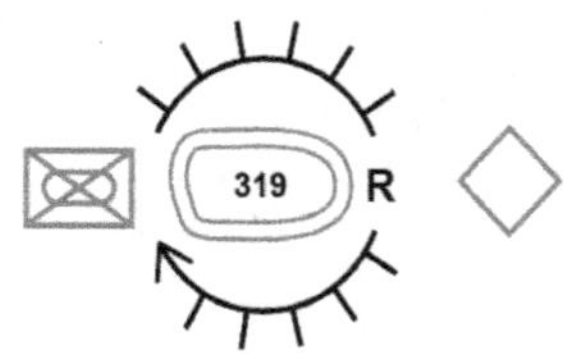 Mechanized infantry unit retains hill 319
secure—A tactical mission task in which a unit prevents the enemy from damaging or destroying a force, facility, or geographical location. (FM 3-90)	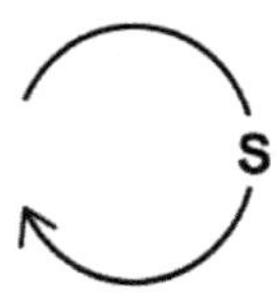	 Secure airport facility
seize—(Army) A tactical mission task in which a unit takes possession of a designated area using overwhelming force. (FM 3-90)	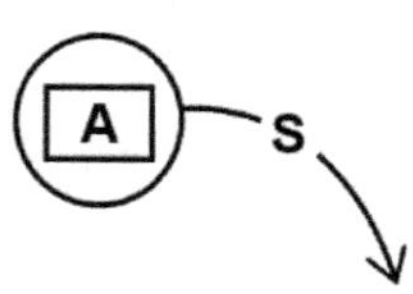	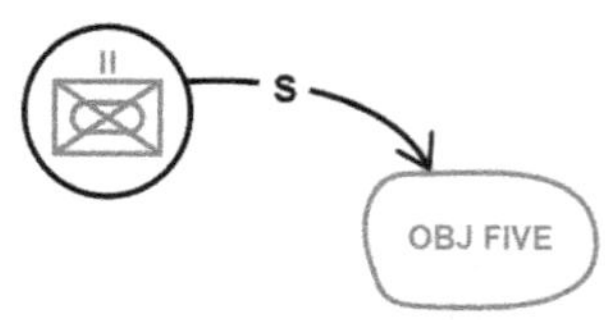 Mechanized infantry battalion seizes objective FIVE

Table 6-1. Tactical mission task symbols (*continued*)

Task	*Symbol*	*Construct Usage Example*
support by fire—A tactical mission task in which a unit engages the enemy by direct fire in support of another maneuvering force. (FM 3-90)		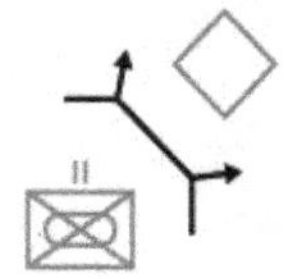 Mechanized infantry battalion support by fire
suppress—A tactical mission task in which a unit temporarily degrades a force or weapon system from accomplishing its mission. (FM 3-90)	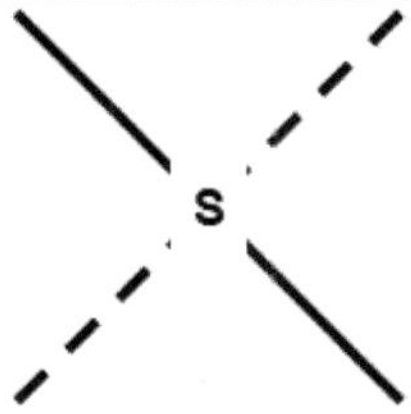	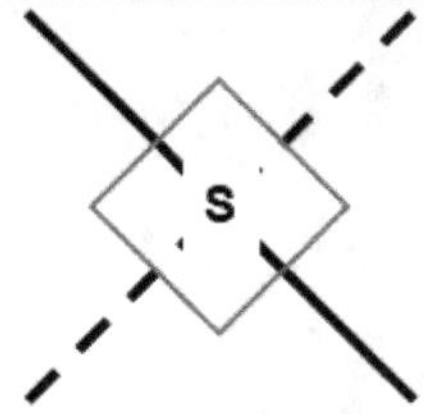 Suppress enemy unit
turn—A tactical mission task in which a unit forces an enemy force from one avenue of approach or mobility corridor to another. (FM 3-90)	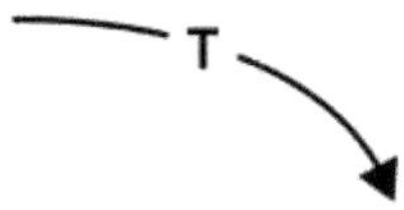	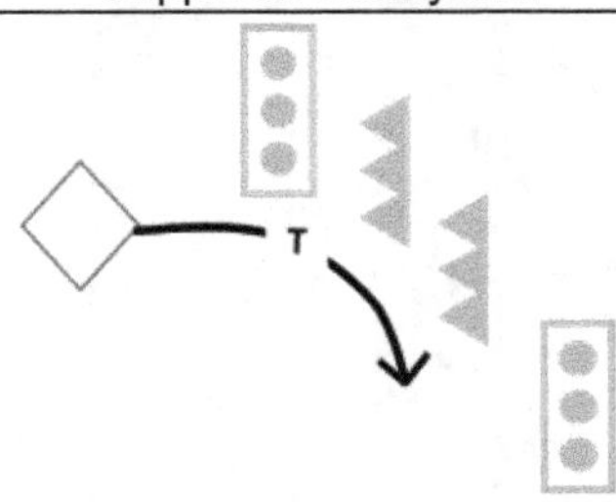 Turn enemy unit

This page intentionally left blank.

Chapter 7
Course of Action Sketch

This chapter discusses the purpose and makeup of a course of action sketch.

PURPOSE OF COURSE OF ACTION SKETCH

7-1. A *course of action* is a scheme developed to accomplish a mission (JP 5-0). Developing, analyzing, and deciding on a course of action for execution is central to planning. Part of course of action development is to produce a course of action statement and sketch. The staff prepares a course of action statement and supporting sketch for each course of action under consideration. A course of action statement clearly describes the array of forces and the sequence of tasks the unit will conduct to accomplish the mission. The statement should be a brief expression of how the combined arms concept will be conducted. The course of action sketch is the graphic portrayal of the course of action statement. The sketch provides a picture of the movement and maneuver aspects of the concept, including positioning of forces. (See FM 6-0 for a detailed discussion of the military decision-making process, including course of action development.)

MAKEUP OF COURSE OF ACTION SKETCH

7-2. The course of action sketch provides a picture of the movement and maneuver aspects of the concept, including the positioning of forces. The course of action sketch becomes the basis for the operation overlay. At a minimum, the course of action sketch includes the array of generic forces and control measures, including—

- Unit and subordinate unit boundaries.
- Unit movement formations (but not subordinate unit formations).
- Reconnaissance and security graphics.
- Ground and air axes of advance.
- Assembly areas, battle positions, strong points, engagement areas, and objectives.
- Obstacle control measures and tactical mission graphics.
- Fire support coordination and airspace coordinating measures.
- Main effort.
- Location of command posts and critical communication nodes.
- Enemy locations, known or template.
- Population concentrations.

7-3. Most symbols for use on the course of action sketch are shown in chapters 2 through 7. However, the unit symbols do not provide decision makers with a quick and easy method of portraying detailed information relating to task organization, composition, or combat effectiveness. Task organization composition symbols portray detailed information for course of action sketches.

TASK ORGANIZATION COMPOSITION SYMBOLS

7-4. Part of course of action development includes determining relative combat power and arraying forces to accomplish the primary tasks envisioned during action. After arraying forces, planners then group these forces underneath a generic headquarters representing the initial task organization. The initial task organization for each generic unit is portrayed in a task organization composition symbol.

7-5. Task organization composition symbols are constructed using a rectangular frame, main and modifier symbols, and amplifiers. Figure 7-1 depicts Main function symbol and amplifier fields. The Main function symbols and amplifiers have specific field placement guidelines that follow current military symbol standards.

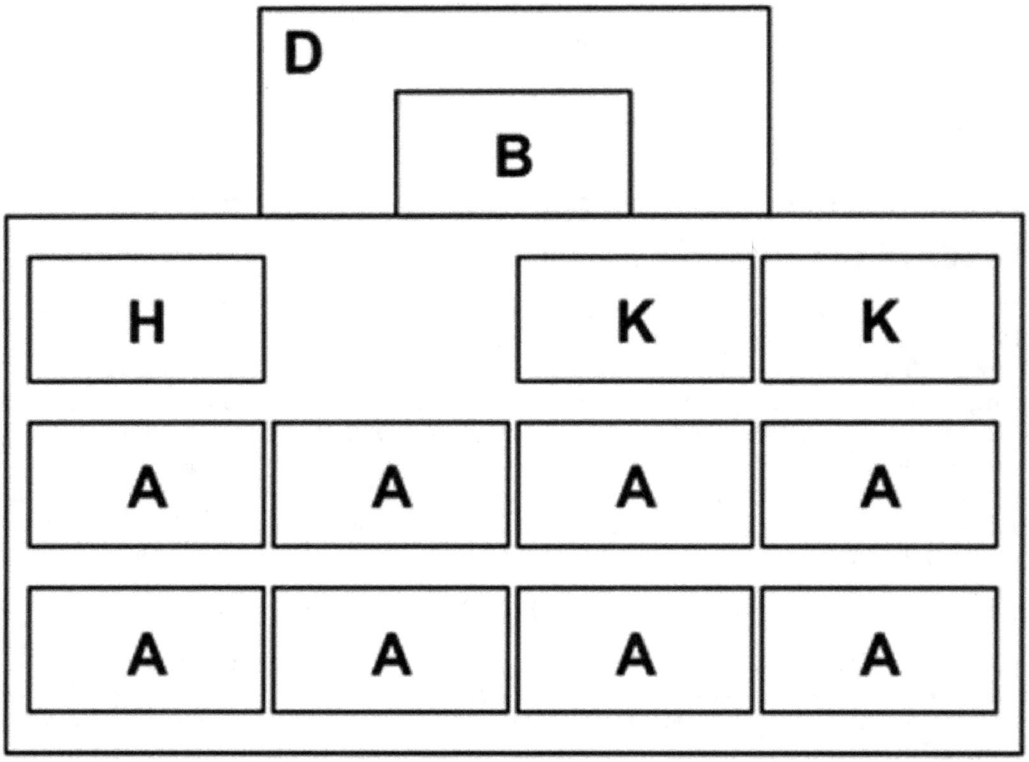

Field A is used to depict subordinate unit symbols using the task organization symbols in table 7-1. While there are four symbols shown in table 7-1, there is no limitation as to the number of symbols in a row or number of rows. If needed, the first line of field A can be reserved to identify only those units that have been reinforced (attached) "(+)", reduced (detached) "(-)", or reinforced and reduced (attached and detached) "(±)" and for units that are one or more echelons lower than the other designated units.

Field B is used to identify the echelon level of the headquarters. See chapter 2, table 2-3 on page 15 for echelon amplifiers.

Field D denotes task force. See chapter 2, table 2-4 on page 19.

Field H is available for including a generic alphanumeric designation of the organization.

Field K is available for including the combat effectiveness using the symbols in table 7-2 on page 249.

Figure 7-1. Task organization Main function symbol and amplifier fields

TASK ORGANIZATION MAIN AND MODIFIER SYMBOLS

7-6. In all cases, task organization symbols are the same as unit main and modifier symbols and amplifiers. Table 7-1 on page 247 shows the most commonly used main and modifier symbols in their appropriate configurations.

Table 7-1. Task organization symbols

Function	*Symbol*	
air assault infantry	 Current usage construct	 Alternate usage construct
airborne infantry		
air defense		
air reconnaissance (cavalry)		
anti-tank		
armor tracked		
armored reconnaissance (cavalry) ***Note.*** Reconnaissance (cavalry) unity that has armored tracked vehicles.	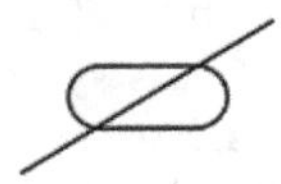	
assault or lift helicopter	 Current usage construct	 Alternate usage construct
attack helicopter		
combined arms		
engineer	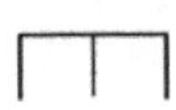	
field artillery		

Table 7-1. Task organization symbols (*continued*)

Function	*Symbol*
infantry	
mechanized infantry ***Note.*** Infantry unit that has armored tracked vehicles.	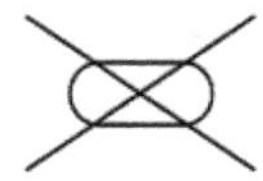
mobile infantry (Stryker)	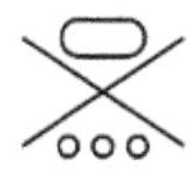
mountain infantry	
reconnaissance (cavalry scout)	

COMBAT EFFECTIVENESS SYMBOLS

7-7. During course of action analysis and war gaming, the staff can track the combat effectiveness of units using combat effectiveness symbols in task organization composition symbols. Combat effectiveness refers to the ability of a unit to perform its mission. Factors such as ammunition, personnel, status of fuel, and weapons systems availability are assessed and rated. Commanders use this information to provide a net assessment of a unit's ability to perform its mission. This assessment can then be expressed graphically using combat effectiveness symbols. Table 7-2 shows two sets of combat effectiveness symbols, which may be also used with task organization composition symbols.

7-8. Table 7-2 depicts combat effectiveness symbols for the overall combat rating of a unit in the center column. Table 7-2 specifies combat effectiveness symbols for the status of selected items of interest in the right column. The four selected items shown in the right column are ammunition; weapons; petroleum, oils, and lubricants; and personnel. Standard operating procedures will specify the items of interest to be reported. Commanders may add to this list for internal reporting and tracking.

Table 7-2. Combat effectiveness symbols

Commander's Assessment of Unit's Ability to Perform Mission	***Effectiveness Pie Charts***	***Selected Status Pie Chart***
	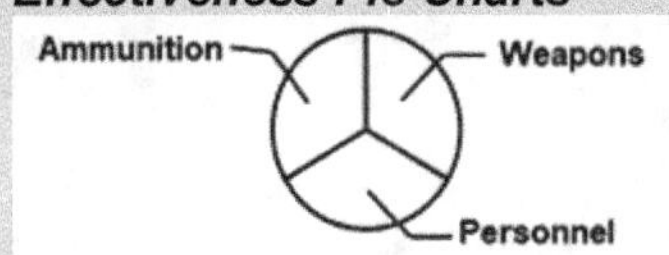	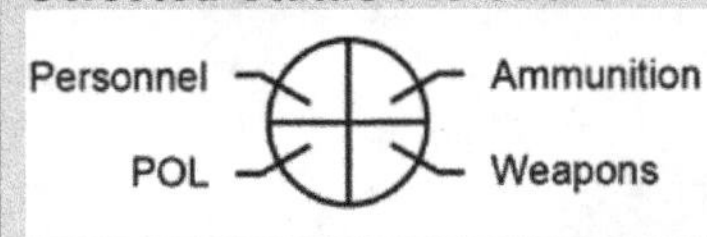
no problems in any area		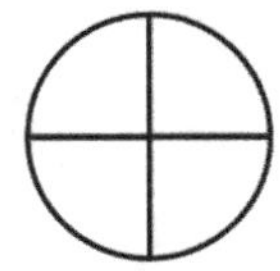
problems in personnel	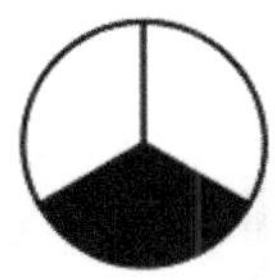	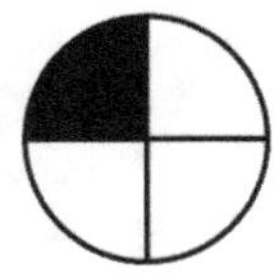
problems in weapons systems	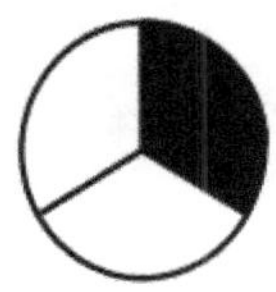	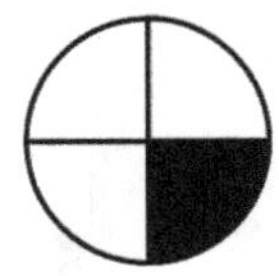
Cannot perform mission: personnel, ammunition, and weapons problems		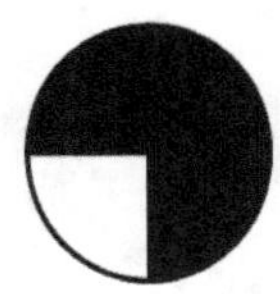
POL: petroleum, oils, and lubricants		

EXAMPLE OF A TASK ORGANIZATION COMPOSITION SYMBOL

7-9. Figure 7-2 depicts a generic example of a task force task organization composition symbol for a battalion task force.

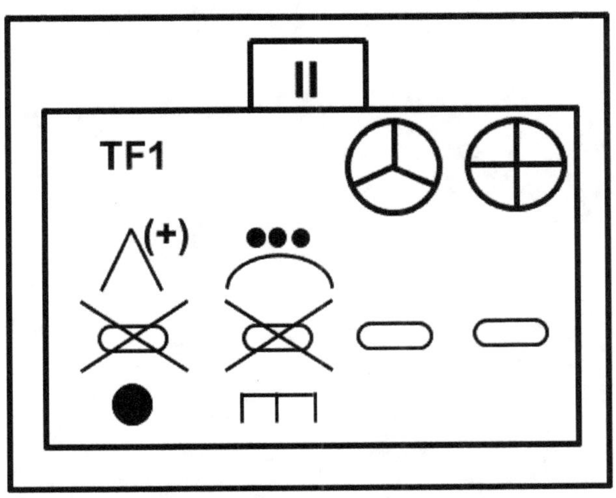

Figure 7-2. Battalion task force example

This page intentionally left blank.

Glossary

The glossary lists acronyms and terms with Army and joint definitions. The proponent publication for terms is listed in parentheses after the definition.

SECTION I – ACRONYMS AND ABBREVIATIONS

ADP	Army doctrine publication
AFTTP	Air Force tactics, techniques, and procedures
APP	Allied procedural publication
ATF	Bureau of Alcohol, Tobacco, Firearms, and Explosives
ATP	Army techniques publication
CBRN	chemical, biological, radiological, and nuclear
DA	Department of the Army
DEA	Drug Enforcement Administration
DOD	Department of Defense
DODD	Department of Defense directive
ED	edition
ENY	enemy
FBI	Federal Bureau of Investigation
FM	field manual
FO	fully operational
JP	joint publication
kph	kilometers per hour
kts	knots per hour
MCRP	Marine Corps reference publication
MCTP	Marine Corps training publication
MCWP	Marine Corps warfighting publication
MIL-STD	military standard
mph	miles per hour
mps	meters per second
NATO	North Atlantic Treaty Organization
NTTP	Navy tactics, techniques, and procedures
POL	petroleum, oil, and lubricants
SIPRNET	SECRET Internet Protocol Router Network
SO	substantially operational
STANAG	standardization agreement
SBCT	Stryker brigade combat team
TC	training circular
TM	technical manual
U.S.	United States

SECTION II – TERMS

aerial delivery

The air transport of cargo, equipment and/or personnel to a desired location on the ground by aircraft. (ATP 4-48)

air and missile defense

Direct [active and passive] defensive actions taken to destroy, nullify, or reduce the effectiveness of hostile air and ballistic missile threats against friendly forces and assets. (JP 3-01)

air assault

The movement of friendly assault forces by rotary-wing or tiltrotor aircraft to engage and destroy enemy forces or to seize and hold key terrain. (JP 3-18)

air corridor

A restricted air route of travel specified for use by friendly aircraft and established for the purpose of preventing friendly aircraft from being fired on by friendly forces. (JP 3-52)

air defense

Defensive measures designed to destroy attacking enemy aircraft or aerodynamic missiles, or to nullify or reduce the effectiveness of such attack. (JP 3-01)

air defense artillery

Weapons and equipment for actively combating air targets from the ground. (JP 3-01)

airfield

An area prepared for the accommodation (including any buildings, installations, and equipment), landing, and takeoff of aircraft. (JP 3-36)

airhead line

A line denoting the limits of the objective area for an airborne assault. (JP 3-18)

airspace coordination area

A three-dimensional block of airspace in a target area, established by the appropriate commander, in which friendly aircraft are reasonably safe from friendly surface fires. (JP 3-09.3)

alternate supply route

A route or routes designated within an area of operations to provide for the movement of traffic when main supply routes become disabled or congested. (FM 4-01)

altitude

The vertical distance of a level, a point or an object considered as a point, measured from mean sea level or height above ellipsoid. (FM 3-09)

ambulance control point

A manned traffic regulating, often stationed at a crossroad or road junction, where ambulances are directed to one of two or more directions to reach loading points and medical treatment facilities. (ATP 4-02.2)

ambulance exchange point

A location where a patient is transferred from one ambulance to another en route to a medical treatment facility. (ATP 4-02.2)

ambulance loading point

This is the point in the shuttle system where one or more ambulances are stationed ready to receive patients for evacuation. (ATP 4-02.2)

ambulance relay point

A point in the shuttle system where one or more empty ambulances are stationed to advance to a loading point or to the next relay post to replace departed ambulances. (ATP 4-02.2)

ambush

A variation of attack from concealed positions against a moving or temporarily halted enemy. (FM 3-90)

ammunition supply point

A retail ammunition support activity where ammunition is received, stored, issued, and accounted for. (ATP 4-35)

ammunition transfer holding point

A designated site operated by a brigade support battalion distribution company, where ammunition is received, transferred, or temporarily stored, or transferred to supported units within a functional or multifunctional brigade. (ATP 4-35)

amphibious warfare ship

A combatant ship having organic capability to embark, land, and support landing forces in amphibious operations and which has characteristics enabling long-duration operations on the high seas. (JP 3-02)

area defense

A type of defensive operation that concentrates on denying enemy forces access to designated terrain for a specific time rather than destroying the enemy outright. (ADP 3-90)

area of operations

An operational area defined by a commander for the land or maritime force commander to accomplish their missions and protect their forces. (JP 3-0)

army group

The largest formation of land forces, normally comprising two or more armies or army corps under a designated commander. (APP-06E)

artillery target intelligence zone

A weapons locating radar search area in enemy territory that the commander monitor closely to detect and report any weapon ahead of all acquisitions other than those from critical friendly zones or call for fire zones. Also called ATIZ. (FM 3-09).

assault

(Army) A short and violent well-ordered attack against a local objective. (FM 3-90)

assault position

A covered and concealed position short of the objective from which final preparations are made to assault the objective. (ADP 3-90)

assembly area

An area a unit occupies to prepare for an operation. (FM 3-90)

attach

1. The placement of units or personnel in an organization where such placement is relatively temporary. (JP 3-0)

attack

A type of offensive operation that defeats enemy forces, seizes terrain, or secures terrain. (FM 3-90)

attack by fire

A tactical mission task using direct and indirect fires to engage an enemy from a distance. (FM 3-90).

attack position

(Army) The last position an attacking force occupies or passes through before crossing the line of departure. (ADP 3-90)

avenue of approach

A path used by an attacking force leading to its objective or to key terrain. Avenues of approach exist in all domains. (ADP 3-90)

axis of advance

The general area through which the bulk of a unit's combat power must move. (ADP 3-90)

base camp

An evolving military facility that supports the military operations of a deployed unit and provides the necessary support and services for sustained operations. (ATP 3-37.10)

base defense zone

An air defense zone established around an air base and limited to the engagement envelope of short-range air defense weapons systems defending that base. (JP 3-52)

battalion

A unit consisting of two or more company-, battery-, or troop-size units and a headquarters. (ADP 3-90)

battery

A company-size unit in a field artillery or air defense artillery battalion. (ADP 3-90)

battle handover line

A designated phase line where responsibility transitions from the stationary force to the moving force and vice versa. (ADP 3-90)

battle position

A defensive location oriented on a likely enemy avenue of approach. (ADP 3-90)

biological agent

A microorganism (or a toxin derived from it) that cause disease in personnel, plants, or animals or causes the deterioration of materiel. (JP 3-11)

block

1. A tactical mission task that denies the enemy access to an area or an avenue of approach. 2. An obstacle effect that integrates fire planning and obstacle effort to stop an attacker along a specific avenue of approach or prevent the attacking force from passing through an engagement area. (FM 3-90)

blue kill box

A fire support and airspace coordination measure that facilitates attacking surface targets with air-to-surface munitions without further coordination with the area of operations commanders' headquarters. (ATP 3-09.34)

boundary

A line that delineates surface areas for the purpose of facilitating coordination and deconfliction of operations between adjacent units, formations, or areas. (JP 3-0)

breach

1. A tactical mission task in which a unit breaks through or establishes a passage through an enemy obstacle. (FM 3-90) A synchronized combined arms activity under the control of the maneuver commander conducted to allow maneuver through an obstacle. (ATP 3-90.4)

bridgehead line

The limit of the objective area in the development of the bridgehead. (ATP 3-90.4)

brigade

A unit consisting of two or more battalions and a headquarters company or detachment. (ADP 3-90)

bypass

A tactical mission task in which a unit deliberately avoids contact with an obstacle or enemy force. (FM 3-90)

call for fire zone

A weapons locating radar search area from which the commander wants to attack hostile firing systems. Also called CFFZ. (FM 3-09)

canalize

(Army) A tactical mission task in which a unit restricts enemy movement to a narrow zone. (FM 3-90)

casualty collection point

A location that may or may not be staffed, where casualties are assembled for evacuation to a medical treatment facility. (ATP 4-02.2)

censor zone

An area from which the weapons locating radar is prohibited from reporting acquisitions. Also called CZ. (FM 3-09)

chemical agent

A chemical substance that is intended for use in military operations to kill, seriously injure, or incapacitate, mainly through physiological effects. (JP 3-11)

civil affairs

Designated Active and Reserve Component forces and units organized, trained, and equipped specifically to conduct civil affairs operations and to support civil-military operations. (JP 3-57)

civil-military cooperation

(NATO) A joint function comprising a set of capabilities integral to supporting the achievement of mission objectives and enabling NATO commands to participate effectively in a broad spectrum of civil-military interaction with diverse non-military actors. (APP 6[D])

clear

A tactical mission task in which a unit eliminates all enemy forces within an assigned area. (FM 3-90)

clearing

A mobility (focused on movement) task that involves the elimination or neutralization of an obstacle that is usually performed by follow-on engineers and explosive ordnance disposal that involves the total elimination or neutralization of an obstacle. (ATP 3-90.4)

combat and operational stress control

A coordinated program of actions taken by military leadership to prevent, identify, and manage reactions to traumatic events that may affect exposed organizations and individuals during unified land operations. (FM 4-02)

combat camera

Specially-trained expeditionary forces from Service-designated units capable of providing high-quality directed visual information during military operations. (JP 3-61)

combat outpost

A reinforced observation post capable of conducting limited combat operations. (FM 3-90)

combatant command

A unified or specified command with a broad continuing mission under a single commander established and so designated by the President, through the Secretary of Defense and with the advice and assistance of the Chairman of the Joint Chiefs of Staff. (JP 1 VOL 2)

command and control

The exercise of authority and direction by a properly designated commander over assigned and attached forces in the accomplishment of the mission. (JP 1 VOL 2)

command post

A headquarters or a portion there of, organized for the exercise of command and control. (FM 6-0)

common sensor boundary

A line depicted by a series of grid coordinates, grid line, phase line, or major terrain feature that divides target acquisition search areas into radar acquisition management areas. Also called CSB. (FM 3-09)

company

A unit consisting of two or more platoons, usually of the same type, with a headquarters and a limited capacity for self-support. (ADP 3-90)

contact point

1. In land warfare, a point on the terrain, easily identifiable, where two or more units are required to make contact. (JP 3-50)

contain

(Army) A tactical mission task in which a unit stops, holds, or surrounds an enemy force. (FM 3-90)

contracting support

The planning, coordination, and execution of contracting authority to legally bind contractors in support of military operations. (JP 4-10)

control

1. The regulation of forces and warfighting functions to accomplish the mission in accordance with the commander's intent. (ADP 6-0) 2. (Army) A tactical mission task in which a unit maintains physical influence over an assigned area. (FM 3-90) 3. (Army) An action taken to eliminate a hazard or reduce its risk. (ATP 5-19)

control measure

The means of regulating forces or warfighting functions. (ADP 6-0).

convoy

2. A group of vehicles organized for the purpose of control and orderly movement with or without escort protection that moves over the same route at the same time and under one commander. (JP 3-02)

coordinated fire line

A line beyond which conventional surface-to-surface direct fire and indirect fire support means may fire at any time within the boundaries of the establishing headquarters without additional coordination but does not eliminate the responsibility to coordinate the airspace required to conduct the mission. (JP 3-09)

coordination point

A point that indicates a specific location for the coordination of tactical actions between adjacent units. (FM 3-90)

cordon and search

A variation of movement to contact where a friendly force isolates and searches a target area. (FM 3-90)

corps

An echelon of command and tactical formation that employs divisions, multi-functional brigades, and functional brigades to achieve objectives on land. (ADP 3-90)

counterattack

A variation of attack by a defending force against an attacking enemy force. (FM 3-90)

course of action

2. A scheme developed to accomplish a mission. (JP 5-0)

cover

(Army) A type of security operation done independent of the main body to protect them by fighting to gain time while preventing enemy ground observation of and direct fire against the main body. (ADP 3-90)

crew

A small military unit that consists of all personnel operating a particular system. (ADP 3-90)

critical friendly zone

A friendly area of coverage employed by weapons locating radar which the maneuver commander designates as critical to the protection of an asset whose loss would seriously jeopardize the mission. (FM 3-09)

cyberspace operations

The employment of cyberspace capabilities where the primary purpose is to achieve objectives in or through cyberspace. (JP 3-0)

decision point

A point in space and time when the commander or staff anticipates making a key decision concerning a specific course of action. (JP 5-0)

decontamination

The process of making any person, object, or area safe by destroying, neutralizing, making harmless, or absorbing and removing chemical or biological agents, or by removing radioactive material clinging to or around it. (JP 3-11)

delay

When a force under pressure trades space for time by slowing down the enemy's momentum and inflicting maximum damage on enemy forces without becoming decisively engaged. (ADP 3-90)

delay line

A phase line over which an enemy is not allowed to cross before a specific date and time or enemy condition. (FM 3-90)

demonstration

(Army) A variation of tactical deception used as a show of force in an area where a unit does not seek a decision and attempts to mislead an adversary. (FM 3-90)

destroy

A tactical mission task that physically renders an enemy force combat ineffective until it is reconstituted. (FM 3-90)

detachment

A tactical element organized on either a temporary or permanent basis for special duties. (ADP 3-90)

detainee holding area

A facility or other location where detainees are administratively processed and provided custodial care pending disposition and subsequent release, transfer, or movement to a theater detention facility. (JP 3-31)

directed energy

An umbrella term covering technologies that relate to the production of a beam of concentrated electromagnetic energy or atomic or subatomic particles. (JP 3-85)

direction finding

A procedure for obtaining bearings of radio frequency emitters by using a highly directional antenna and a display unit on an intercept receiver or ancillary equipment. (JP 3-85)

direction of attack

A specific direction or assigned route a force uses and does not deviate from when attacking. (ADP 3-90)

direction of fire

The direction on which a fire unit is laid to the most significant threat in the target area, to the chart direction to the center of the zone of fire, or to the target. (ATP 3-09.50)

disengage

A tactical mission task in which a unit breaks contact with an enemy to conduct another mission or to avoid becoming decisively engaged. (FM 3-90)

disrupt

1. A tactical mission task in which a unit upsets an enemy's formation or tempo and causes the enemy force to attack prematurely or in a piecemeal fashion. 2. An obstacle effect that focuses fire planning and obstacle effort to cause the enemy force to break up its formation and tempo, interrupt its timetable, commit breaching assets prematurely, and attack in a piecemeal effort. (FM 3-90)

division

An echelon of command and tactical formation that employs brigade combat teams, multi-functional brigades, and functional brigades to achieve objectives on land. (ADP 3-90)

drop zone

A specific area upon which airborne troops, equipment, or supplies are airdropped. (JP 3-36)

early-entry command post

A lead element of a headquarters designed to control operations until the remaining portions of the headquarters are deployed and operational. (FM 6-0)

electromagnetic warfare

Military action involving the use of electromagnetic and directed energy to control the electromagnetic spectrum or to attack the enemy. (JP 3-85)

enabling operation

An operation that sets the friendly conditions required for mission accomplishment. (FM 3-90)

encirclement

Where one force loses its freedom of maneuver because an opposing force is able to isolate it by controlling all ground lines of communications and reinforcement. (FM 3-90)

engagement area

An area where the commander masses effects to contain and destroy an enemy force. Also called EA. (FM 3-90)

engineer regulating point

A location used to ensure that vehicles do not exceed the capacity of the crossing means and to give drivers final instructions on site-specific procedures and information. Also called ERP. (ATP 3-90.4)

engineer work line

A coordinated boundary or phase line used to compartmentalize an area of operations to indicate where specific engineer units have primary responsibility for the engineer effort. (FM 3-34)

envelopment

A form of maneuver in which an attacking force avoids an enemy's principal defense by attacking along an assailable flank. (FM 3-90)

exfiltrate

A tactical mission task in which a unit removes Soldiers or units from areas under enemy control by stealth, deception, surprise, or clandestine means. (FM 3-90)

exploitation

(Army) A type of offensive operation following a successful attack to disorganize the enemy in depth. (FM 3-90)

explosive ordnance disposal

(Army) The detection, identification, on-site evaluation, rendering safe, exploitation, recovery, and final disposal of explosive ordnance. (FM 4-30)

feint

(Army) A variation of tactical deception that makes contact solely to deceive the adversary as to the location, time of attack or both. (FM 3-90)

field artillery

(Army) The equipment, supplies, ammunition, and personnel involved in the use of cannon, rocket, or surface-to-surface missile launchers. (FM 3-09)

final coordination line

A phase line close to the enemy position used to coordinate the lifting or shifting of supporting fires with the final deployment of maneuver elements. (ADP 3-90)

final protective fire

An immediately available prearranged barrier of fire designed to impede enemy movement across defensive lines or areas. (JP 3-09.3)

fire direction center

That element of a command post, consisting of gunnery and communications personnel and equipment, by means of which the commander exercises fire direction and/or fire control. (JP 3-09.3)

fire support area

An appropriate maneuver area assigned to fire support ships by the naval force commander from which they can deliver gunfire support to an amphibious operation. (JP 3-09)

fire support coordination line

A fire support coordination measure established by the land or amphibious force commander to support common objectives within an area of operation, beyond which all fires must be coordinated with affected commanders prior to engagement and, short of the line, all fires must be coordinated with the establishing commander prior to engagement. (JP 3-09)

fire support station

An exact location at sea within a fire support area within which a fire support ship delivers fire. (JP 3-02)

fix

1. A tactical mission task in which a unit prevents the enemy from moving from a specific location for a specific period. 2. An obstacle effect that focuses fire planning and obstacle effort to slow an attacker's movement within a specified area, normally an engagement area. (FM 3-90-1)

follow and assume

A tactical mission task in which a committed force follows a lead force conducting an offensive operation and continues the mission if the lead force cannot continue. (FM 3-90)

follow and support

A tactical mission task in which a committed force follows and supports a lead force conducting an offensive operation. (FM 3-90)

forms of maneuver

Distinct tactical combinations of fire and movement within a unique set of doctrinal characteristics that differs primarily in the relationship between the maneuvering force and the enemy. (ADP 3-90)

forward arming and refueling point

A temporary facility, organized, equipped, and deployed, to provide fuel and ammunition necessary for the employment of aviation maneuver units in combat. (JP 3-09.3)

forward edge of the battle area

The foremost limits of a series of areas in which ground combat units are deployed to coordinate fire support, the positioning of forces, or the maneuver of units, excluding areas in which covering or screening forces are operating. (JP 3-09.3)

forward line of own troops

A line which indicates the most forward positions of friendly forces in any kind of military operation at a specific time. (FM 3-90)

forward observer

An individual operating with front line troops trained to adjust ground or naval gunfire and pass back battlefield information. (JP 3-09)

forward passage of lines

Occurs when a unit passes through another unit's positions while moving toward the enemy. (ADP 3-90)

free-fire area

A specific area into which any weapon system may fire without additional coordination with the establishing headquarters. (JP 3-09)

frontal attack

A form of maneuver in which the attacking force seeks to destroy a weaker enemy force or fix a larger enemy force in place over a broad front. (FM 3-90)

gap

1. An area free of obstacles that enables forces to maneuver in a tactical formation. (FM 3-90) 2. A ravine, mountain pass, river, or other terrain feature that presents an obstacle that may be bridged. (ATP 3-90.4)

general engineering

Those engineering capabilities and activities, other than combat engineering, that provide infrastructure and modify, maintain, or protect the physical environment. (JP 3-34)

guard

A type of security operation done to protect the main body by fighting to gain time while preventing enemy ground observation of and direct fire against the main body. (ADP 3-90)

guerrilla

An irregular, predominantly indigenous member of a guerrilla force organized similar to military concepts and structure in order to conduct military and paramilitary operations in enemy-held, hostile, or denied territory. (ATP 3-05.1)

high-altitude missile engagement zone

A defined airspace in which high-altitude surface-to-air missiles engage air and missile threats. (JP 3-01)

high-density airspace control zone

Airspace designated in an airspace control plan or airspace control order in which there is a concentrated employment of numerous and varied weapons and airspace users. (JP 3-52)

improvised explosive device

A weapon fabricated or emplaced in an unconventional manner incorporating destructive, lethal, noxious, pyrotechnic, or incendiary chemicals. Also called IED. (JP 3-42)

infiltration

(Army) A form of maneuver in which an attacking force conducts undetected movement through or into an area occupied by enemy forces. (FM 3-90)

infiltration lane

A control measure that coordinates forward and lateral movement of infiltrating units and fixes fire planning responsibilities. (FM 3-90)

information operations force

A force consisting of units, staff elements, and individual military professionals in the Active and Reserve Components, and Department of Defense civilian employees who conduct or directly support the integration of information-related capabilities against adversaries and potential adversaries during military operations, as well as those who train these professionals. (DODD 3600.01)

intelligence operations

(Army) The tasks undertaken by military intelligence units through the intelligence disciplines to obtain information to satisfy validated requirements. (ADP 2-0)

interdict

A tactical mission task in which a unit prevents, disrupts, or delays the enemy's use of an area or route in any domain. (FM 3-90)

intermodal operations

The process of using multiple modes (air, sea, highway, rail) and conveyances (i.e. truck, barge, containers, pallets) to move troops, supplies and equipment through expeditionary entry points and the network of specialized transportation nodes to sustain land forces. (ADP 4-0)

isolate

A tactical mission task in which a unit seals off an enemy, physically and psychologically, from sources of support and denies it freedom of movement. (FM 3-0)

isolated personnel

United States military, Department of Defense civilians, and contractor personnel (and others designated by the President or Secretary of Defense) who are separated from their unit (as an individual or a group) while participating in a United States sponsored military activity or mission and are, or may be, in a situation where they must survive, evade, resist, or escape. (JP 3-50)

joint engagement zone

A defined airspace in which surface-to-air missiles and aircraft are simultaneously employed to engage air and missile threats. (JP 3-01)

key terrain

(Army) An identifiable characteristic whose seizures or retention affords a marked advantage to either combatant. (ADP 3-90)

kill zone

The location where fires are concentrated in an ambush. (FM 3-90)

landing zone

Any specified zone used for the landing of aircraft. (JP 3-36)

lateral boundary

A boundary defining the left or right limit of a unit's assigned area. (FM 3-90)

liaison

That contact or intercommunication maintained between elements of military forces or other agencies to ensure mutual understanding and unity of purpose and action. (FM 6-0)

limit of advance

A phase line used to control forward progress of the attack. (ADP 3-90)

line of contact

A general trace delineating the location where friendly and enemy forces are engaged. (FM 3-90)

line of departure

1. In land warfare, a line designated to coordinate the departure of attack elements. (JP 3-31)

linkup point

A designated place where two forces are scheduled to meet. (FM 3-90)

low-altitude missile engagement zone

A defined airspace in which low- to medium altitude surface-to-air missiles engage air and missile threats. (JP 3-01)

low-level transit route

A temporary air corridor of defined dimensions established to minimize the risk to friendly aircraft from friendly air defenses. (JP 3-52)

main command post

A portion of a unit headquarters containing the majority of the staff designed to command and control current operations, conduct detailed analysis, and plan future operations. (FM 6-0)

main supply route

The route or routes designated within an operational area upon which the bulk of traffic flows in support of military operations. (FM 1-02.1)

maintenance collection point

A temporary location established within the battalion echelon for the collection of equipment needing or undergoing field maintenance. (ATP 4-33)

maneuver

The employment of forces in the operational area, through movement in combination with fires and information, to achieve a position of advantage in respect to the enemy. (DOD Dictionary)

medical evacuation

The timely and effective movement of the wounded, injured, or ill to and between medical treatment facilities on dedicated and properly marked medical platforms with en route care provided by medical personnel. (ATP 4-02.2)

medical treatment facility

(Army) Any facility established for the purpose of providing medical treatment. This includes battalion aid stations, Role 2 facilities, dispensaries, clinics, and hospitals. (FM 4-02)

military deception

Actions executed to deliberately mislead adversary military, paramilitary, or violent extremist organization decision makers, thereby causing the adversary to take specific actions (or inactions) that will contribute to the accomplishment of the friendly mission. (JP 3-13.4)

minimum-risk route

A temporary air corridor of defined dimensions used by aircraft that presents the minimum known hazards to low-flying aircraft transiting the combat zone. Also called MRR. (JP 3-52)

missile defense

Defensive measures designed to destroy attacking enemy missiles or to nullify or reduce the effectiveness of such attack. (JP 3-01)

mobile defense

A type of defensive operation that concentrates on the destruction or defeat of the enemy through a decisive attack by a striking force. (ADP 3-90)

mobility

A quality or capability of military forces which permits them to move from place to place while retaining the ability to fulfill their primary mission. (JP 3-36)

mortuary affairs

Provides for the search, recovery, identification, preparation, and disposition of human remains of persons for whom the Services are responsible by status and executive order. (JP 4-0)

movement

The positioning of combat power to establish the conditions for maneuver. (ADP 3-90)

movement to contact

(Army) A type of offensive operation designed to establish or regain contact to develop the situation. (FM 3-90)

multidomain operations

The combined arms employment of joint and Army capabilities to create and exploit relative advantages to achieve objectives, defeat enemy forces, and consolidate gains on behalf of joint force commanders. (FM 3-0)

multinational operations

A collective term to describe military actions conducted by forces of two or more nations, usually undertaken within the structure of a coalition or alliance. (JP 3-16)

named area of interest

The geospatial area or systems node or link against which information that will satisfy a specific information requirement can be collected, usually to capture indications of adversary courses of action. (JP 2-0)

neutralize

A tactical mission task in which a unit renders the enemy incapable of interfering with an operation. (FM 3-90)

no-fire area

An area designated by the appropriate commander into which fires or their effects are prohibited. (JP 3-09.3)

nongovernmental organization

(DOD) A private, self-governing, not-for-profit organization dedicated to alleviating human suffering; and/or promoting education, health care, economic development, environmental protection, human rights, and conflict resolution; and/or encouraging the establishment of democratic institutions and civil society. Also called NGO. (JP 3-08)

objective area

A geographical area, defined by competent authority, within which is located an objective to be captured or reached by the military forces. (JP 3-06)

observation post

A position from which observations are made, or fires directed and adjusted. Also called OP. (FM 3-90)

obstacle

Any barrier designed or employed to disrupt, fix, turn, or block the movement and maneuver, and to impose additional losses in personnel, time, and equipment. (JP 3-15)

obstacle belt

A brigade-level command and control measure, normally depicted graphically, to show where within an obstacle zone the ground tactical commander plans to limit friendly obstacle employment and focus the defense. (JP 3-15)

obstacle control measures

Specific measures that grant obstacle-emplacing authority. (FM 3-90)

obstacle group

One or more individual obstacles that provide a specific obstacle effect. (FM 3-90)

obstacle restricted area

A command and control measure used to limit the type or number of obstacles within an area. (JP 3-15)

obstacle zone

A division-level command and control measure to designate specific land areas where lower echelons are allowed to employ tactical obstacles. (JP 3-15)

occupy

A tactical mission task in which a unit moves into an area to control it without enemy opposition. (FM 3 90)

passage lane

A lane through an obstacle that provides safe passage for a passing force. (FM 3-90)

passage of lines

An operation in which a force moves forward or rearward through another force's combat positions with the intention of moving into or out of contact with the enemy. (JP 3-18)

passage point

A designated place where passing units pass through the stationary unit. Also called PP. (FM 3-90)

penetration

A form of maneuver in which a force attacks on a narrow front. (FM 3-90)

personnel recovery

The sum of military, diplomatic, and civil efforts to prepare for and execute the recovery and reintegration of isolated personnel. Also called PR. (JP 3-50)

phase line

An easily identified feature in the operational area utilized for control and coordination of military operations. (JP 3-09)

platoon

A subdivision of a company or troop consisting of two or more squads or sections. (ADP 3-90)

point of departure

The point where the unit crosses the line of departure and begins moving along a direction of attack. (ADP 3-90)

position area for artillery

An area assigned to an artillery unit to deliver surface to surface fires. Also called PAA. (FM 3-90)

probable line of deployment

A phase line that designates the location where the commander intends to deploy the unit into assault formation before beginning the assault. (ADP 3-90)

psychological actions

Lethal and nonlethal actions planned, coordinated, and conducted to produce a psychological effect in a foreign individual, group, or population. (FM 3-53)

public affairs

Communication activities with external and internal audiences. (JP 3-61)

purple kill box

A fire support and airspace coordination measure that facilitates attacking surface targets with subsurface-to-surface, surface-to-surface, and air-to-surface munitions without further coordination with the area of operation commander's headquarters. (ATP 3-09.34)

pursuit

A type of offensive operation to catch or cut off a disorganized hostile force attempting to escape, with the aim of destroying it. (FM 3-90)

raid

An operation to temporarily seize an area to secure information, confuse an enemy, capture personnel or equipment, or to destroy a capability culminating with a planned withdrawal. (JP 3-0) (Army) A variation of attack to temporarily seize an objective with a planned withdrawal. (FM 3-90)

rally point

(Army) An easily identifiable point on the ground at which units can reassemble and reorganize if they become dispersed. (FM 3-90)

Rangers

Rapidly deployable airborne light infantry organized and trained to conduct highly complex joint direct action operations in coordination with or in support of other special operations units of all Services. (JP 3-05)

rearward passage of lines

Occurs when a unit passes through another unit's positions while moving away from the enemy. (ADP 3-90)

recovery

4. Actions taken to extricate damaged or disabled equipment for return to friendly control or repair at another location. (JP 3-34)

release point

A designated place on a route where elements are released from centralized control. Also called RP. (FM 3-90)

relief in place

An operation in which, by direction of higher authority, all or part of a unit is replaced in an area by the incoming unit and the responsibilities of the replaced elements for the mission and the assigned zone of operations are transferred to the incoming unit. (JP 3-07.3)

restricted operations zone

Airspace reserved for specific activities in which the operations of one or more airspace users is restricted. (JP 3-52)

restrictive fire area

A location in which specific restrictions are imposed and into which fires that exceed those restrictions will not be delivered without coordination with the establishing headquarters. (JP 3-09)

restrictive fire line

A specific boundary established between converging, friendly surface forces that prohibits fires or their effects from crossing. (JP 3-09)

retain

A tactical mission task in which a unit prevents enemy occupation or use of terrain. (FM 3-90)

retirement

When a force out of contact moves away from the enemy. (ADP 3-90)

retrograde

The process for the movement of non-unit equipment and materiel from a forward location to a reset (replenishment, repair, or recapitalization) program or to another directed area of operations to replenish unit stocks, or to satisfy stock requirements. (JP 4-09) (Army) A type of defensive operation that involves organized movement away from the enemy. (ADP 3-90)

route

The prescribed course to be traveled from a point of origin to a destination. (FM 3 90)

screen

A type of security operation that primarily provides early warning to the protected force. (ADP 3-90)

SEAL team

United States Navy forces organized, trained, and equipped to conduct special operations with an emphasis on maritime, coastal, and riverine environments. (JP 3-05)

search

In personnel recovery, a systematic reconnaissance of a defined area. (JP 3-50)

search and rescue

The use of aircraft, surface craft, submarines, and specialized rescue teams and equipment to search for and rescue distressed persons on land or at sea in a permissive environment. (JP 3-50)

search and rescue point

A predesignated specific location, relative to which isolated personnel provide their position to recovery forces. (JP 3-50)

section

(Army) A tactical unit of the Army and Marine Corps smaller than a platoon and larger than a squad. (ADP 3-90)

secure

A tactical mission task in which a unit prevents the enemy from damaging or destroying a force, facility, or geographical location. (FM 3-90)

security

1. Measures taken by a military unit, activity, or installation to protect itself against all acts designed to, or which may, impair its effectiveness. (JP 3-10)

security operations

Those operations performed by commanders to provide early and accurate warning of enemy operations, to provide the forces being protected with time and maneuver space within which to react to the enemy, and to develop the situation to allow commanders to effectively use their protected forces. (ADP 3-90)

seize

(Army) A tactical mission task in which a unit takes possession of a designated area using overwhelming force. (FM 3-90)

short-range air defense engagement zone

A defined airspace in which short-range air defense weapons engage air threats and which may be established within a low- or high-altitude missile engagement zone. (JP 3-01)

space forces

The space and terrestrial systems, equipment, facilities, organizations, and personnel necessary to access, use and, if directed, control space for national security. (JP 3-14)

special forces

United States Army forces organized, trained, and equipped to conduct special operations with an emphasis on unconventional warfare capabilities. (JP 3-05)

special operations forces

Those Active and Reserve Component forces of the Services designated by the Secretary of Defense and specifically organized, trained, and equipped to conduct and support special operations. (JP 3-05)

squad

A small military unit typically containing two or more fire teams. (ADP 3-90)

standard use Army aircraft flight route

Route established below the coordination level to facilitate the movement of Army aviation assets. (JP 3-52)

start point

A designated place on a route where elements fall under the control of a designated march commander. Also called SP. (FM 3-90)

strong point

A heavily battle position tied to a natural or reinforcing obstacle to create an anchor for the defense or to deny the enemy decisive or key terrain. (ADP 3-90)

supply

(Army) The process of providing all items necessary to equip, maintain, and operate a military command. (FM 1-02.1)

support

1. The action of a force that aids, protects, complements, or sustains another force in accordance with the directive requiring such action. 2. A unit that helps another unit in battle. 3. An element of a command that assists, protects, or supplies other forces in combat. (JP 1 VOL 2)

support area

The portion of the commander's area of operations that is designated to facilitate the positioning employment, and protection of base sustainment assets required to sustain, enable, and control operations. (ADP 3-0)

support by fire

A tactical mission task in which a unit engages the enemy by direct fire in support of another maneuvering force. (FM 3-90)

suppress

A tactical mission task in which a unit temporarily degrades a force or weapon system from accomplishing its mission. (FM 3-90)

surveillance

The systematic observation of aerospace, cyberspace, surface or subsurface areas, places, persons, or things by visual, aural, electronic, photographic, or other means. (JP 3-0)

sustainment

(Army) The provision of logistics, financial management, personnel services, and health service support necessary to maintain operations until successful mission completion. (ADP 4-0)

tactical command post

A portion of a unit headquarters designed to command and control operations as directed. (FM 6-0)

tactical mission task

The specific activity a unit performs while executing a tactical operation or form of maneuver. (FM 3-90)

target acquisition

The detection, identification, and location of a target in sufficient detail to permit the effective employment of capabilities that create the required effects. (JP 3-60)

target area of interest

The geographical area where high value targets can be acquired and engaged by friendly forces. (JP 2-0)

target number extension

A sequentially assigned number identifying the individual elements in a target. (MIL-STD 6017)

target reference point

A predetermined point of reference, normally a permanent structure or terrain feature that can be used when describing a target location. (JP 3-09.3)

task organization

(Army) A temporary grouping of forces designed to accomplish a particular mission. (ADP 5-0)

theater army

An echelon of command designated as the Army Service component command responsible for recommendations of allocations and employment of Army forces to the geographic combatant commander. (JP 3-31)

traffic control post

A manned post that is used to preclude the interruption of traffic flow or movement along a designated route. (FM 3-39)

transportation

A logistics function that includes movement control and associated activities to incorporate military, commercial, and multinational motor, rail, air, and water mode assets in the movement of units, personnel, equipment, and supplies in support of the concept of operations. (FM 1-02.1)

troop

A company-size unit in a cavalry organization. (ADP 3-90)

turn

1. A tactical mission task in which a unit forces an enemy force from one avenue of approach or mobility corridor to another. 2. A tactical obstacle effect that integrates fire planning and obstacle effort to divert an enemy formation from one avenue of approach to an adjacent avenue of approach or into an engagement area. (FM 3-90)

turning movement

(Army) A form of maneuver in which the attacking force seeks to avoid the enemy's principal defensive positions by attacking to the rear of their current positions forcing them to move or divert forces to meet the threat. (FM 3-90)

unit

Any military element whose structure is prescribed by a competent authority. (JP 3-33)

unmanned aircraft

(DOD) An aircraft that does not carry a human operator and is capable of flight with or without human remote control. (JP 3-30)

unmanned aircraft system

That system whose components include the necessary equipment, network, and personnel to control an unmanned aircraft. (JP 3-30)

weapon engagement zone

A defined airspace in which a particular weapon system engages air and missile threats. (JP 3-01)

weapons free zone

An air defense zone established for the protection of key assets or facilities, other than air bases, where weapon systems may be fired at any target not positively recognized as friendly. (JP 3-01)

withdraw

To disengage from an enemy force and move in a direction away from the enemy. (ADP 3-90)

DO NOT DELETE THIS SECTION BREAK—delete this sentence after adding all chapters.

References

All websites accessed on 30 January 2024.

REQUIRED PUBLICATIONS

These documents must be available to intended users of this publication.

Army Dictionary online. Available at https://jdeis.js.mil/jdeis/index.jsp?pindex=207.

DOD Dictionary of Military and Associated Terms. January 2024.

FM 1-02.1. *Operational Terms*. 09 March 2021.

MIL-STD 2525E. *Department of Defense Interface Standard Joint Military Symbology*. 31 December 2022. https://quicksearch.dla.mil/qsDocDetails.aspx?ident_number=114934.

RELATED PUBLICATIONS

These documents contain relevant supplemental information.

DEPARTMENT OF DEFENSE AND JOINT PUBLICATIONS

Unless otherwise indicated, most Department of Defense publications are available online at https://www.esd.whs.mil/DD/. Unless otherwise indicated, most joint publications are available online at https://www.jcs.mil/doctrine.

DOD Law of War Manual. June 2015.

DODD 3600.01. *Information Operations (IO)*. 02 May 2013.

JP 1 Volume 1. Joint Warfighting. 27 August 2023.

JP 1 Volume 2. *The Joint Force*. 19 June 2020.

JP 2-0. *Joint Intelligence*. 26 May 2022.

JP 3-0. *Joint Campaigns and Operations*. 18 June 2022.

JP 3-01. *Countering Air and Missile Threats*. 06 April 2023.

JP 3-02. *Amphibious Operations*. 04 January 2019.

JP 3-05. *Joint Doctrine for Special Operations*. 22 September 2020.

JP 3-06. *Joint Urban Operations*. 20 November 2013.

JP 3-07.3. *Peace Operations*. 01 March 2018.

JP 3-08. *Interorganizational Cooperation*. 12 October 2016.

JP 3-09. *Joint Fire Support*. 10 April 2019.

JP 3-09.3. *Close Air Support*. 10 June 2019.

JP 3-10. *Joint Security Operations in Theater*. 25 July 2019.

JP 3-11. *Operations in Chemical, Biological, Radiological, and Nuclear Environments*. 29 October 2018.

JP 3-13.4. *Military Deception*. 14 February 2017.

JP 3-14. *Joint Space Operations*. 23 August 2023.

JP 3-15. *Barriers, Obstacles, and Mine Warfare for Joint Operations*. 26 May 2022.

JP 3-16. *Multinational Operations*. 01 March 2019.

JP 3-18. *Joint Forcible Entry Operations*. 27 June 2018.

JP 3-22. *Foreign Internal Defense*. 17 August 2018.

JP 3-30. *Joint Air Operations*. 25 July 2019.

JP 3-31. *Joint Land Operations*. 03 October 2019.

JP 3-33. *Joint Force Headquarters*. 09 June 2022.

JP 3-34. *Joint Engineer Operations*. 06 January 2016.

JP 3-36. *Joint Air Mobility and Sealift Operations*. 04 January 2021.

JP 3-42. *Joint Explosive Ordnance Disposal*. 14 September 2022.

JP 3-50. *Personnel Recovery*. 14 August 2023.

JP 3-52. *Joint Airspace Control*. 22 October 2022.

JP 3-57. *Civil-Military Operations*. 09 July 2018.

JP 3-60. *Joint Targeting*. 28 September 2018.

JP 3-61. *Public Affairs*. 17 November 2015.

JP 3-85. *Joint Electromagnetic Spectrum Operations*. 22 May 2020.

JP 4-0. *Joint Logistics*. 20 July 2023.

JP 4-02. *Joint Health Services*. 29 August 2023.

JP 4-09. *Distribution Operations (U//FOUO)*. 14 March 2019.

JP 4-10. *Operational Contract Support*. 04 March 2019.

JP 4-18. *Joint Terminal and Joint Logistics Over-the-Shore Operations (CUI)*. 05 December 2022.

JP 5-0. *Joint Planning*. 01 December 2020.

JP 6-0. *Joint Communications*. 04 December 2023.

ARMY PUBLICATIONS

Unless otherwise indicated, most Army doctrinal publications are available online at https://armypubs.army.mil/.

ADP 2-0. *Intelligence*. 31 July 2019.

ADP 3-0. *Operations*. 31 July 2019.

ADP 3-37. *Protection*. 31 July 2019.

ADP 3-90. *Offense and Defense*. 31 July 2019.

ADP 4-0. *Sustainment*. 31 July 2019.

ADP 5-0. *The Operations Process*. 31 July 2019.

ADP 6-0. *Mission Command: Command and Control of Army Forces*. 31 July 2019.

ATP 1-19. *Army Bands*. 28 July 2021.

ATP 2-22.7. *Geospatial Intelligence*. 26 March 2015.

ATP 3-04.1. *Aviation Tactical Employment*. 07 May 2020.

ATP 3-05.1. *Unconventional Warfare at the Combined Joint Special Operations Task Force Level*. 09 April 2021. Common access card required.

ATP 3-05.60. *Special Operations Communications System*. 30 November 2015.

ATP 3-09.02. *Field Artillery Survey*. 16 February 2016.

ATP 3-09.12. *Field Artillery Counterfire and Weapons Locating Radar Operations*. 26 October 2021.

ATP 3-09.23. *Field Artillery Cannon Battalion*. 24 September 2015.

ATP 3-09.30. *Observed Fires*. 28 September 2017.

ATP 3-09.34/MCRP 3-31.4/NTTP 3-09.2.1/AFTTP 3-2.59. *Kill Box: Multi-Service Tactics, Techniques, and Procedures for Kill Box Planning and Employment*. 07 October 2022.

ATP 3-09.50. *The Field Artillery Cannon Battery*. 04 May 2016.

ATP 3-09.60. *Techniques for Multiple Launch Rocket System (MLRS) and High Mobility Artillery Rocket System (HIMARS) Operations*. 29 July 2020.

ATP 3-11.36/MCRP 10-10E.1/NTTP 3-11.34/AFTTP 3-2.70. *Multi-Service Tactics, Techniques, and Procedures for Chemical, Biological, Radiological, and Nuclear Planning*. 24 September 2018.

ATP 3-11.37/MCRP 10-10E.7/NTTP 3-11.29/AFTTP 3-22.44). *Multi-Service Tactics, Techniques, and Procedures for Chemical, Biological, Radiological, and Nuclear Reconnaissance and Surveillance*. 31 March 2021.

ATP 3-18.13. *Special Forces Use of Pack Animals*. 30 October 2014.

ATP 3-20.15/MCRP 3-10B.1. *Tank Platoon*. 03 July 2019.

ATP 3-20.96. *Cavalry Squadron*. 12 May 2016.

ATP 3-21.8. *Infantry Platoon and Squad.* 12 April 2016.

ATP 3-21.10. *Infantry Rifle Company.* 14 May 2018.

ATP 3-21.11. *Stryker Brigade Combat Team Infantry Rifle Company*. 25 November 2020.

ATP 3-21.20. *Infantry Battalion*. 28 December 2017.

ATP 3-21.51. *Subterranean Operations*. 01 November 2019.

ATP 3-21.90/MCTP 3-01D. *Tactical Employment of Mortars*. 09 October 2019.

ATP 3-21.91. *Stryker Brigade Combat Team Weapons Troop*. 11 May 2017.

ATP 3-27.5. *AN/TPY-2 Forward Based Mode System Operations*. 11 February 2022. Common access card required.

ATP 3-34.45/MCRP 3-40D.17. *Electric Power Generation and Distribution*. 06 July 2018.

ATP 3-34.80. *Geospatial Engineering*. 22 February 2017.

ATP 3-37.10/MCRP 3-40D.13. *Base Camps*. 27 January 2017.

ATP 3-37.11. *Chemical, Biological, Radiological, Nuclear and Explosives Command.* 28 August 2018.

ATP 3-39.12. *Law Enforcement Investigations*. 19 August 2013.

ATP 3-39.30. *Security and Mobility Support*. 21 May 2020.

ATP 3-39.34. *Military Working Dogs*. 19 May 2022.

ATP 3-52.1/MCRP 3-20F.4/NTTP 3-56.4/AFTTP 3-2.78. *Multi-Service Tactics, Techniques, and Procedures for Airspace Control.* 21 June 2023.

ATP 3-90.1. *Armor and Mechanized Infantry Company Team*. 24 October 2023.

ATP 3-90.4/ MCTP 3-34A/MCWP 3-17.8. *Combined Arms Mobility*. 10 June 2022.

ATP 3-90.5. *Combined Arms Battalion*. 15 July 2021.

ATP 3-90.8/MCTP 3-34B. *Combined Arms Countermobility*. 30 November 2021.

ATP 3-90.97. *Mountain Warfare and Cold Weather Operations*. 29 April 2016.

ATP 3-91. *Division Operations*. 17 October 2014.

ATP 3-96.1. *Security Force Assistance Brigade*. 02 September 2020.

ATP 4-02.1. *Army Medical Logistics*. 29 October 2015.

ATP 4-02.2. *Medical Evacuation*. 12 July 2019.

ATP 4-11. *Army Motor Transport Operations*. 14 August 2020.

ATP 4-14. *Expeditionary Railway Center Operations*. 22 June 2022.

ATP 4-15. *Army Watercraft Operations*. 03 April 2015.

ATP 4-33. *Maintenance Operations*. 09 July 2019.

ATP 4-35. *Munitions Operations*. 31 January 2023.

ATP 4-41. *Army Field Feeding and Class 1 Operations*. 31 December 2015.

ATP 4-42. *Materiel Management, Supply, and Field Services Operations*. 02 November 2020.

ATP 4-43. *Petroleum Supply Operations*. 18 April 2022.

ATP 4-44/ MCRP 3-40D.14/MCRP 3-17.7Q. *Water Support Operations*. 16 December 2022.

ATP 4-46/MCRP 3-40G.3/NTTP 4-06/AFTTP 3-2.51. *Multi-Service Tactics, Techniques, and Procedures for Mortuary Affairs in Theaters of Operations*. 03 August 2022.

ATP 4-48. *Aerial Delivery*. 28 August 2023.

ATP 4-90. *Brigade Support Battalion*. 18 June 2020.

ATP 5-19. *Risk Management*. 09 November 2021.

ATP 6-0.5. *Command Post Organization and Operations*. 01 March 2017.

ATP 6-02.54. *Techniques for Satellite Communications*. 05 November 2020.

ATP 6-02.60. *Tactical Network Techniques for Corps and Below*. 09 August 2019.

ATP 6-02.70. *Techniques for Spectrum Management Operations*. 16 October 2019.

DA PAM 385-64. *Ammunition and Explosives Safety Standards*. 24 July 2023.

FM 1-0. *Human Resources Support*. 25 August 2021.

FM 1-05. *Religious Support*. 21 January 2019.

FM 1-06. *Financial Management Operations*. 15 April 2014.

FM 2-0. *Intelligence*. 01 October 2023.

FM 3-0. *Operations*. 01 October 2022.

FM 3-04. *Army Aviation*. 06 April 2020.

FM 3-07. *Stability*. 02 June 2014.

FM 3-09. *Fire Support and Field Artillery Operations*. 30 April 2020.

FM 3-11. *Chemical, Biological, Radiological, and Nuclear Operations*. 23 May 2019.

FM 3-18. *Special Forces Operations*. 28 May 2014.

FM 3-34. *Engineer Operations*. 18 December 2020.

FM 3-39. *Military Police Operations*. 09 April 2019.

FM 3-50. *Army Personnel Recovery*. 02 September 2014.

FM 3-53. *Military Information Support Operations*. 04 January 2013.

FM 3-57. *Civil Affairs Operations*. 28 July 2021.

FM 3-63. *Detainee Operations*. 02 January 2020.

FM 3-81. *Maneuver Enhancement Brigade*. 09 November 2021.

FM 3-84. *Legal Support to Operations*. 01 September 2023.

FM 3-90. *Tactics*. 01 May 2023.

FM 3-94. *Armies, Corps, and Division Operations*. 23 July 2021.

FM 3-96. *Brigade Combat Team*. 19 January 2021.

FM 3-98. *Reconnaissance and Security Operations*. 10 January 2023.

FM 3-99. *Airborne and Air Assault Operations*. 06 March 2015.

FM 4-0. *Sustainment Operations*. 31 July 2019.

FM 4-01. *Army Transportation Operations*. 03 April 2014.

FM 4-02. *Army Health System*. 17 November 2020.

FM 4-30. *Ordnance Operations*. 01 April 2014.

FM 4-40. *Quartermaster Operations*. 22 October 2013.

FM 6-0. *Commander and Staff Organization and Operations*. 16 May 2022.

FM 6-02. *Signal Support to Operations*. 13 September 2019.

TC 3-21.75. *The Warrior Ethos and Soldier Combat Skills*. 13 August 2013.

TC 3-22.10. *Sniper*. 07 December 2017.

TM 3-11.32/MCRP 10-10E.5/NTRP 3-11.25/AFTTP 3-2.56. *Multi-Service Reference for Chemical, Biological, Radiological, and Nuclear Warning and Reporting and Hazard Prediction Procedures*. 15 May 2017.

TM 3-23.25. *Shoulder-Launched Munitions*. 15 September 2010.

TM 3-34.85/MCRP 3-17A. *Engineer Field Data*. 17 October 2013.

INTERNATIONAL STANDARDIZATION AGREEMENTS

Most Standardization Agreements are available on the NATO Standardization Office website (user registration required): https://nso.nato.int/nso/.

STANAG 1059 (ED. 8). *Letter Codes for Geographical Entities*. 19 February 2004.

STANAG 1241 (ED. 6). *NATO Standard Identity Description Structure for Tactical Use*. 26 January 2023.

STANAG 2019 (ED. 8)/APP-06(E). *NATO Joint Military Symbology*. 11 October 2023.

WEBSITES

Bureau of Alcohol, Tabacco, Firearms and Explosives website. https://www.atf.gov/

Federal Bureau of Investigation website. https://www.fbi.gov/

United States Coast Guard website. https://www.uscg.mil/

United States Drug Enforcement Administration website. https://www.dea.gov/

United States Secret Service website. https://www.secretservice.gov/

U.S. Marshals Service website. https://www.usmarshals.gov/

PRESCRIBED FORMS

This section contains no entries.

REFERENCED FORMS

Unless otherwise indicated, DA forms are available on the Army Publishing Directorate (APD) website at https://armypubs.army.mil/.

DA Form 2028. *Recommended Changes to Publications and Blank Forms.*

This page intentionally left blank.

Index

Entries are by paragraph number.

Entries are by paragraph number.

Entries are by paragraph number.

This page intentionally left blank.

FM 1-02.2

28 February 2024

By Order of the Secretary of the Army:

RANDY A. GEORGE
General, United States Army
Chief of Staff

Official:

MARK F. AVERILL
Administrative Assistant
to the Secretary of the Army
2405402

DISTRIBUTION:
Active Army, Army National Guard, and United States Army Reserve. To be distributed in accordance with the initial distribution number (IDN) 116180, requirements for FM 1-02.2.

www.ingramcontent.com/pod-product-compliance
Lightning Source LLC
LaVergne TN
LVHW080925210926
848485LV00004B/727